Second Edition

Kellie Ploeger Cox, PhD

New York Chicago San Francisco Athens London Madrid
Mexico City Milan New Delhi Singapore Sydney Toronto

Copyright © 2022, 2019 by McGraw Hill. All rights reserved. Printed in the United States of America. Except as permitted under the United States Copyright Act of 1976, no part of this publication may be reproduced or distributed in any form or by any means, or stored in a database or retrieval system, without the prior written permission of the publisher.

1 2 3 4 5 6 7 8 9 LCR 27 26 25 24 23 22

ISBN 978-1-264-28579-2
MHID 1-264-28579-5

e-ISBN 978-1-264-28580-8
e-MHID 1-264-28580-9

Interior design by Steve Straus of Think Book Works.
Cover and letter art by Kate Rutter.

McGraw Hill books are available at special quantity discounts to use as premiums and sales promotions or for use in corporate training programs. To contact a representative, please visit the Contact Us pages at www.mhprofessional.com.

McGraw Hill is committed to making our products accessible to all learners. To learn more about the available support and accommodations we offer, please contact us at accessibility@mheducation.com. We also participate in the Access Text Network (www.accesstext.org), and ATN members may submit requests through ATN.

Dedication

I need to thank all of my students. The words on these pages have been formed by years of their thoughtful and clever questions.

I would like to thank my better half, John. When I thought I'd run out of words, he assured me there were more.

I want to thank my dear daughters, Raina and Kenley, whose adherence to homework schedules inspired me to stick to my own.

And finally, to the cats, whose frequent walks across my keyboard kept things interessssssssssmku8jhewq3wwgggggggggggggg.

Contents

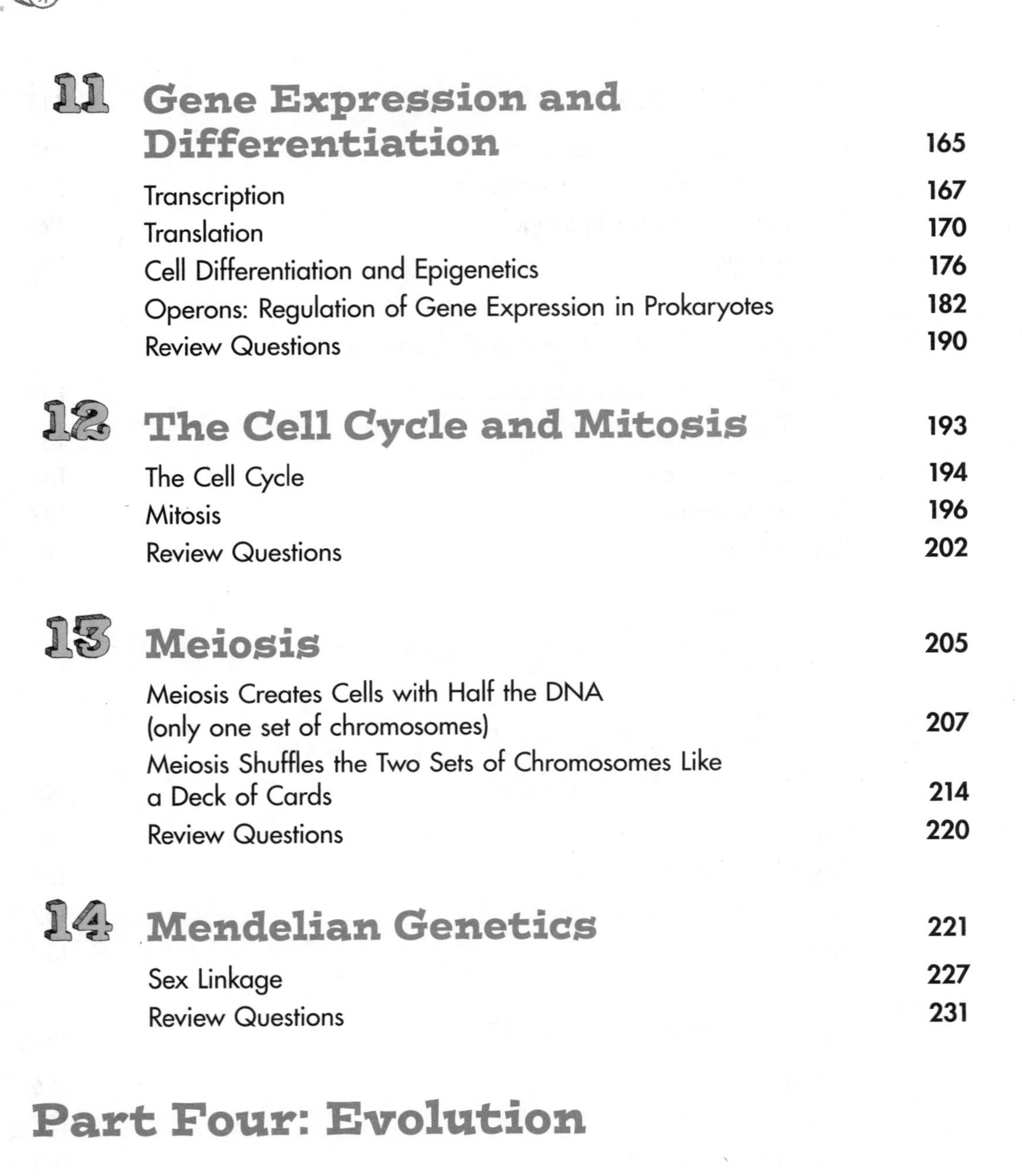

Introduction

Welcome to your new biology book! Let us explain why we believe you've made the right choice with this new edition. This probably isn't your first go-round with either a textbook or other kind of guide to a school subject. You've probably had your fill of books asking you to memorize lots of terms. This book isn't going to do that—although you're welcome to memorize anything you take an interest in. You may also have found that a lot of books make a lot of promises about all the things you'll be able to accomplish by the time you reach the end of a given chapter. In the process, those books can make you feel as though you missed out on the building blocks that you actually need to master those goals.

With *Must Know High School Biology*, we've taken a different approach. When you start a new chapter, right off the bat you will immediately see one or more **must know** ideas. These are the essential concepts behind what you are going to study, and they will form the foundation of what you will learn throughout the chapter. With these **must know** ideas, you will have what you need to hold it together as you study, and they will be your guide as you make your way through each chapter.

To build on this foundation, you will find easy-to-follow discussions of the topic at hand, accompanied by comprehensive examples that show you how to apply what you're learning to solving typical biology questions. Each chapter ends with review questions—more than 300 throughout the book—designed to instill confidence as you practice your new skills.

This book has other features that will help you on this biology journey of yours. It has a number of "sidebars" that will both help provide information or just serve as a quick break from your studies. The BTW sidebars ("by the way") point out important information as well as study tips and exceptions to the rule. Every once in a while, an IRL sidebar ("in real life") will tell you what you're studying has to do with the real world; other IRLs may just be interesting factoids.

But that's not all—this new edition has taken it a step further. We know our Biology students well and we want to make sure you're getting the most out of this book. We added new EASY MISTAKE sidebars that point out common mistakes and things *not* to do. For those needing a little assistance, we have our EXTRA HELP A+ feature, where more challenging concepts, topics, or questions are given some more explanation. And finally, one special note for the teachers (because we didn't forget about you!)—a **Teacher's Guide** section at the back of the book is a place where you can go to find tips and strategies on teaching the material in the book, a behind-the-scenes look at what the author was thinking when creating the material, and resources curated specifically to make your life easier!

In addition, this book is accompanied by a flashcard app that will give you the ability to test yourself at any time. The app includes more than 100 "flashcards" with a review question on one side and the answer on the other. You can either work through the flashcards by themselves or use them alongside the book. To find out where to get the app and how to use it, go to the next section, The Flashcard App.

Before you get started, though, let me introduce you to your guide throughout this book. Kellie—I mean Dr. Cox—has taught biology for more than 20 years and also teaches AP Biology. It's been a pleasure to see how enthusiastic she is about her subject as well as how adept she is at explaining complex subjects. I don't know whether she'd consider herself an artist, but we thought her sketches were so terrific we decided to leave them in, as we're confident they will help you picture some of the trickier topics.

Dr. Cox has a clear idea of what you should get out of a biology course and has developed strategies to help you get there. She has seen the kinds of pitfalls that students get into, and she is experienced at solving those difficulties. In this book, she applies that experience both to showing you the most effective way to learn a given concept—as well as advising you how to extricate yourself from traps you may have fallen into. She will be a trustworthy guide as you expand your biology knowledge and develop new skills.

Before we leave you to your author's surefooted guidance, let us give you one piece of advice. While we know that saying something "is the *worst*" is a cliché, if anything is the worst in biology, it's the concept of energy transformation. (Running into photosynthesis or cellular respiration for the first time can be intimidating.) Let Dr. Cox introduce you to the concept and show you how to apply it confidently to your biology work. Take our word for it, mastering the topic of energy transformation will leave you in good stead throughout your biology career.

Good luck with your studies!

The Editors at McGraw Hill

The Flashcard App

This book features a bonus flashcard app. It will help you test yourself on what you've learned as you make your way through the book (or in and out). It includes 100-plus "flashcards," both "front" and "back." It gives you two options on how to use it. You can jump right into the app and start from any point that you want. Or you can take advantage of the handy QR Codes near the end of each chapter in the book; they will take you directly to the flashcards related to what you're studying at the moment.

To take advantage of this bonus feature, follow these easy steps:

Search for ***McGraw Hill Must Know*** App from either Google Play or the App Store.

↓

Download the app to your smartphone or tablet.

↓

Once you've got the app, you can use it in either of two ways.

↙ ↘

Just open the app and you're ready to go.	Use your phone's QR Code reader to scan any of the book's QR codes.
You can start at the beginning, or select any of the chapters listed.	You'll be taken directly to the flashcards that match your chapter of choice.

↘ ↙

Be ready to test your biology knowledge!

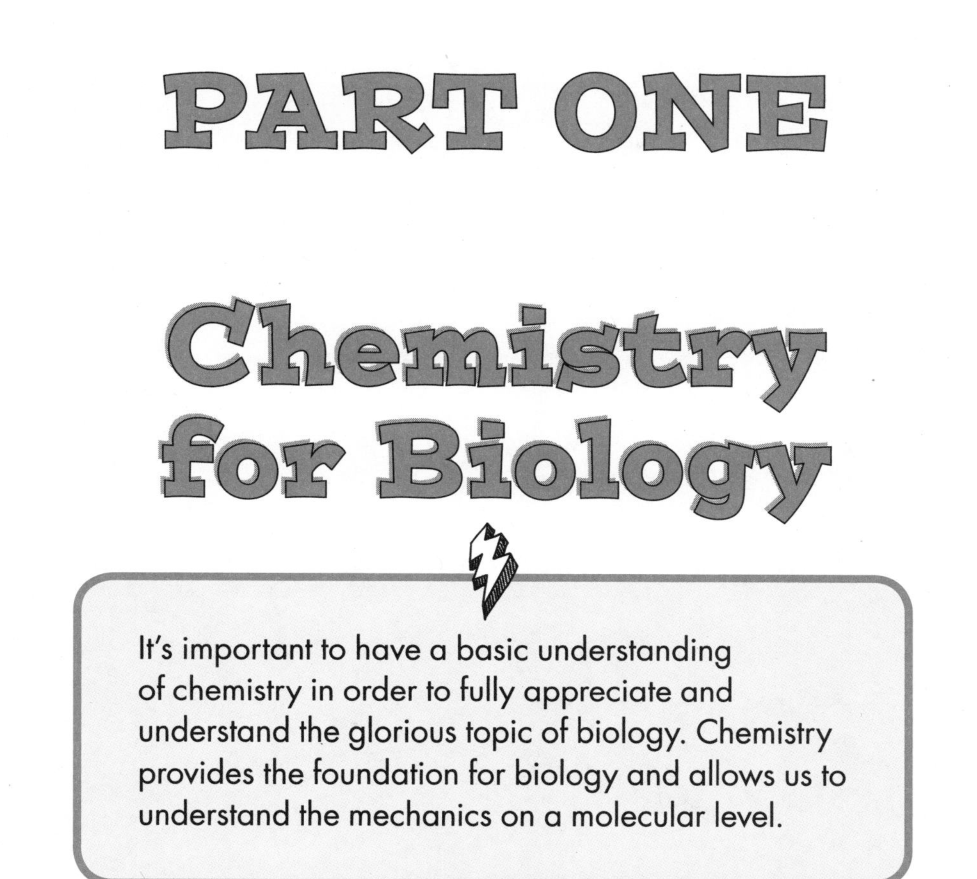

PART ONE

Chemistry for Biology

It's important to have a basic understanding of chemistry in order to fully appreciate and understand the glorious topic of biology. Chemistry provides the foundation for biology and allows us to understand the mechanics on a molecular level.

1 Chemical Bonds and Reactions

MUST KNOW

- Atoms are rearranged and bonds are broken and formed in chemical reactions.

- Interactions between molecules are pivotal to proper biological function.

Biochemical reactions, in the simplest terms, are based on changing one thing into another. In order to do that, the covalent bonds within the starting molecule(s) must be broken and new bonds formed to create the new molecule(s).

The rearrangement of bonds is significant, because it is related to whether the reaction is spontaneous or not, or if energy is needed or released. From a biological point of view, this aspect of chemistry is really important because cellular metabolism is based on breaking and rearranging bonds in order to build molecules and release energy.

BTW

*Yes, I am oversimplifying. But in regards to what I want my biology students to know, this is what we should focus on. It is, simply speaking, the **must know** concept!*

Covalent, Ionic, and Intermolecular Bonds

Think back to chemistry class and you may remember the different types of bonds that occur within and between molecules: covalent, ionic, and intermolecular. Our **must know** concept will focus on the rearrangement of covalent bonds, but all three of these bond types are important.

An **ion** is a charged atom, and an **ionic bond** occurs between a positive and a negative ion. Whether an atom is an ion is based on the number of electrons it has. Normally, the number of electrons in an atom equals the number of protons; since each electron has a negative charge, and each proton has a positive charge, the overall charges will balance each other out and the atom will be neutral.

If a neutral atom gains an extra electron, it becomes a negatively charged ion (**anion**).

BTW

The atoms depicted on a periodic table are the neutral variety, with equal numbers of electrons and protons. An element's atomic number equals the number of protons; the atomic number, therefore, also indicates the numbers of electrons. An element's atomic mass is the sum of the number of protons and the number of neutrons. All the weight (mass) of the atom is in the nucleus, where the protons and neutrons reside. So, you add them all up!

If a neutral atom loses an electron, it becomes a positively charged atom (**cation**). The saying "opposites attract" holds true here, because that negative anion and the positive cation seek out and hold on to one another. Table salt—NaCl—is created through an ionic bond formed between the Na^+ and the Cl^-. A **covalent bond**, unlike the ionic bond, is formed between atoms that are *sharing* electrons. A single covalent bond is created when two electrons are shared between two atoms; a double bond involves four shared electrons. Both ionic bonds and covalent bonds are called intramolecular bonds (*intra* = within) because they occur *within* a single molecule. An intermolecular bond (*inter* = between) is different because it occurs between two *separate* molecules. We will talk about two types of intermolecular forces in this book: van der Waals and hydrogen bonding. Hydrogen bonding occurs in molecules that contain a hydrogen atom bonded to an oxygen, nitrogen, or fluorine atom. We will talk more about hydrogen bonding once we get to the water section of this chapter.

Intermolecular Forces: van der Waals and Hydrogen Bonding

Entire molecules can have positively and negatively charged regions. That doesn't mean they are ionic (having lost or gained an electron). It is simply that electrons like to hang out in some regions of a molecule more than other regions, creating an unequal charge distribution. Because one area has a slightly more positive charge (because the electrons are not hanging out there) and one area has a slightly more negative charge, separate molecules will stick to one another, just like magnets. These sorts of attractions are called **van der Waals interactions**, and they only occur if the molecules are really, really close to one another. I have a totally cool example of van der Waals forces in nature. You know what a gecko is? Geckos are those little lizards that live in all sorts of habitats, and they particularly like to hang out on rocks and the sides of trees.

If a gecko happens to wander into somebody's house, they are just as happy clinging to the walls as a rock. A gecko can hang upside down on a ceiling, even if it's made of glass! These remarkable little lizards' climbing abilities are so stunning, they have even been the subject of Aristotle's praise. For years, scientists have studied this phenomenon to figure out how, exactly, this lizard can seemingly defy gravity and cling to almost any surface. Was it exuding some sort of sticky glue-like substance from its feet? Were there tiny little barbs that physically clung to the surface like Velcro? Maybe miniscule suction cups? Nope; it's because of van der Waals forces between their toes and the wall! And there needs to be a *ton* of individual van der Waals interactions in order to support the weight of an entire gecko. This is possible because the gecko's feet are flat and splayed out, and the entire surface is covered by a dizzying amount of tiny, microscopic folds and hairs. This creates a super-high surface area, thus increasing the potential number of van der Waals interactions. Behold! The amazing gecko and his glorious feet:

The common house gecko (*Hemidactylus frenatus*)

Author: Firos AK. https://commons.wikimedia.org/wiki/File:Asian_House_Gecko_close_up_from_bangalore.jpg

Close-up of the underside of a gecko's foot as it walks on a glass wall

Author: Bjørn Christian Tørrissen. https://commons.wikimedia.org/wiki/File:Gecko_foot_on_glass.JPG

Gecko toes (and intermolecular interactions in general) are perfect examples of our **must know** concept that interactions between molecules are an important aspect of biology.

Energy Transformation (and How It Must Follow the Rules of the Universe)

Thermodynamics is the study of energy transformations, and it's a perfect application of our **must know** concept. There are different forms of energy (e.g., chemical, thermal, mechanical) and transformation refers to switching one form to another, a very helpful ability for cells. On a small scale, this transformation occurs because the atoms and bonds of chemicals are broken up and rearranged in new ways. On a big scale, energy is transformed and transferred between us and the environment because organisms are referred to as "open systems" (a system in which mass or energy can be lost to or gained from the environment). For example: you eat food and it is broken down and transformed into the chemical form of energy called ATP (thank you, cellular respiration). During the process, some of the energy is lost as heat, radiating from our bodies into the surroundings. Two laws of thermodynamics explain the scenario above.

- **First Law of Thermodynamics: Energy cannot be created or destroyed.** It's impossible to create energy from scratch; it can only be transformed from one form to another. This transformation is due to our **must know** concept of breaking bonds and rearranging atoms to create new things. Does your electric company *create* the energy that flows throughout your house? Nope. It *converts* it from one form (oil, coal, solar) into the electricity that travels through the power lines. In an upcoming chapter, we will cover two very important metabolic processes that transform energy from one form to another: photosynthesis and cellular respiration. Photosynthesis grabs photons of light from the sun (solar energy) and transforms them into chemical energy (glucose). Cellular respiration takes the chemical energy of glucose and converts it into another chemical form, ATP. The first law of thermodynamics aligns perfectly with our **must know** concept. Chemical reactions involve the breaking, reforming, and rearranging of bonds; we rely on breaking, reforming, and rearranging bonds in order to convert energy from one form to another.
- **Second Law of Thermodynamics: Energy transformations increase the disorder of the universe.** If you think your life is always tending toward chaos, then congratulations, you're fully aligned

with the second law of thermodynamics: things tend toward disorder. Specifically, every energy transformation will increase the disorder of the universe. The term *entropy* is used as a measure of disorder and randomness. When energy is transformed, some of the potentially usable energy is lost as heat. For example, when your cells break and rearrange the bonds of glucose in order to transform it into the chemical energy of ATP, not all of the potential energy in glucose is used; some is lost as heat. This heat is radiating off your body, into the universe, increasing the overall disorder. Never realized you had such a huge impact on the universe, did you?

IRL

Pseudoscience alert: There cannot be a perpetual motion machine. The two laws of thermodynamics are ironclad, and if a machine is converting one form of energy into motion, there needs to be a net input of energy because as per the second law, every transformation will result in a loss of some energy into the surroundings. A machine that is 100% efficient and no energy is ever lost? Nope, can't be done. If you hear of someone claiming to invent such a machine, they have either made a grievous error in their results, or they are trying to scam someone.

Catabolic and Anabolic Reactions

Our **must know** focus is that chemical reactions involve the breaking and rearrangement of covalent bonds within a molecule. When creating a new bond and two atoms are glued together, energy is required. When that bond is broken, it releases the energy stored within it.

A **catabolic** reaction breaks larger, more complex molecules into smaller, simpler molecules. Covalent bonds are cleaved apart and energy is released. This is an example of an **exergonic** reaction: the products end up with less energy than the starting reactants and energy is released over the course of the reaction. Any reaction that has net release of energy has a *negative* ΔG.

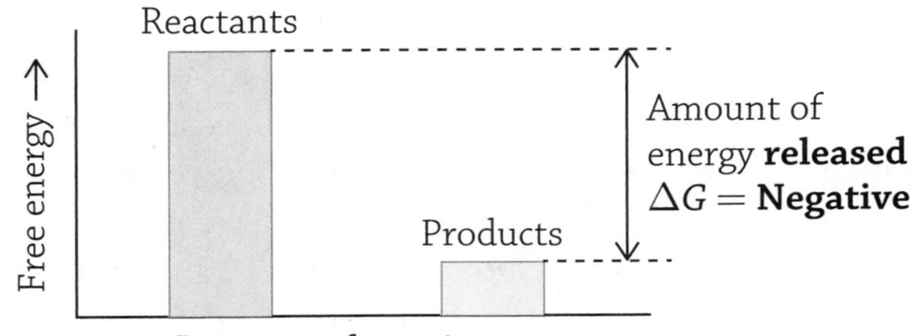

Graph of a catabolic reaction

An **anabolic** reaction is a "building-up" reaction (to help remember this, think about "anabolic steroids" and how they dangerously build up muscle mass). In order to form a new, complex molecule from smaller parts, new covalent bonds are created and energy is used; this is called an **endergonic** reaction. This means that the final product of an anabolic reaction stores more energy than the reactants it was made from. Any reaction that absorbs free energy has a positive ΔG.

What is this strange "delta gee" (ΔG) I speak of? The letter G stands for Gibbs free energy (or just ***free energy****), and it means the amount of a reaction's energy that can be used to do some sort of "work." When you use ΔG to describe a reaction, you are talking about the change in free energy from the start of the reaction (the amount of energy in the reactants) to the end of the reaction (the amount of energy in the products). If you have more energy stored at the start (in the reactants), then energy will be released once the products are formed; this indicates a $-\Delta G$. Reactions with a $-\Delta G$ are spontaneous, meaning it is energetically favorable for the reaction to occur . . . it just wants to happen! If instead energy needs to be added in order to create the products, the reaction is said to have a $+\Delta G$; this reaction is not spontaneous and needs a push (energy added) in order for it to occur.*

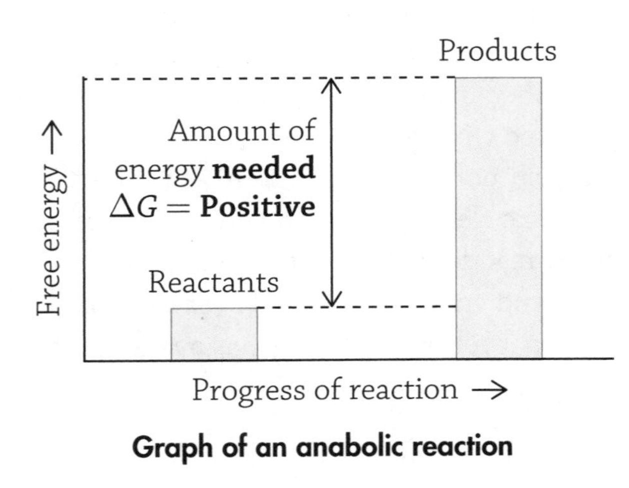

Graph of an anabolic reaction

A perfect example of energy-releasing ($-\Delta G$) catabolic reactions and energy-storing (ΔG) anabolic reactions relates to something we will learn more about later: photosynthesis and cellular respiration.

This is the summary equation for the process of **photosynthesis**:

$$6CO_2 + 6H_2O \text{ (+ sun energy)} \rightarrow C_6H_{12}O_6 + 6O_2$$

And this is the summary equation for **cellular respiration:**

$$C_6H_{12}O_6 + 6O_2 \rightarrow 6CO_2 + 6H_2O \text{ (+ chemical energy)}$$

Even though each of these equations is actually the composite of many, many individual reactions, they summarize the biochemical pathways quite nicely. Photosynthesis is the process of combining six carbon dioxide molecules with six water molecules and, using the energy of the sun, rearranging the bonds to create a high-energy molecule of glucose and six molecules of diatomic oxygen (O_2). The sugar molecule contains more energy in its bonds than found in the reactants (carbon dioxide and water) and the energy input of the sun was needed to make the final product ($+\Delta G$); this is an anabolic reaction. The glucose molecule stores the energy of the sun within it! When this sugar molecule is subjected to the catabolic pathways of cellular respiration, it is broken apart and the stored energy is released ($-\Delta G$). The small resulting molecules of carbon dioxide float away into the atmosphere, and the only way they will once again reform a molecule of glucose is if there is again an input of sun energy to once again rearrange the atoms of six carbon dioxides and six water molecules into a single energy-storing glucose molecule.

Enzyme Function

Your metabolism consists of trillions of reactions occurring in your body, in every cell, all the time. Some of these reactions, if left to their own devices, would occur so slowly as to not help the cell in its pursuit of a happy and metabolically productive life. Luckily, there are enzymes to facilitate the process. Enzymes help us understand both **must know** concepts: the interactions they participate in (with their substrate) are pivotal to their function (breaking and rearranging bonds).

Enzymes are biological catalysts that speed up chemical reactions but are themselves not used up in the reaction (they are reusable). The term *biological catalyst* is a fancy way of saying they are made by cells. Most enzymes are made of proteins, though there are RNA-based enzymes (called ribozymes). Since enzymes come from cells, it makes sense that they tend to prefer conditions that remind them of home: body temperature, neutral pH. There are definitely exceptions to this rule, and they're cool to consider. You have enzymes in your stomach, for example, and conditions there are harshly acidic (a pH of 2!). There are certain species of archaebacteria (a type of prokaryotic cell) that enjoy living in extreme environments where no other form of life would have a chance of surviving. These single-celled critters' enzymes have evolved to thrive in such punishing environments as 100°C hot springs, 0°C arctic waters, super-salty bodies of water, and solutions with a pH ranging from 2 to 11. These crazy "extremozymes" are, obviously, the exception to the rule.

Enzymes are very specific and will catalyze only one reaction. Generally speaking, a reaction can either be a building (anabolic) reaction or a breaking (catabolic) reaction. The molecule(s) the enzyme grabs onto is called the **substrate**, and it fits in a specific location called the enzyme's **active site**. Once the enzyme grabs onto its target, it forms the **enzyme-substrate complex**, a temporary pairing of the substrate before the reaction actually occurs. If this is a breaking-apart reaction, the enzyme will squeeze a bit, putting stress on the covalent bonds within the substrate. The enzyme is making the substrate easier to break apart! Once it does, the enzyme releases the products and will happily await another substrate molecule.

An anabolic enzyme will work a bit differently, because it needs to bring together two reactants and create a larger molecule. The enzyme will help this process along by grabbing the reactants and making sure they line up perfectly for them to bond together.

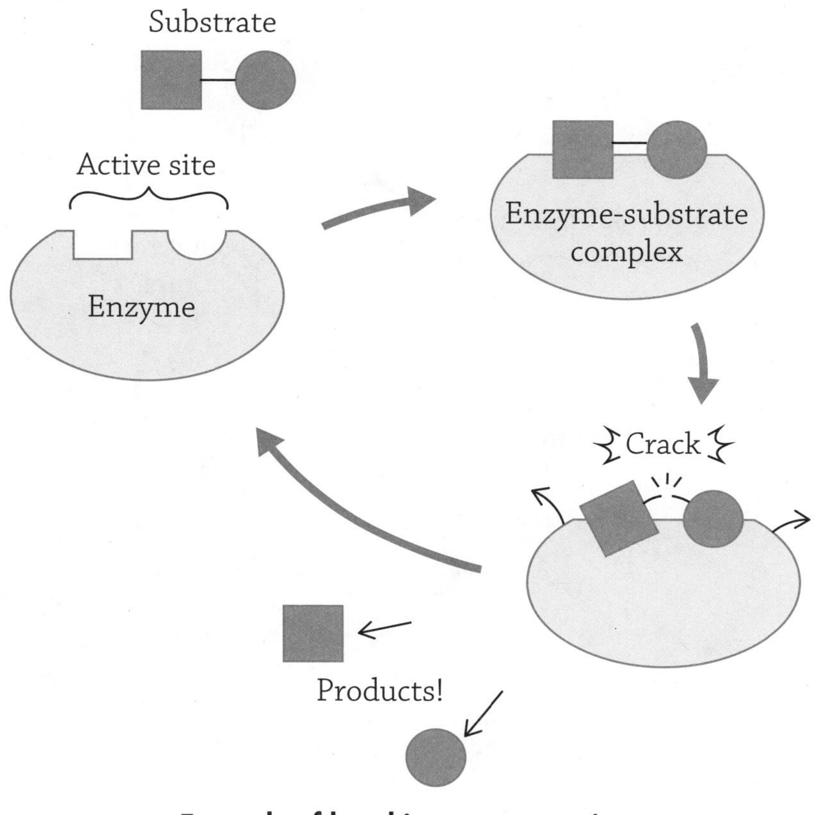

Example of breaking-apart reaction

Enzyme Inhibition

Enzymes are good at their job, but sometimes the cell needs to put on the metabolic brakes. The presence of inhibitors in a cell can help regulate enzyme activity. There are two different categories of enzyme inhibition, depending on where, exactly, this little inhibitor binds. As you know, the substrate must fit into the enzyme's active site in order for catalysis to occur. If, for some reason, the substrate cannot get into that active site, then the reaction cannot happen! A **competitive inhibitor** does just what its name implies: it competes for the active site and blocks the actual substrate. It is

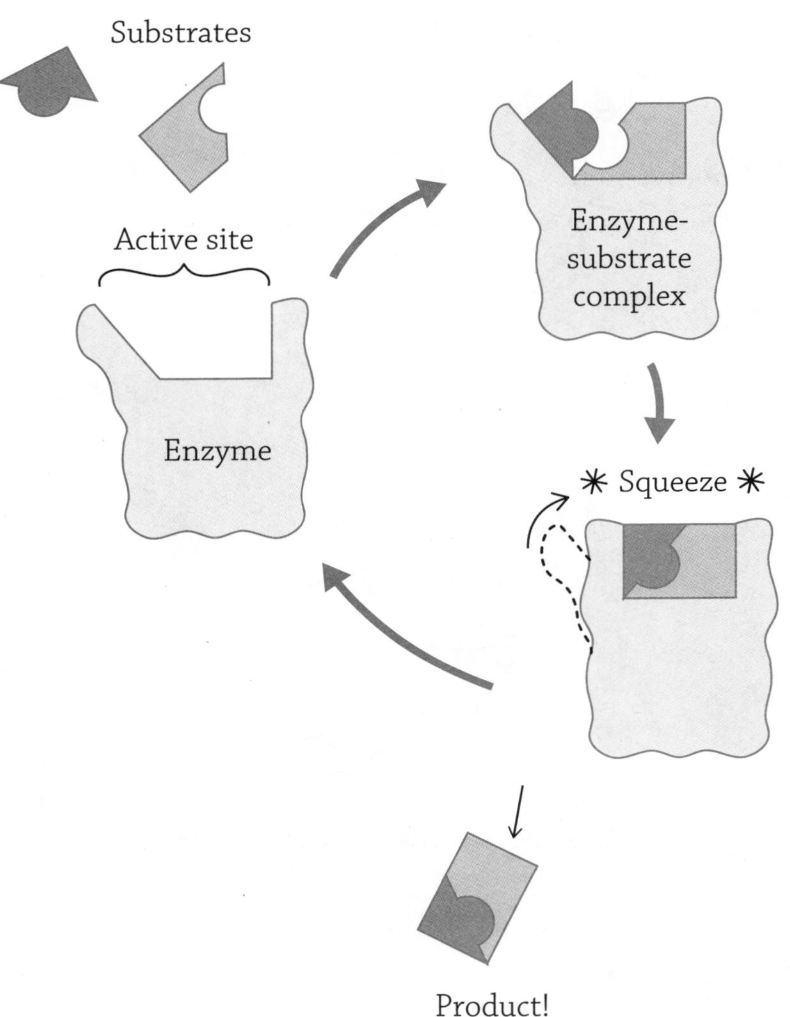

Example of building reaction

a rude sort of molecule, shouldering its way into the space that is supposed to be for the substrate. This pairing, however, is not permanent (luckily, otherwise that particular enzyme molecule would be kaput). The competitive enzyme hangs out for a bit, then is eventually let go. The enzyme will then grab another molecule, whether it be the actual substrate or another competitive inhibitor. If there was a relatively small concentration of inhibitor, the chances of the next molecule being a substrate molecule are relatively high; if there was a high concentration of inhibitor, the enzyme might once again grab the wrong molecule.

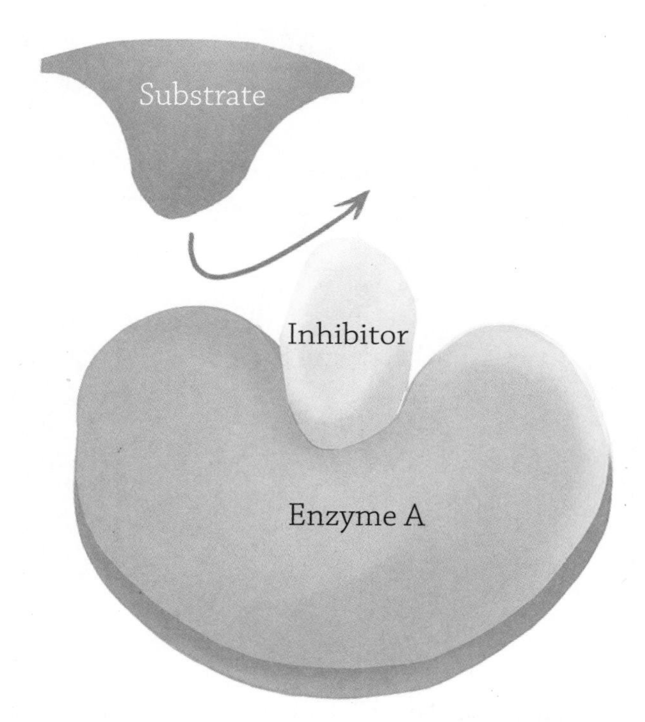

Competitive enzyme inhibition. If the competitive inhibitor binds the active site, it blocks the substrate from entering.

Author: California16. https://commons.wikimedia.org/wiki/File:Enzyme_inhibition.png

There is this other kind of inhibitor that is a bit sneakier. A **noncompetitive inhibitor** binds at a site *other* than the active site, but once it binds, it makes the active site change shape so the substrate no longer fits. As was the case in the competitive inhibitor, the noncompetitive inhibitor is not permanently bound to the enzyme. But if you increase the concentration of the actual substrate, would that increase the chance of the reaction occurring (as it did with the competitive inhibitor)? Nope. Since this particular passive-aggressive inhibitor is casually latching into a different location, it couldn't care less how much substrate is around; it's still going to screw up enzyme activity.

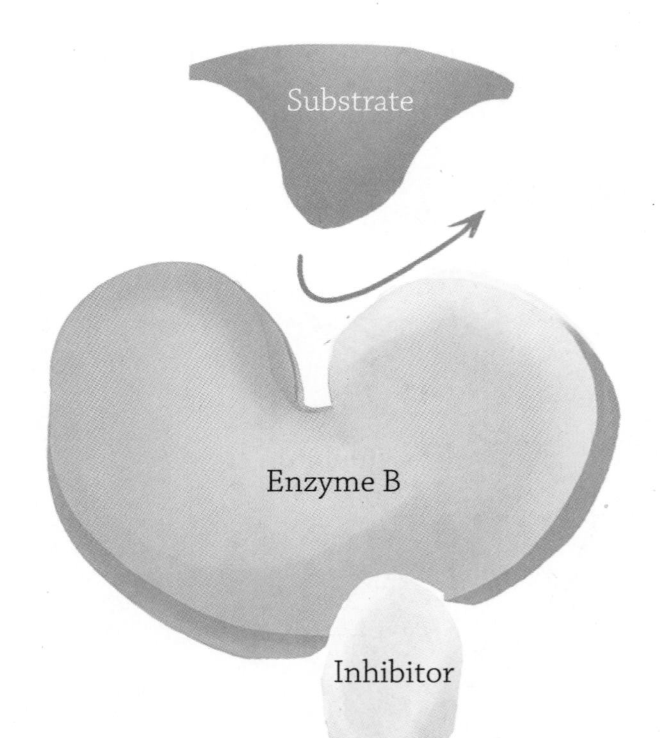

Noncompetitive enzyme inhibition. The inhibitor binds a region other than the active site, so it is not in direct competition with the substrate.

Author: California16. https://commons.wikimedia.org/wiki/File:Competitive%26NonCompetitive_Enzyme_Inhibition.jpg

Allosteric Enzymes

Some enzymes have evolved clever means to finely control their activity. It's important that enzymes aren't running at full speed, all the time . . . that would be wasteful and inefficient. Enzymes that can tweak their reaction speeds are called **allosteric enzymes**, and they are regulated in a way very similar to the noncompetitive inhibition example from earlier.

Allosteric enzymes usually are composed of multiple protein subunits, each with their own active site. An allosteric enzyme vacillates between two forms: active and inactive. There are also other binding sites to which a **regulator protein** can bind.

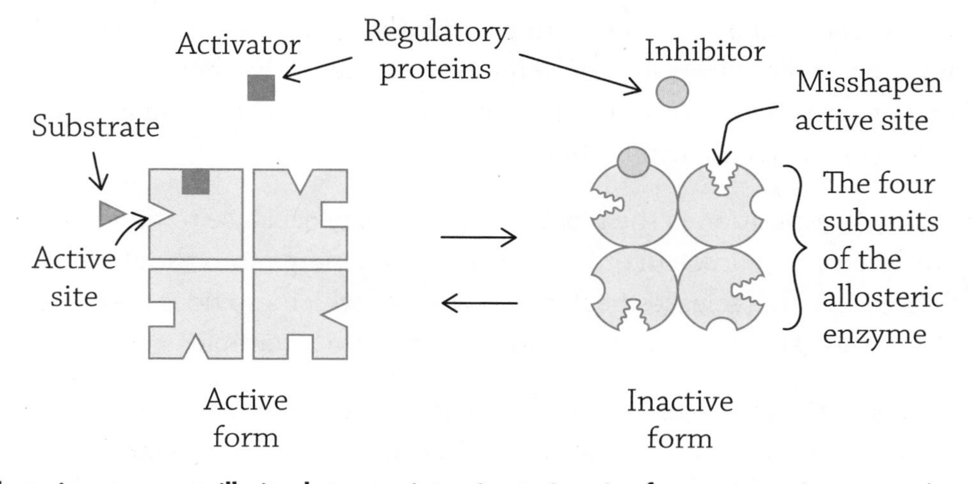

Allosteric enzyme oscillating between its active or inactive forms. An activator regulatory protein binds to an allosteric site and stabilizes the active form, allowing the substrate to fit into the active site.

Allosteric enzymes and noncompetitive inhibitors are very similar in function, the difference being that allosteric enzymes can fine-tune reaction rates by both decreasing AND increasing the speed of reaction.

There are two different kinds of regulator molecules: activators and inhibitors. If an activator binds to the regulatory site, the subunits are stabilized in the active shape. It is "active" because the shape of the active site perfectly fits the substrate. This speeds up the reaction. If, however, the inhibitor binds to the regulatory site, then all the subunits are stabilized in the inactive form. This allows the cell to fine-tune enzyme activity through the presence of activators and inhibitors. One important allosteric enzyme has a major role in adjusting your metabolic production of ATP to fit the current energy needs. The enzyme phosphofructokinase (PFK) is an allosteric enzyme that regulates the speed of glycolysis, and we'll learn more about PFK in Chapter 6. Enzyme function is rooted in our **must know** concept of crucial interactions between molecules being a foundation of proper biological function.

REVIEW QUESTIONS

1. Molecules of water have an uneven distribution of electrons, resulting in one end of the molecule having a slightly negative charge. The slightly negative end of one water molecule will adhere to the slightly positive end of a second water molecule. What type of chemical bond is occurring between these two water molecules?

2. Cellular respiration is the biochemical pathway cells use to break up a molecule of glucose in order to convert the energy into the chemical form, ATP. This releases heat in the process, which provides body heat. Which law(s) of thermodynamics applies to this example?

3. Choose the right term from each of the follow pairs: A reaction that breaks bonds is called a(n) **anabolic/catabolic** reaction because energy is **needed/released**.

4. You run an experiment with an enzyme that catalyzes the reaction of turning A (reactant) into B (the product). Your friend plays a joke on you by adding a noncompetitive inhibitor to the reaction, slowing the reaction down significantly.
   a. Where on the enzyme does the inhibitor bind?
   b. Can the reaction be sped up again by adding more of the enzyme's substrate (A)?

5. Choose the correct answer: A(n) **ionic/covalent/intermolecular** bond involves sharing two electrons between two atoms.

6. Referring to the two reaction diagrams below, indicate which of the two reactions: Requires an input of energy? Creates products that contain more energy than the reactants? Has a negative delta *G*?

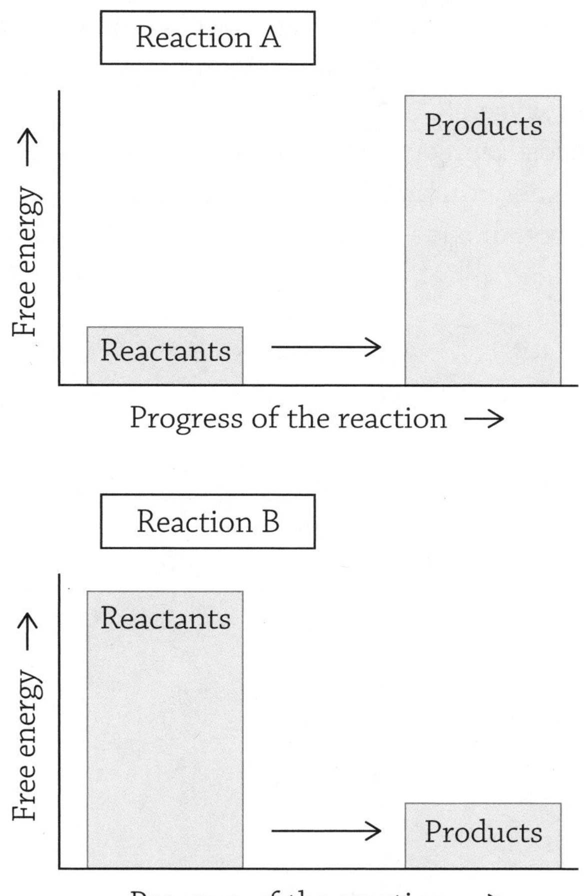

7. Why are enzymes called biological catalysts?

8. How does an enzyme help a catabolic reaction occur?

9. How is an allosteric enzyme similar to a noncompetitive inhibitor? How is it different?

10. Which of the following is an example of an anabolic reaction?

a. Hydrogen peroxide (H_2O_2) being broken down into water (H_2O) and oxygen (O_2)

b. Glucose being used by the mitochondrion to create ATP

c. The chemical processes of food digestion

d. Carbon dioxide (CO_2) and water (H_2O) combined to form glucose

Properties of Water

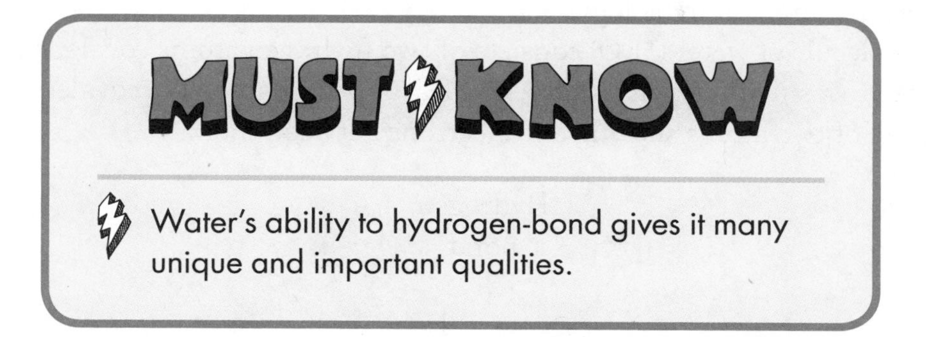

Water's ability to hydrogen-bond gives it many unique and important qualities.

ater is really important. You know that, because you've probably heard that about 70% of Earth's surface is covered with water, 60% of your body is composed of water, and you need to drink plenty of water (otherwise you'll become dehydrated!). What I want to talk about is why water is important on a molecular level. This little three-atom molecule is a key component of many life functions.

As we progress through this chapter, we will keep mentioning a special bond that occurs between water molecules. This bond (a hydrogen bond) is going to allow water to do many special things that contribute to its important function in biology.

A molecule of water (H_2O) consists of two hydrogen atoms covalently bonded to a single oxygen atom. As we learned in Chapter 1, a covalent bond consists of two atoms sharing two electrons between them.

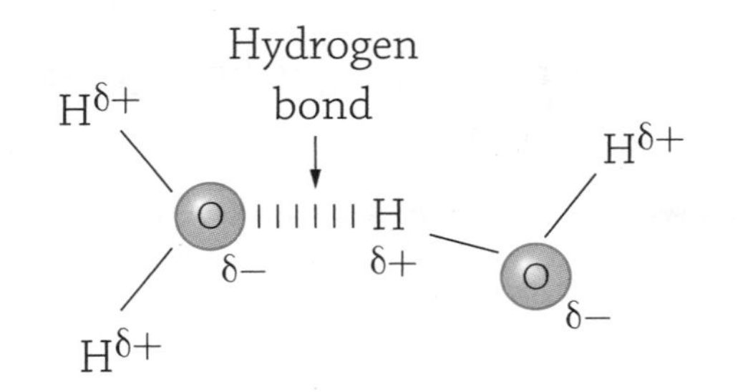

Hydrogen bonding between two water molecules

The figure above shows two water molecules involved in a specific type of intermolecular bond called a **hydrogen bond**. Each water molecule (circled) consists of two hydrogen atoms and an oxygen atom held together by covalent bonds (the straight lines). Because the covalent bonds are occurring within a single water molecule, they are examples of intramolecular bonds. Now, you notice those strange symbols (δ) next to the hydrogens and oxygen? That means "slightly negative" or "slightly positive." Recall that a covalent bond involves the sharing of two electrons. The truth is, the atoms involved in that covalent bond don't

Remember to draw a dashed line to represent a hydrogen bond. A solid line means a covalent bond!

necessarily share equally. When a hydrogen and an oxygen participate in a covalent bond, the oxygen tends to pull the shared electrons closer to it, and away from the hydrogen. That gives the oxygen atom a slightly negative change; the electrons are hanging out closer to the oxygen atom. Because the negative charges are farther away from the hydrogen, the hydrogen atom is left with a slightly positive charge. The "slightly" adjective is really important because it is not a fully negative or positive charge . . . that would mean it was an ion (and it's not because no electrons were fully lost or gained). Molecules that have this difference in charge distribution are called **polar** molecules.

The slightly negative oxygen of one water molecule is attracted to the slightly positive hydrogen of a different water molecule. It's as if these two water molecules stick to each other like magnets, forming an intermolecular bond (remember that term from Chapter 1?). The reason all of this is important is because it relates to our **must know** concept. The ability of water molecules to form hydrogen bonds is way more important than you may think. Why will a closed bottle of water burst if left in the freezer (or even worse, why will water pipes in your house break if they freeze)? Why does a droplet of water make a dome-like shape? And why does it tend to be cooler along the coastline in the summer and a bit warmer in the winter? Read on for the answers to these mysteries!

Unique Properties of Water

Water is special. It is the liquid basis of all life. All living organisms are made mostly of water, and individual cells are about 80% water. Water has many unique properties that make life possible, including:

- Water is involved in cohesion and adhesion.
- Water expands when freezes and becomes less dense.
- Water has a high heat capacity.

Let's go through each of these items in detail. As you do, keep in mind the **must know** idea that hydrogen bonding is the reason these things can occur.

- **Water is involved in both cohesion and adhesion** When a molecule of water hydrogen-bonds with itself, it is called **cohesion**. We learned about hydrogen bonds (and other intermolecular bonds) earlier in this chapter. Now, what we didn't mention is that water can form hydrogen bonds with things other than water molecules; this is called **adhesion**. If you measured water in a graduated cylinder and noticed how the edge of the water sticks a bit to the glassware, you have seen an example of adhesion. The water is hydrogen-bonding to the glass (or plastic) of the graduated cylinder, forming a **meniscus**. Cohesion and adhesion work together in one very important example we will visit again later in Chapter 20 (Plant Structure and Transport): it allows a "chain" of water molecules to be pulled up a plant's vascular tissue.

  When water molecules hang around together, they act quite sticky. The surface of a body of water forms a "skin" of molecules hydrogen-bonded together. This gives water its tremendous surface tension. A rain droplet forms a lovely dome-shaped structure when it hits your car. Water striders can walk on water because the surface tension resists their tiny little legs punching through.

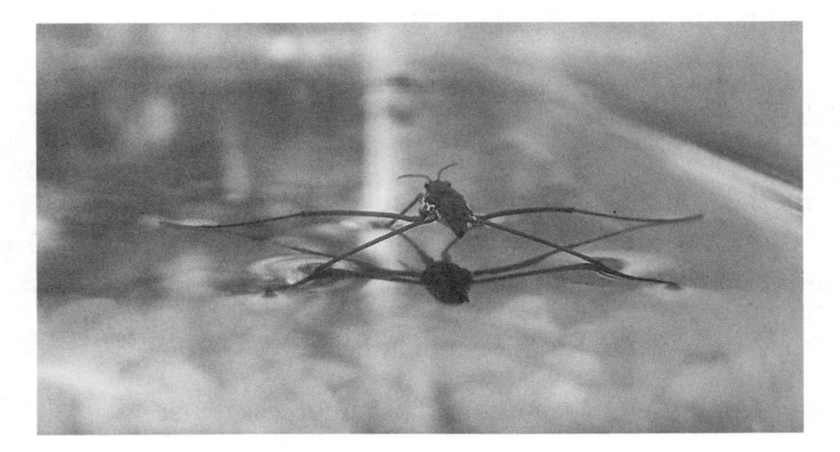

Water strider

Author: TimVickers. https://commons.wikimedia.org/wiki/File:Water_strider.jpg

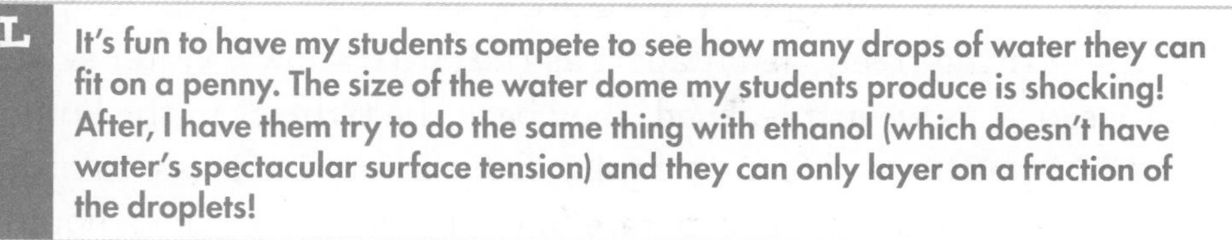

IRL It's fun to have my students compete to see how many drops of water they can fit on a penny. The size of the water dome my students produce is shocking! After, I have them try to do the same thing with ethanol (which doesn't have water's spectacular surface tension) and they can only layer on a fraction of the droplets!

- **Water expands when frozen and becomes less dense** The hydrogen bonding that occurs causes the water molecules to align in such a way that water is less dense as a solid than as a liquid. That's really weird! When water turns solid, the hydrogen-bonded molecules form what is called a "crystal lattice." This arrangement gives the molecules quite a bit of elbow room. If you compare two equal volumes—one of liquid water and one of solid ice—there are fewer water molecules in the ice than the liquid! When water freezes, it expands and gets bigger—that's why ice cubes sort of "pop" out of the ice cube tray. Furthermore, ice is less dense that its liquid counterpart . . . and that explains why those ice cubes float on the top of your drink and don't sink to the bottom. This floating-ice fact has serious significance. First of all, in a large body of water (think oceans and lakes) floating ice insulates any water that's below it. Secondly, if ice sank, those large bodies of water would easily freeze solid in the winter. Then, in the summer, only the top of the water would melt! How could anything live in lakes or oceans when their habitat freezes solid on a yearly basis? For that matter, would life exist AT ALL? The first forms of life evolved in the oceans. If they froze solid . . . NO LIFE. No us. Nothing. Whoa!

  Next time you swirl your ice cubes in your drink, ruminate on the significance the lowly ice cube has to the existence of life.

- **Water has a high heat capacity.** Take a moment to consider the three basic states of matter. In a solid, molecules pack the tightest; in a liquid, they are a little farther apart; and in a gas, they are floating freely, as far apart as the space allows. Water is weird because, as we just learned, they pack the tightest in the liquid form. But let us instead focus on going from a liquid to a gas. In order for any substance to evaporate (meaning it turns from a liquid to a gas), enough energy must be supplied to break the intermolecular bonds between the molecules. As we know, water molecules have quite a grip on each other, thanks to copious hydrogen bonding. In order to transition a liquid to a gas, those hydrogen bonds must be overcome (and broken apart) by heat. Consider a ring of water left on the table by your sweaty soda can. It will eventually absorb enough heat from the environment to evaporate and dry up. If it was instead a little puddle of another solvent, such as ethanol, hexane, or benzene, the liquid would evaporate and disappear in a fraction of the time that water took. So, what is the significance of this, since we're learning biology right now? Water has the ability to stabilize temperatures because it resists temperature change, due to all that hydrogen bonding (our **must know** concept proving important once again). A large body of water will keep stable and resist wild swings in temperature. This is good for the life that lives in aquatic environments. Furthermore, because it resists a temperature change, oceans help to even out temperatures along the coastline—a bit warmer in the winter, and a bit cooler in the summer.

REVIEW QUESTIONS

1. Draw two water molecules involved in a hydrogen bond. Label a covalent bond and a hydrogen bond.
2. When a water molecule forms a hydrogen bond with another water molecule, this is called ________________; when a water molecule hydrogen-bonds to another substance, it is called ________________.
3. What happens to water's density when it freezes?
4. Water resists quick changes in temperature and doesn't easily evaporate (or freeze); this is due to water's ________________.
5. Why is it important to use the term *partial* when describing charges on a water molecule?
6. How does the polarity of a water molecule play a role in its ability to form hydrogen bonds?
7. What property of water is responsible for the creation of surface tension?
8. When water freezes, it becomes less dense and floats on the surface of bodies of water. Explain why this is helpful to aquatic life.
9. When water changes state—either from a solid to a liquid, or a liquid to a gas—what bonds are being broken?

Macromolecules

MUST KNOW

Large macromolecules are created with dehydration reactions and broken apart by hydrolysis reactions.

Glucose is the monomer used to create many large carbohydrate polymers.

Phospholipids are key to the formation of cell membranes (and all life).

A protein's structure is based on its sequence of amino acids.

The four different bases of DNA and RNA provide the variability for the genetic code.

arth is populated by organic life-forms, meaning we are all composed of a bunch of carbon-based molecules (organic chemistry focuses on the study of carbon compounds). Even though the term *life-form* could mean a ton of wildly different things (a speck of lichen, a piece of kelp, a colony of *E. coli*), they all have the same basic building block: the cell. We will talk about the structural brick of life in detail when we get to Part Two (Cells), but for now, let's take a closer look at some of the key categories of molecule that make up a cell: carbohydrates, lipids, proteins, and nucleic acids. It is a **must know** to recognize that each macromolecule performs a very specific function for the cell.

A **molecule** is defined as a group of atoms held together by covalent bonds. Water (H_2O), glucose ($C_6H_{12}O_6$), and caffeine ($C_8H_{10}N_4O_2$) are all molecules. Now, if I use the term **macromolecule**, hopefully it makes you think of something that's big; many of biology's important molecules are big macromolecules. Furthermore, if this gigantic macromolecule is created by covalently linking together smaller, repeating molecules, it's referred to as a **polymer** (the repeating subunit of which is a **monomer**). If this all makes your head spin, I understand. I think it's easier if I show you specific examples. First, however, let's talk about the chemical reactions needed to create these big ol' macromolecules. As the other **must know** states, the creation (and destruction) of these macromolecules all use the same reactions: dehydration and hydrolysis.

Dehydration and Hydrolysis Reactions

The process of creating macromolecules from smaller components relies on a type of reaction called a **dehydration reaction**. If you are dehydrated, it means you are low on water, and if you dehydrate foods, it means you remove water from them. It makes sense that a dehydration reaction involves the removal of water, specifically, the removal of a water molecule between the two monomers that are being connected. Hanging off of one monomer there is a hydrogen (—H), and on the other monomer there is a hydroxyl group (—OH). A specific enzyme catalyzes the removal of the water molecule and leaves in its place a brand-new covalent bond:

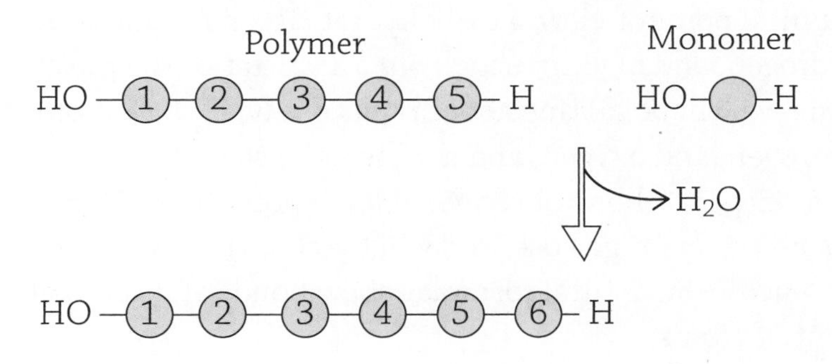

If you were to reverse the process, individual monomers would be cleaved from the larger polymer. Since a water molecule was removed to form the link, the cell needs to add a water molecule back to sever the link. Using water (hydro-) to break (-lysis) a covalent bond is called a **hydrolysis reaction**:

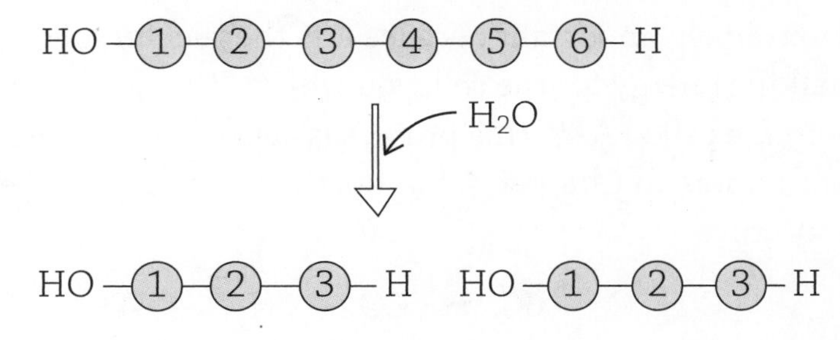

The reactions that occur in your digestive tract are mostly hydrolysis reactions, breaking up polymers in your food into individual monomers for your body to use!

Carbohydrates

The term *carbohydrate* may bring to mind pasta and bread and other starchy deliciousness. Indeed, you are correct—these foods are categorized as carbohydrates. In this case, you are using the term in the dietary sense. From a chemical point of view, a carbohydrate is a molecule containing carbon, hydrogen, and oxygen, usually in a 1:2:1 ratio. The most important carbohydrate we will be talking about is glucose ($C_6H_{12}O_6$). It only contains carbon, hydrogen, and oxygen, and it adheres to the ratio of twice as many hydrogens as either carbons or oxygen. Please note that this → $C_6H_{12}O_6$ is the *molecular formula* for glucose, and what you see below is the *structural formula* for glucose. Structural formulas show bonds and the spatial arrangement of atoms.

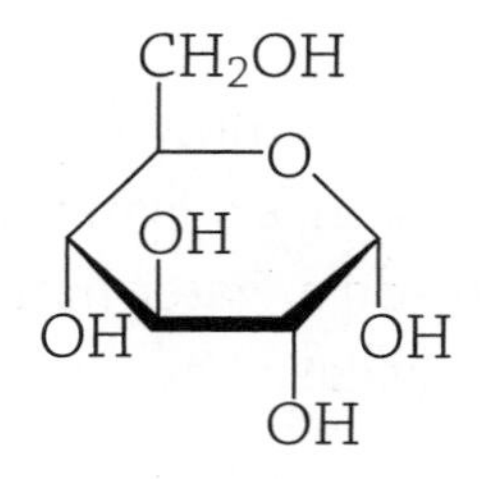

Structural formula of glucose

Glucose is extremely important because it is the fuel for all cells. When we learn about making energy for the cell, you will see that glucose is "burned" to create a molecule called ATP; this process is called cellular respiration, and we'll talk about it more in Chapter 6. For now, this is the overall equation for cellular respiration:

$$C_6H_{12}O_6 + 6O_2 \rightarrow 6CO_2 + 6H_2O + ATP$$

The molecule ATP is the true energy currency of all cells. It's like a battery that powers all life-forms. Unfortunately, ATP isn't very stable and a cell can't stock up on ATP and expect it to stay "charged." But guess what is super stable and can be easily stored? Glucose! The best way for a cell to store energy is to stockpile glucose, and then when it needs the useable form of energy (ATP), it sends the glucose through cellular respiration. Here's another thing to consider—how does a cell efficiently store all that glucose? The answer is by covalently linking a bunch of glucose molecules together into a larger structure that can be tucked away for later use. The covalent link formed between glucose monomers is called a **glycosidic link**. Just like the **must know** concept suggested, the resulting large macromolecule was created by a dehydration reaction between adjacent simple sugar molecules.

If you are a plant, you store a huge amount of glucose as the polymer **starch**; if you are an animal, you store it as the polymer **glycogen**. To summarize, a cell stores the monomer glucose by covalently bonding thousands of them together using a dehydration reaction. The glucose-storage polymer for plants is starch, and for animals it is glycogen. Even though both of these carbohydrate polymers (also called **polysaccharides**) are made up of the same monomer (glucose), they have different structures when you look at the overall shape. Plants store their starch in organelles called amyloplasts; animals store their glycogen in their liver and muscle cells.

BTW

The root word "glyco-" means sugar! Also, the term saccharine refers to something sweet.

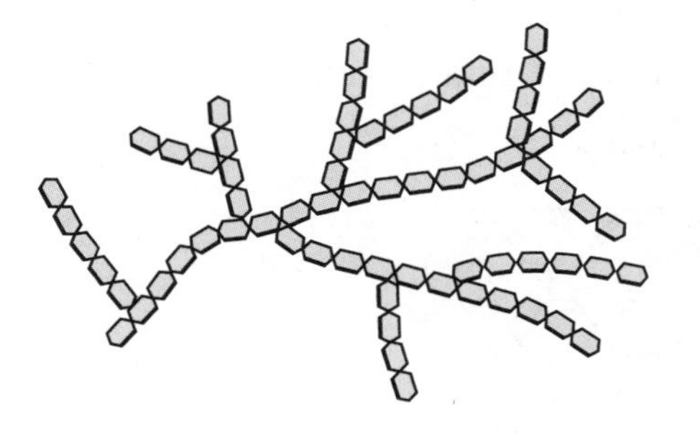

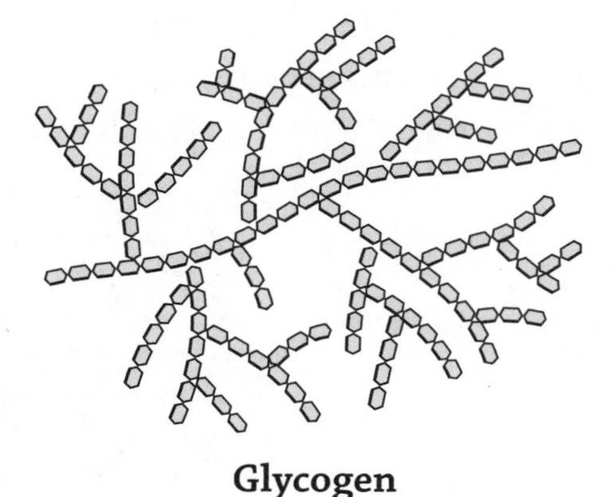

Overall structures of starch and glycogen. Each individual hexagon is a single glucose monomer.

We eat plants and pillage their storage of starch. Referring back to our **must know** concept, recall that a hydrolysis reaction is going to break up a polymer into individual monomers. Animals have an enzyme called **amylase** to hydrolyze the glycosidic bonds between the glucose monomers of both starch and glycogen. One thing to note that we'll talk about later: the specific type of glycosidic bond found in starch is called an *alpha-glycosidic link*. Once the starch polymer is broken down and glucose is freed by our digestive system, the glucose jumps into our circulatory system and is transported throughout the body to be used in cellular respiration to make ATP. But what if after eating (and digesting) a huge starchy meal, we end up with extra glucose? Do we then re-form the polymer starch and store it in our own bodies? Nope. Animals *eat* starch, but we don't *store* starch. Instead we have the enzymes for dehydration reactions to covalently link the glucose monomers into glycogen, our own energy storage polymer! The glycogen can then be hydrolyzed to release its glucose whenever we need a boost of energy. Whether we are referring to plants' starch or animals' glycogen, they both perform the same **must know** function for the cell of energy storage polymer.

Interestingly enough, there is another polymer of glucose called **cellulose**. Unlike starch and glycogen, its function is NOT to store glucose to use for energy. The **must know** function of cellulose is instead for structure and support, and it is found in plant cell walls. Cellulose, unlike starch and glycogen, forms straight chain structures that stack one on top of another and hydrogen-bond together to form a nice, strong layer:

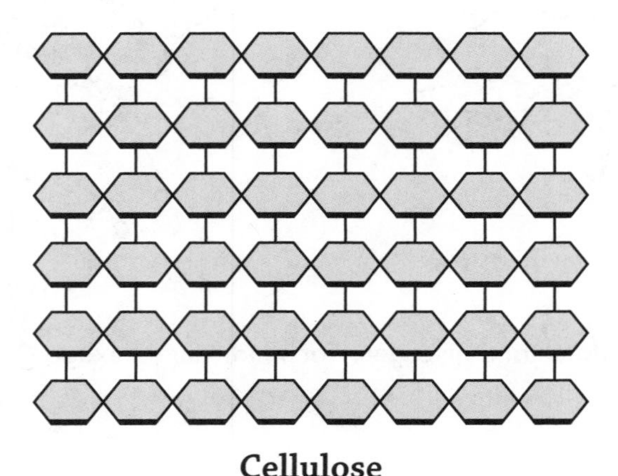

Cellulose

Cellulose. Once again, each small hexagon is a glucose monomer.

Just like starch, the glucose monomers in cellulose are linked together by dehydration reactions. However, the resulting bond is a *beta-glycosidic link* (unlike starch's alpha-glycosidic link). This is significant! The only reason we can't use cellulose as a source of glucose in our diet is because we lack the specific hydrolytic enzymes to break the beta-glycosidic link. Our amylase enzymes (produced in our salivary glands and by our pancreas) target starch's alpha links, but they can't do a thing with those silly beta links. That means we can't digest the cellulose in the plants we eat. That's doesn't mean it's bad . . . on the contrary, dietary cellulose (also called fiber) is important for your health. A high-fiber diet helps regulate your blood glucose levels, makes you feel full, and keeps you "regular" because the fiber travels intact (and quickly) through your digestive system!

There's another carbohydrate polymer that, like cellulose, is important for structure. It is called **chitin** (pronounced KI'-tin) and it is the structural polymer for the exoskeletons of arthropods (insects and crustaceans) and the cell walls of fungi (mushrooms and mold). It is a little weird because it is the only carbohydrate to contain something other than carbon, hydrogen, and oxygen . . . can you see what element is also found in chitin?

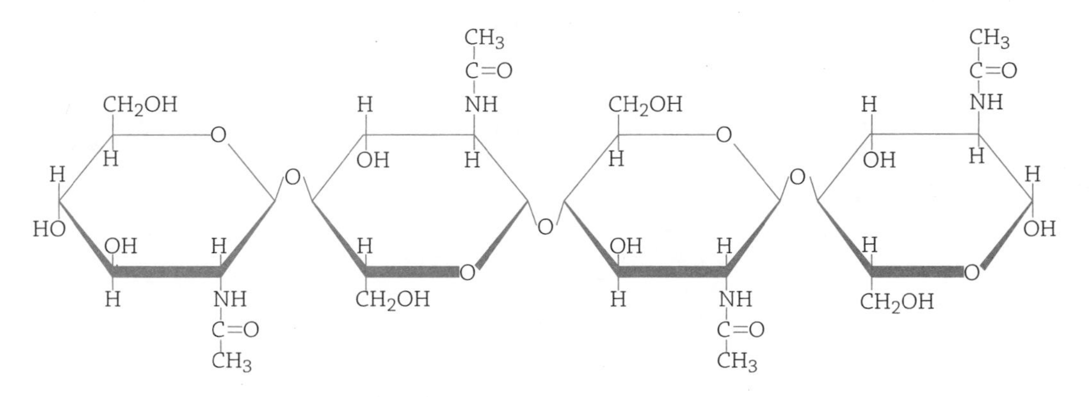

Structural formula for chitin. Notice that there is nitrogen in this carbohydrate (which is not found in any of the others).

Lipids

The next class of macromolecules we'll talk about is the lipids. These are indeed big molecules, but don't expect to see the repeating-monomer-structure of the other large carbohydrates: lipids are *not* polymers. They are, however, big molecules with significant structural components. They are also created using dehydration reactions. The three types of lipids that we'll cover are *steroids, fats,* and *phospholipids*. Generally speaking, all lipids are hydrophobic, meaning they hate water. You may already know that if you've ever poured some olive oil into water—the blob of oil just sits there and doesn't mix well. This is because molecules that contain mostly carbon and hydrogen are nonpolar; nonpolar molecules are hydrophobic.

EASY MISTAKE! Don't refer to this entire class of macromolecules as "fats." Fats are a specific subset of the larger category "lipids."

Steroids

The general structure of a steroid is a four-ringed molecule, like this:

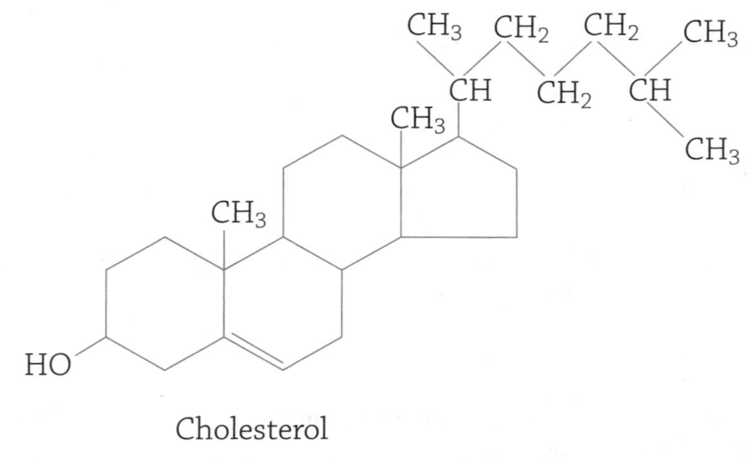

Cholesterol (a type of steroid)

Though cholesterol gets a bad rap because many folks have high levels in their blood (which isn't healthy) and we should avoid foods with high cholesterol, it is also a **must know** concept that cholesterol plays a very important role in the formation of cell membranes.

Fats

In your daily life, if you mention "fats" you are most likely referring to food. Fat is, indeed, the stuff in your diet like butter, cooking oil, lard, and the like. A fat molecule is composed of two components: one glycerol molecule with three fatty acids attached to it:

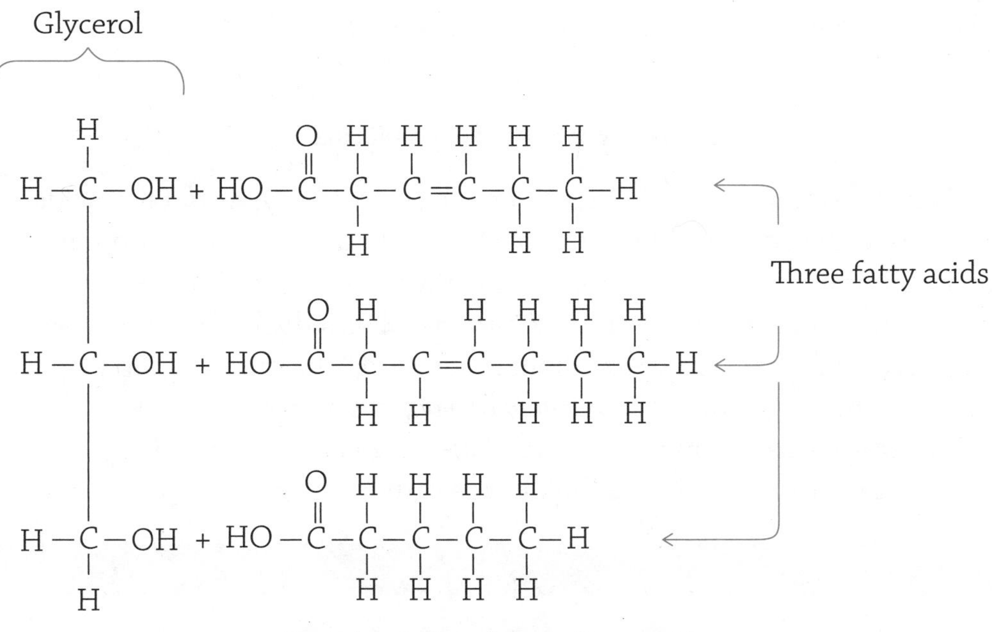

One glycerol and three fatty acids

Each fatty acid molecule is attached to the glycerol through—you guessed it—a dehydration reaction! The resulting bond is called an ester linkage, and the final fat molecule looks like this:

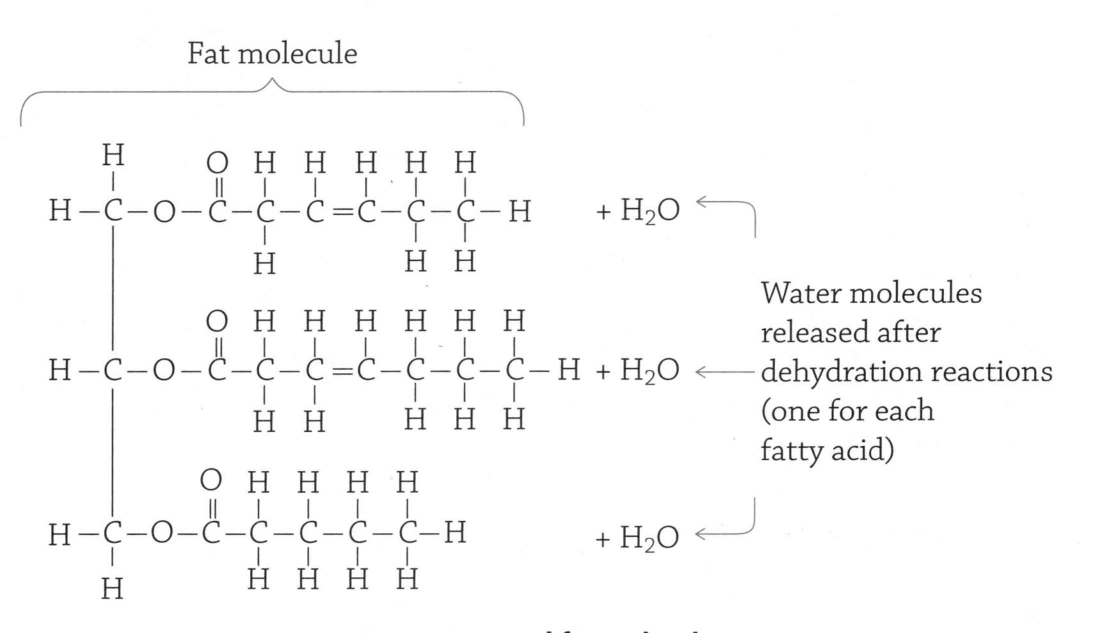

An unsaturated fat molecule

There are different kinds of fats, depending on the types of fatty acids attached to the glycerol. More than 20 kinds of fatty acids are found in foods, but it is not important for you to know them. You do, however, need to pay attention to whether or not a fatty acid has a double bond between two carbons. If there is a double bond, that is an "unsaturated" fatty acid. If there are only single bonds between carbons, it is a "saturated" fatty acid.

The reason this is important is because the double bonds in an unsaturated fatty acid make a kink in the carbon chain.

EXTRA HELP

For my students, it helps if I explain why, exactly, a fatty acid is described using the terms saturated and unsaturated. It has to do with the fact that carbons like to form four covalent bonds. If a carbon in the fatty acid chain has two hydrogens stuck on to it, that's as many as it could possibly hold (the last carbon in the chain doesn't count). Therefore, it is "saturated" with hydrogens. If, instead, there's a carbon-carbon double bond, each of those carbons can only afford to have one hydrogen attached (because each carbon can only have four total covalent bonds). That is why it is referred to as "unsaturated" . . . it isn't saturated with the maximum number of hydrogens!

A weird-shaped fatty acid means that the fat molecules cannot pack tightly together. There are three main states of matter, right? Solid, liquid, and gas. If molecules of a fat can't pack super-close, it is going to be a liquid at room temperature, such as olive oil and canola oil.

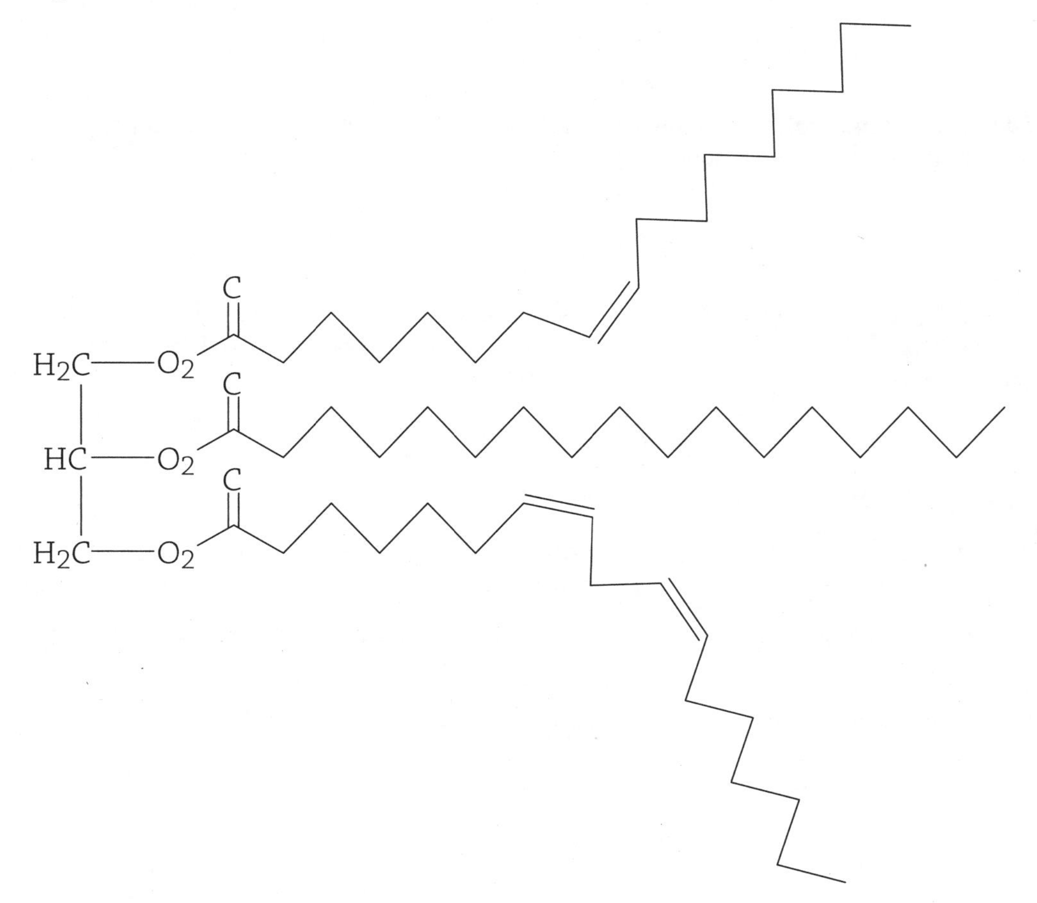

Unsaturated fat molecule. The double bond within the fatty acid tails make an irregular "kink" in the chain.

If, instead, each of the three fatty acids attached to the glycerol are saturated, there are only single bonds and the chains lie relatively flat. These saturated fats can pack tightly together, creating a solid at room temperature—butter and lard, for example.

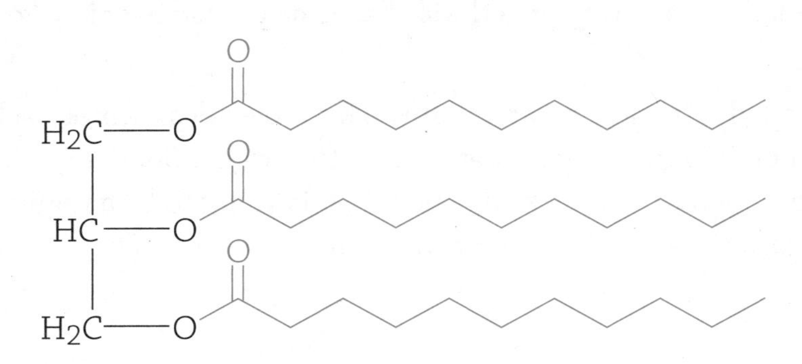

Saturated fat molecule. Since there are no double bonds within the fatty acid tails, they are straight and can pack tightly together.

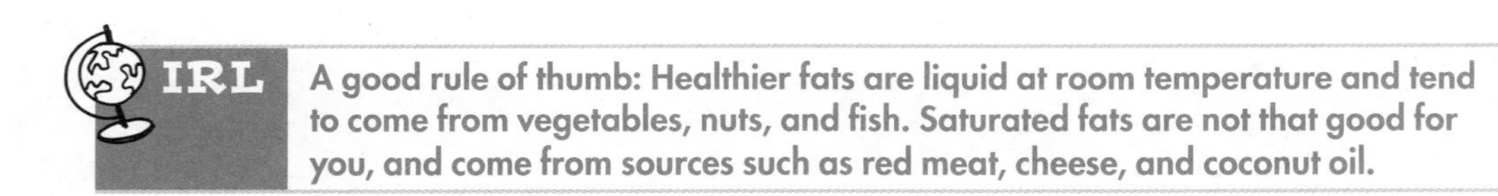

A good rule of thumb: Healthier fats are liquid at room temperature and tend to come from vegetables, nuts, and fish. Saturated fats are not that good for you, and come from sources such as red meat, cheese, and coconut oil.

Phospholipids

Now we will discuss possibly the most important lipid in the subject of biology. Indeed, if it wasn't for phospholipids, life would not exist! The basic building block of life is the cell, and a cell is made of a cell membrane that separates the surrounding environment from the inside of the cell. The major component of all cell membranes is the phospholipid:

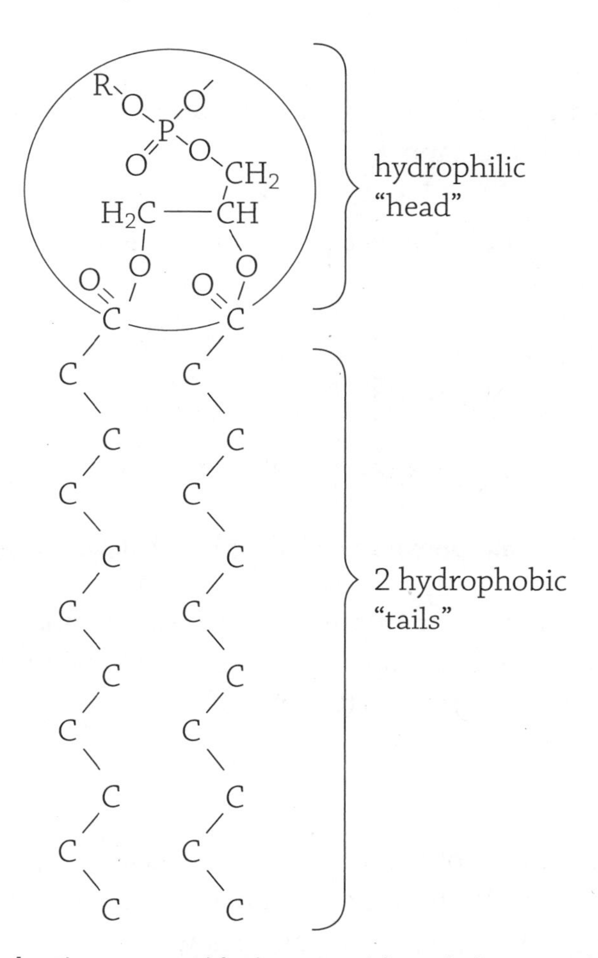

Phospholipid molecule. This is a simplified structural formula because the hydrogens on the two fatty acid tails have been omitted.

Notice that a phospholipid looks very similar to a fat: there is a glycerol backbone onto which two fatty acids are attached by a dehydration reaction. The major difference, however, is instead of a third fatty acid, there is a *phosphate group*. This is super significant! Look closely and you will notice that the phosphate group is charged. Any sort of positive or negative charge makes that part of the molecule polar, and polar things love water. The two nonpolar fatty acid chains remain hateful of water. So, unlike the fat molecule that is entirely hydrophobic, this molecule is half hydrophobic (the long carbon tails) and half hydrophilic (the phosphate group). The dual-nature of phospholipids is referred to as **amphipathic**. Something cool happens when a ton of these amphipathic molecules are dumped into water—they spontaneously arrange themselves in such a way that both parts of the molecule are happy!

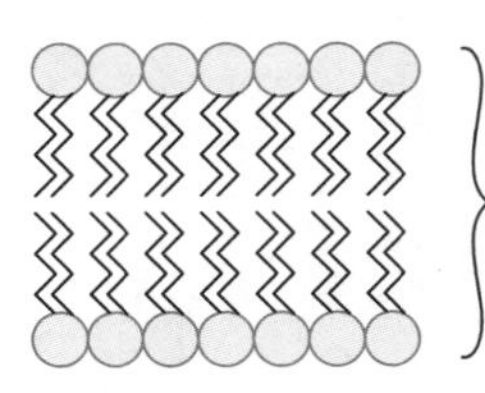

Cell membrane; also called a phospholipid bilayer

This allowed the formation of cell membranes, the **must know** major and important function of phospholipids. You will learn much more about phospholipid bilayers in Part Two (Cells).

Proteins

Proteins are important. Not too surprising, considering the genes in your DNA code for proteins. Proteins perform many **must know** functions for the cell, including the examples provided below:

Protein's Function	Example
Enzymes	Catalase is an enzyme that breaks hydrogen peroxide into water and oxygen gas.
Hormones	Insulin is a hormone that regulates blood sugar concentration.
Immune system	Antibodies are tiny proteins that stick onto viruses and bacteria and mark them for destruction.
Transport proteins	Facilitated diffusion and active transport rely on these transport proteins to provide passage through the cell membrane.
Structure	Hair, fingernails, and feathers are made of the protein keratin. The protein collagen makes up connective tissues and bones.

Proteins (also called polypeptides) are polymers made of amino acids (the monomer). Each amino acid is attached to a growing chain of protein by the lovely dehydration reaction we keep referring to. The resulting covalent bond between amino acid monomers is called a *peptide bond*:

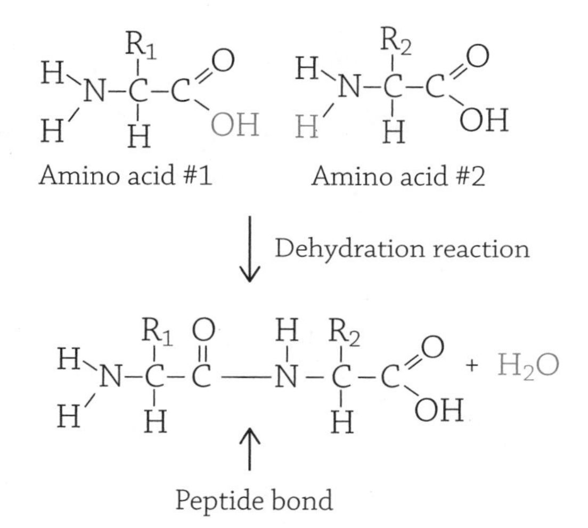

Two amino acid monomers being linked through a dehydration reaction

If you look at the image above, you can see the basic structure of an amino acid. The "R" means the "rest" of the molecule, which can be one of 20 different things (there are 20 different amino acids). Each amino acid's R group determines its chemical properties, such as polar, nonpolar, or charged. This matters because the properties of the amino acid have an impact on the protein's overall function.

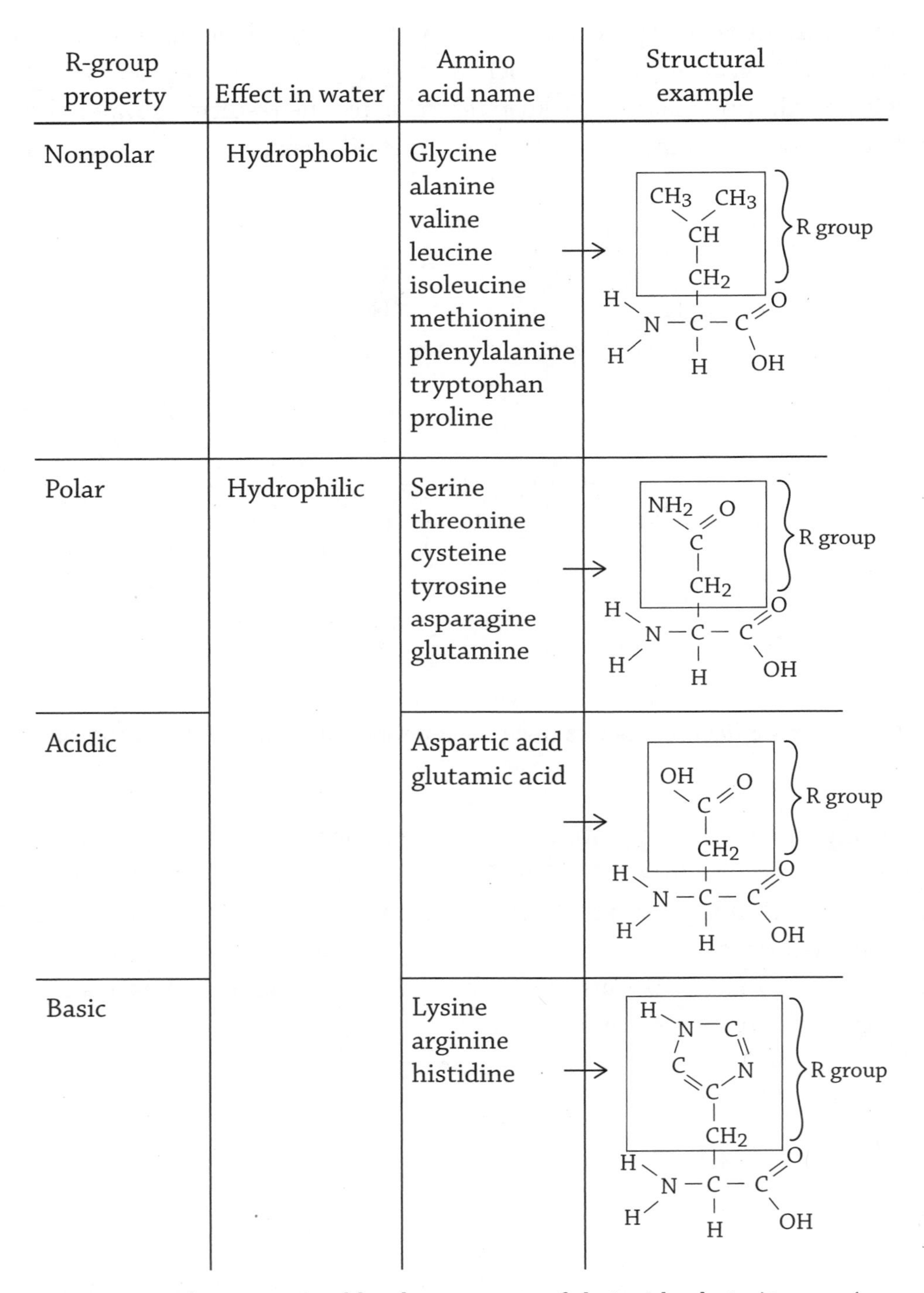

R-group property	Effect in water	Amino acid name	Structural example
Nonpolar	Hydrophobic	Glycine alanine valine leucine isoleucine methionine phenylalanine tryptophan proline	R group: CH_3 CH_3 / CH / CH_2; H, H, N – C – C (=O, OH), H
Polar	Hydrophilic	Serine threonine cysteine tyrosine asparagine glutamine	R group: NH_2 O / C / CH_2; H, H, N – C – C (=O, OH), H
Acidic		Aspartic acid glutamic acid	R group: OH O / C / CH_2; H, H, N – C – C (=O, OH), H
Basic		Lysine arginine histidine	R group: H, N – C, C, N, C / CH_2; H, H, N – C – C (=O, OH), H

Amino acids categorized by the property of their side chain (R group)

The sequence of amino acids that makes up a protein is the protein's **primary structure**. The human muscle protein titin, for example, holds the crown for longest primary structure: titin is composed of a whopping 30,000 amino acids (give or take a few thousand).

A long protein doesn't remain a boring straight chain of amino acids, however—instead, the repeating amino acid backbones (not the R groups) hydrogen-bond to one another in a repeated, regular fashion. A protein's **secondary structure** can be, for example, a helix:

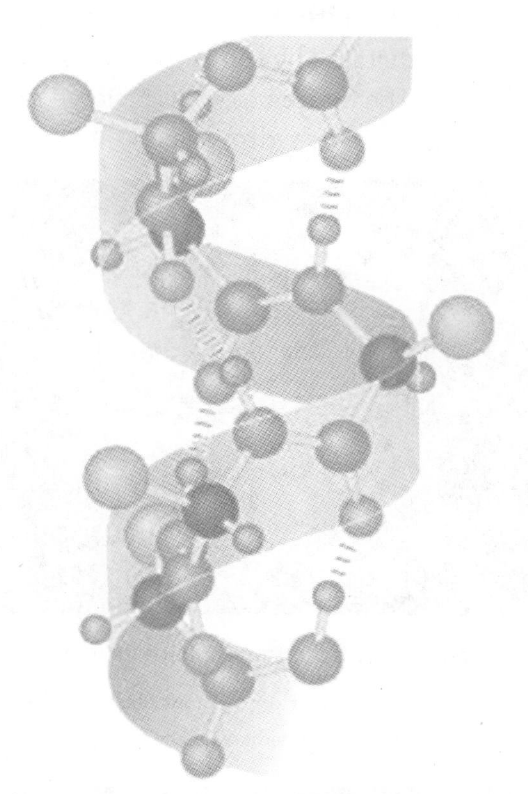

Protein secondary structure (α helix). The dashed lines indicate hydrogen bonds occurring between an oxygen atom and a hydrogen hanging off of a nitrogen atom.

Author: National Institutes of Health. https://commons.wikimedia.org/wiki/File:AlphaHelixProtein.jpg

It gets interesting at the **tertiary structure** level. Here, those individual amino acid R groups really start to matter. The protein will fold into a seemingly random blob dictated by interactions between R groups. For example, if there's a number of nonpolar amino acids in the protein, they will cluster together at the core of the protein to try and "hide" from water that surrounds the protein. The 3D structure of a protein is extremely important to its function—just consider an enzyme. Enzymes are proteins that catalyze (speed up) reactions by grabbing onto other molecules (the enzyme's **substrate**). An enzyme has a specific location called an **active site** that perfectly fits its substrate like a key fits a lock, and this shape is dictated by the amino acids that are part of the protein's primary structure. The figure below is the large, blobby, tertiary shape of the bacterial enzyme isocitrate dehydrogenase:

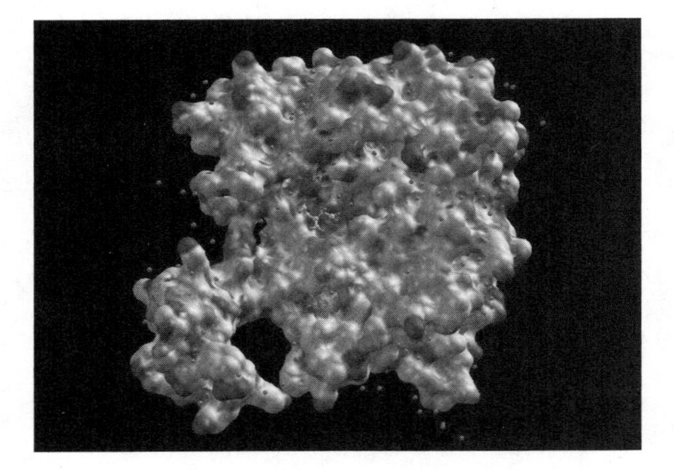

Isocitrate dehydrogenase from *E. coli.* Each small sphere indicates an individual amino acid.

Author: Haynathart. https://commons.wikimedia.org/w/index.php?search=enzyme+structure&title=Special%3ASearch&profile=default&fulltext=1#/media/File:Ecoli_IDH_with_surface_pocket.jpg

EXTRA HELP

If enzymes becomes too hot they often stop working because they unravel and lose their substrate-specific 3-dimensional shape. This unraveling is referred to as **denaturation**.

My favorite example for the levels (and significance) of protein structure is the protein hemoglobin, responsible for carrying oxygen in red blood cells. Hemoglobin forms a **quaternary structure**, meaning there are two or more *individual proteins* that come together to form the functional unit. Hemoglobin consists of two alpha subunits and two beta subunits, and they fit together in happy functional harmony:

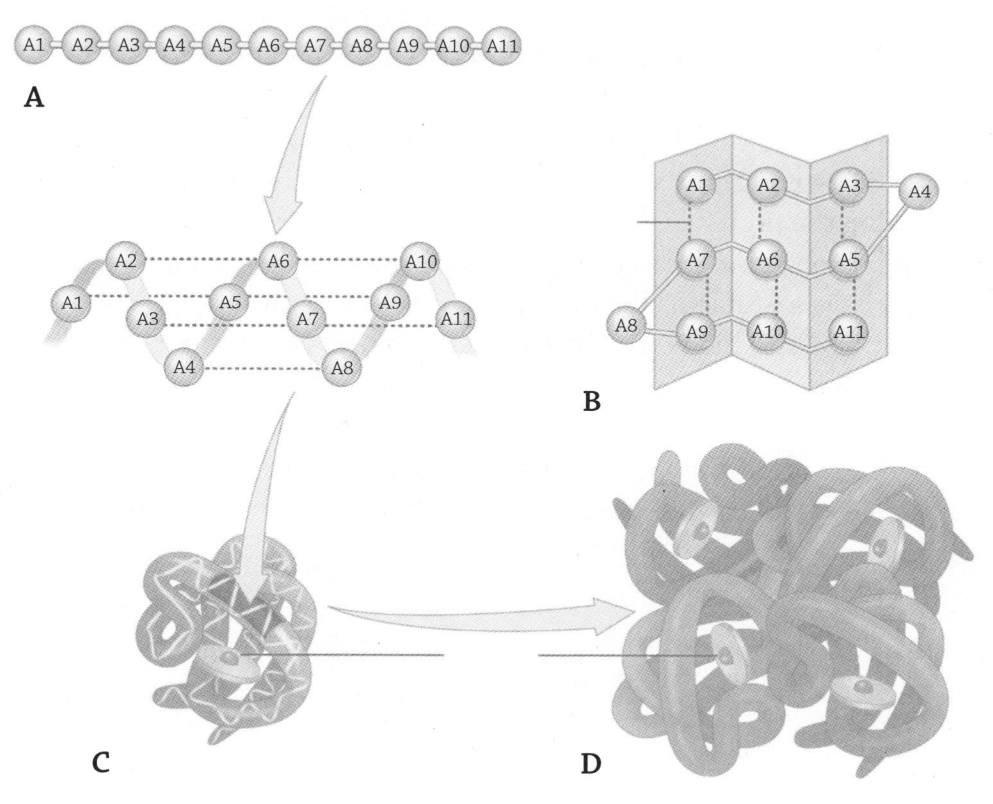

The four levels of hemoglobin structure. The primary level (A) indicates the sequence of amino acids that make up the protein, which then folds into a regulated, repeated folding pattern (second structure, B). The protein forms a particular 3D shape (tertiary structure, C), based on the amino acids' R groups. Finally, four individual proteins come together to create hemoglobin's large quaternary structure (D).

Author: OpenStax. https://commons.wikimedia.org/w/index.php?search=hemoglobin+quaternary+structure&title=Special:Search&profile=default&fulltext=1#/media/File:225_Peptide_Bond-01_labeled.jpg

To really stress the importance of primary structure, consider the seemingly insignificant difference between a healthy hemoglobin molecule and one that leads to the disease sickle cell anemia. A mutation occurs that effects only *one* amino acid: a glutamic acid is changed into a valine. Think back to those different categories of amino acids: valine is hydrophobic, whereas the original glutamic acid is hydrophilic. This minor change in the primary sequence doesn't have any effect on the secondary structure, because the regular folding doesn't have a thing to do with the different side chains. But now in the tertiary structure, there's a new hydrophobic region that's uncomfortably exposed to the water surrounding the protein. Get a bunch of these mutated and misfolded hemoglobin molecules into a red blood cell, and then you really have a problem. The hemoglobin molecules clump together in order to hide those hydrophobic bits, and the protein ends up crystallizing into a fiber that not only cannot carry oxygen very well, but it also causes the red blood cells to become deformed and sickle-shaped. These unfortunately shaped little cells clog tiny arteries and cause big circulation problems.

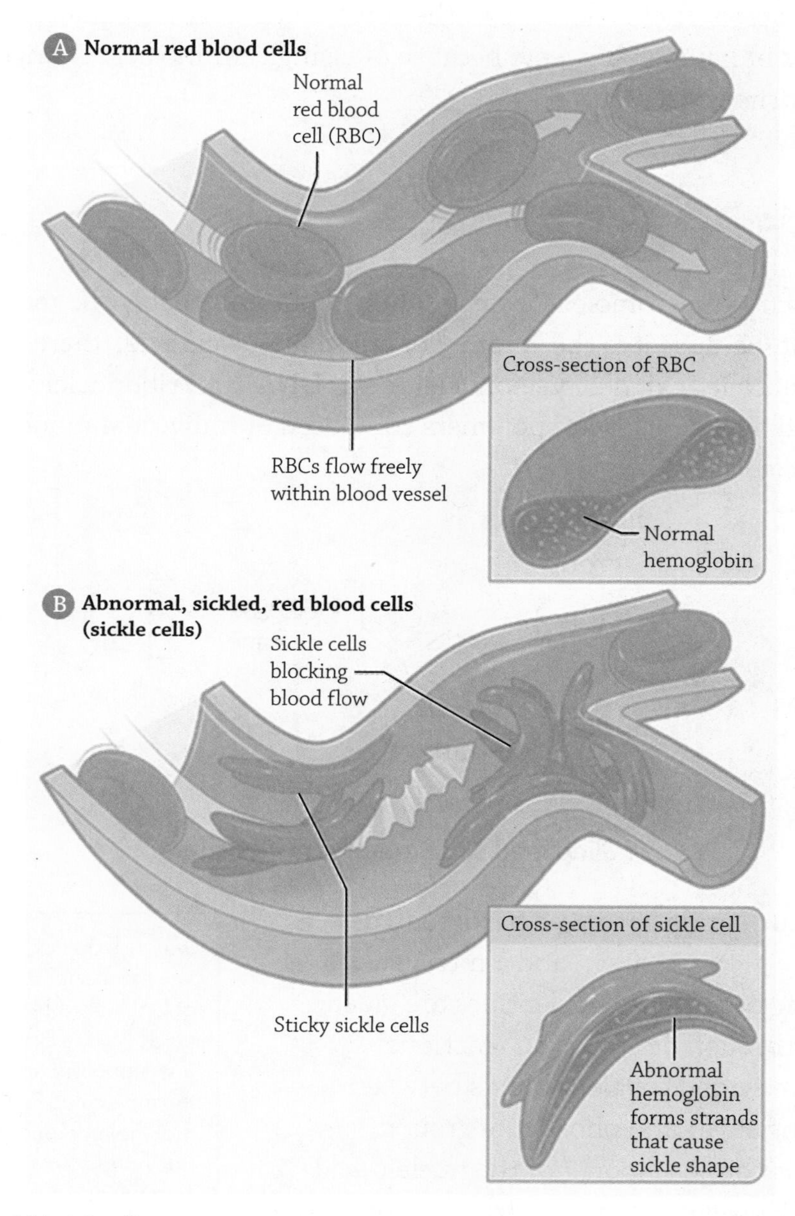

Sickled, red blood cells

Author: The National Heart, Lung, and Blood Institute (NHLBI). https://commons.wikimedia.org/w/index.php?search=hemoglobin+sickle+cell&title=Special:Search&profile=default&fulltext=1#/media/File:Sickle_cell_01.jpg

All of these structural issues in the complex tertiary and quaternary structures of hemoglobin were because of a single amino acid change in the protein's primary structure!

Nucleic Acids

The final large macromolecule we will talk about is arguably the most important, because it is the "brains" of any cell: nucleic acid. There are two kinds of nucleic acid: deoxyribonucleic acid (DNA) and ribonucleic acid (RNA). All nucleic acids are polymers composed of individual monomers called nucleotides:

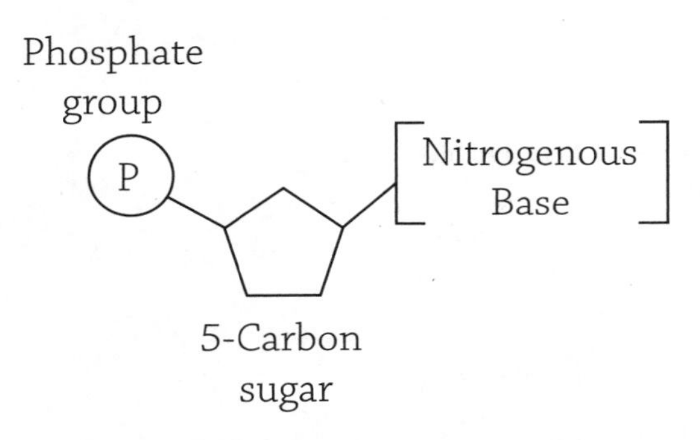

Simplified structure of a nucleotide

Each nucleotide has three parts to it: a phosphate group, a sugar, and a nitrogen-containing base. These nucleotides are glued together through dehydration reactions that create strong covalent bonds between alternating sugar and phosphate groups. This forms the "backbone" of the nucleic acid polymer. However, the part of the monomer that matters the most is the nitrogenous (nitrogen-containing) base. DNA has four different kinds of bases: adenine, thymine, guanine, and cytosine. RNA has three of the same,

If asked, do not say the monomer of DNA or RNA is a "nucleic acid" (you're accidentally just repeating the name of the molecule). Remember that the monomer is called a *nucleotide*. Furthermore, don't make the mistake of saying the monomer of DNA or RNA is a nitrogenous base; that is just one component of the monomer (along with the 5-carbon sugar and the phosphate group).

except instead of thymine it has uracil. All of these bases fit into one of two categories: purines or pyrimidines. Purines are double-ringed structures and include adenine and guanine, whereas pyrimidines are single-ringed structures and include thymine, cytosine, and uracil. You will learn more detail about this later, when we study genetics.

There are different types of RNA and they perform different tasks for the cell. Messenger RNA (mRNA), for example, carries the directions to make a protein from the nuclear-imprisoned genes to the protein-building ribosomes in the cytoplasm. Some RNA molecules play an important role in regulating gene expression, such as little micro RNAs or small interfering RNAs. Others carry amino acids over to the ribosome for protein synthesis (transfer RNA) and help the ribosome synthesize the final protein product (ribosomal RNA). Each of these interesting RNAs will be discussed later.

In general, all RNA nucleotides contain a sugar called ribose and are single-stranded molecules, unlike their chemical cousin, DNA, a large double-stranded molecule constructed with the sugar deoxyribose. The double-stranded nature of DNA is extremely important, something we will discuss later. For now, notice that the two strands of a DNA molecule are complementary to one another. They are both composed of nucleotides, the upright portion of which is made of alternating sugar-phosphate groups bonded strongly together by covalent bonds. The nitrogenous base portion points toward the middle of the two strands, and is the source of the hydrogen bonding that holds the two strands together. If one strand has an adenine nucleotide, it must pair with a thymine nucleotide on the complementary strand; a guanine nucleotide pairs with a cytosine nucleotide. Furthermore, these two strands are called **antiparallel** because they run in opposite directions. The end with the free phosphate group is referred to as the 5′ (five-prime) end, and the free hydroxyl (—OH) group is the 3′ (three-prime) end. These numbers indicate which carbon of the sugar the phosphate or hydroxyl group is hanging off of:

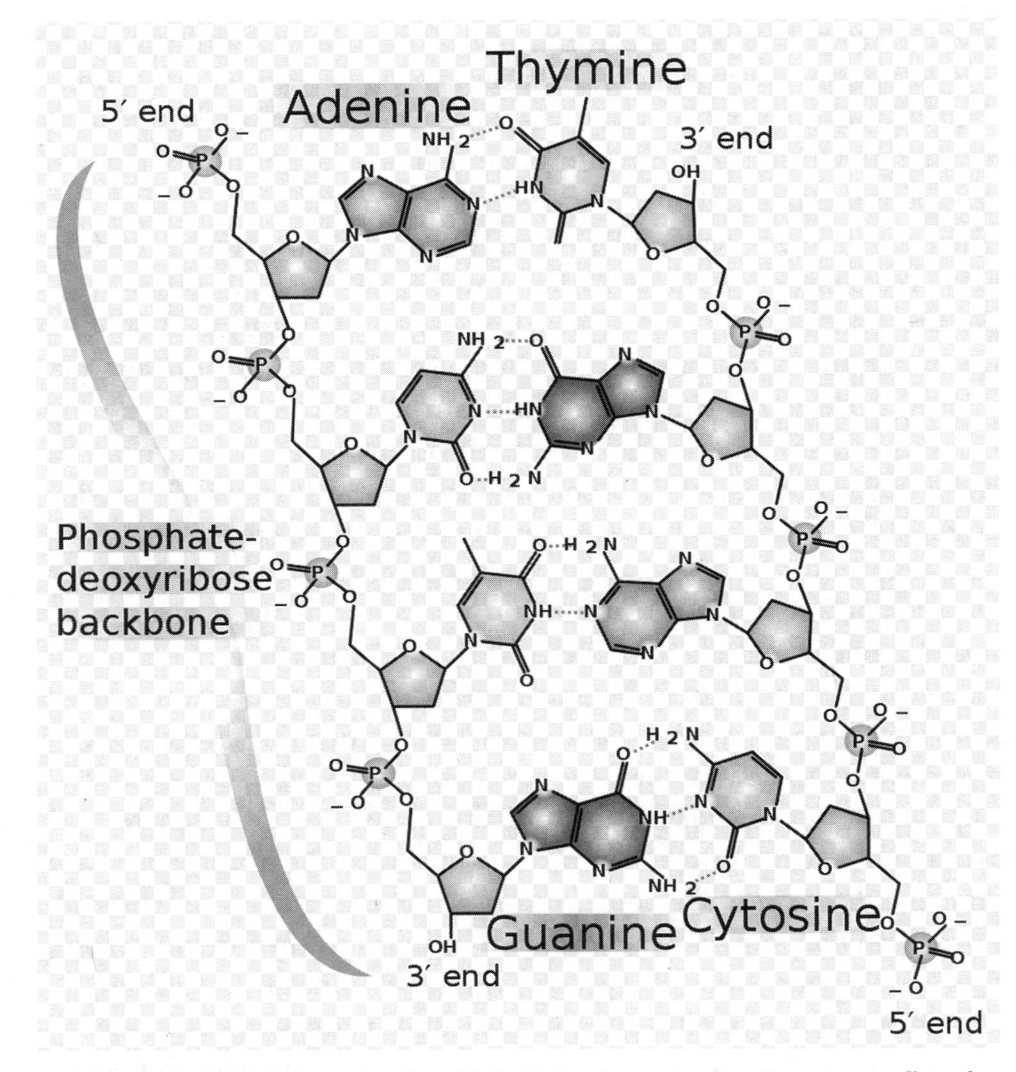

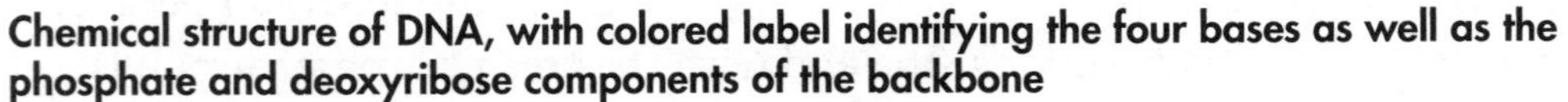

Chemical structure of DNA, with colored label identifying the four bases as well as the phosphate and deoxyribose components of the backbone

Author: Madeleine Price Ball. https://commons.wikimedia.org/wiki/File:DNA_chemical_structure.svg

All of these macromolecules work together in different ways to create both the structure of the cell and to perform its various functions. As we progress through the book, proteins, nucleic acids, lipids, and carbohydrates will be mentioned time and time again in their various roles and functions.

REVIEW QUESTIONS

1. Fill in the blanks: A polymer is a large ______________ (term for large molecule) that is composed of repeating subunits called ______________.

2. Given these two amino acid monomers, show the resulting peptide bond that would form after a dehydration reaction.

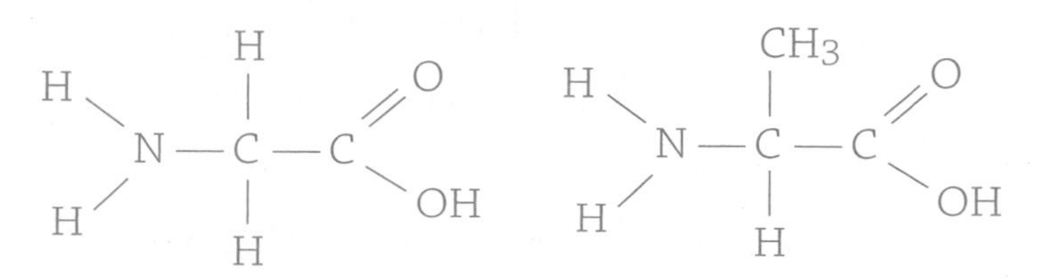

3. Match the macromolecule with its function:

Macromolecule	Function
a. Glycogen	1. Dietary component that is liquid at room temperature
b. Phospholipid	2. Structural polymer in plant cell walls
c. Unsaturated fat	3. Energy storage for animals
d. Antibodies (protein)	4. Dietary component that is solid at room temperature
e. Saturated fat	5. The main component of cell membranes
f. Keratin (protein)	6. Helps your immune system mark invaders for destruction
g. Cellulose	7. Structure component of hair, nails, and feathers

4. List three differences between DNA and RNA.

5. Explain why the base-pairing rules exist. Why, exactly, must adenine base-pair with thymine, and guanine base-pair with cytosine?

6. Match the macromolecule with its function:

Macromolecule	Function
a. Catalase (protein)	1. Energy storage for plants
b. Chitin	2. An enzyme that breaks down hydrogen peroxide
c. Insulin (protein)	3. A type of lipid in your diet
d. Fat	4. Important component of cell membranes
e. Starch	5. Structural polymer in animal exoskeletons
f. Cholesterol	6. A hormone that helps regulate blood sugar levels
g. Transport proteins	7. Help move things through the cell membrane

7. Choose the correct term: If a protein's primary sequence had a bunch of hydrophilic amino acids, you would expect them to **hide from/face toward** water.

8. Choose the correct term: When you digest starch, **dehydration/ hydrolysis** enzymes work at breaking up the starch polymer.

9. Hemoglobin is the protein in red blood cells that carries oxygen throughout the circulatory system. The hemoglobin molecule is made up of four total protein chains (four *subunits*) clumped together. Two of the protein chains (the alpha subunits) are made of 141 amino acids each. The other two proteins (the beta subunits) are made of 146 amino acids each. Based on this information, which of the four levels of protein structure is described?

10. Which of the following statements is correct?
    a. The sugar component of an RNA nucleotide is deoxyribose.
    b. Cellulose is an energy storage polymer.
    c. The monomer of a protein is an amino acid.
    d. The hydrophobic portion of a phospholipid bilayer is composed of phosphate groups.

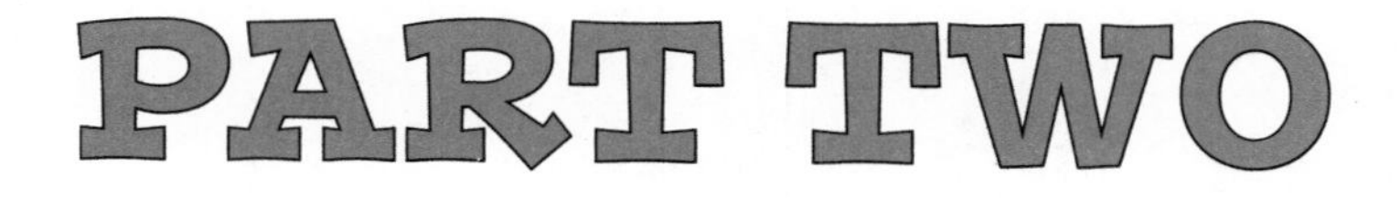

Cells are the building blocks of all life, and nothing below the organizational level of a cell can be qualified as being alive (sorry, viruses, you don't cut it). That doesn't mean a single cell is a simple sort of structure—there's a lot going on in each and every one of our little building blocks!

4 Overview of the Cell

MUST KNOW

- Cells are the building blocks of life.
- The structure of the cell membrane is very important to its function.

When you say something is "alive," what do you really mean? That it has eyes and can see? That it is able to move? That it eats food? The truth is, for something to be alive, it must be made of at least one cell. As our **must know** idea points out, the cell is the building block of life; something smaller than a cell (such as a virus) does not qualify as being "alive." An organism may be composed of a single cell (unicellular), such as a bacterium, an amoeba, or a yeast cell. Just as bricks can be adhered together to form larger structures, so too can cells. A multicellular organism is one that is composed of more than one cell. Multicellularity allows cells to begin to specialize and form specific tissues, each with their own functions. The concept of specialization will be addressed further in Chapter 11 (Gene Expression and Differentiation). By understanding that a single cell is the building block for all types of life, you will be able to understand the processes shared by all living organisms on Earth.

Cell Membranes

All cells have an outer boundary called a cell membrane. This leads us directly into our second **must know** concept: the structure of the membrane is important to its function as a selective permeable barrier (meaning it chooses what is able to pass through). The membrane consists of two layers of a special macromolecule called a **phospholipid**. This "phospholipid bilayer" (get it? It is two [bi-] layers of phospholipids!) is key to the evolution of cells because it is picky about what it lets pass through. The selective permeability of the cell membrane allows the cell to have an environment on the inside that is different from the environment surrounding the cell. The inside of the cell is the cytoplasm, and it contains dissolved substances such as ions, proteins, glucose molecules, and many other chemicals. The cell actively collects substances it wants (food! molecular building materials!) and exports wastes.

Membranes Are Composed of Lipids

The main component of a cell membrane is the phospholipid. Phospholipids are special because they are **amphipathic**, meaning the molecule has both **hydrophilic** (water loving) and **hydrophobic** (water avoiding) parts. If you look closely, you will notice that the phosphate group has some charges associated with it, yet those two long hydrocarbon tails are very nonpolar:

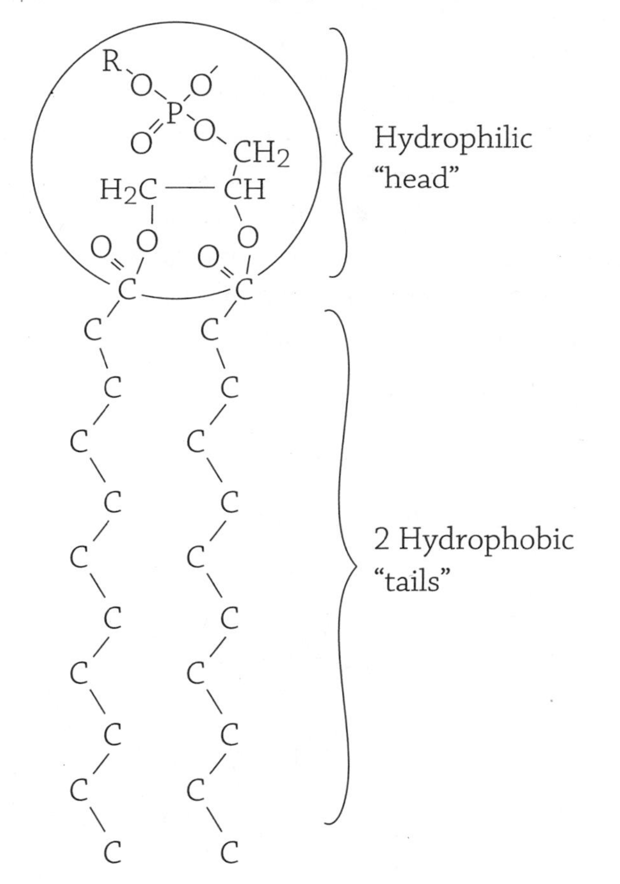

A single phospholipid molecule. Please note that "R" means the "rest" of the molecule (but for our purposes, it isn't significant). The structure of the "tails" is also simplified by omitting the hydrogens.

Charged areas are polar, and since water is also polar, they "mix" well. The long carbon-carbon tails are very nonpolar, so they do whatever they can to hide from water.

BTW

There's a saying: "Like dissolves like." This means that polar molecules dissolve in polar solvents (such as water), whereas nonpolar molecules dissolve in nonpolar solvents (such as oil). If you mix a polar and a nonpolar substance, they will not mix well at all! If you've ever poured oil into water, you've seen this firsthand.

A cell's membrane is happy to be composed of two layers of phospholipids. Look at the image below and notice how the polar phosphate "head" groups face outward toward the water surrounding the cell and the water inside the cell. The hydrophobic "tails," meanwhile, are tucked into the middle of the membrane, away from any chance of touching anything aqueous. This creates the semipermeable barrier around the cell. There are also embedded proteins that play an important role in transport across the cell membrane. Specifically, they help in either active transport (serving as tiny little pumps) or facilitated diffusion (tunnels through which molecules will passively flow). You'll learn more about this in the cell transport chapter.

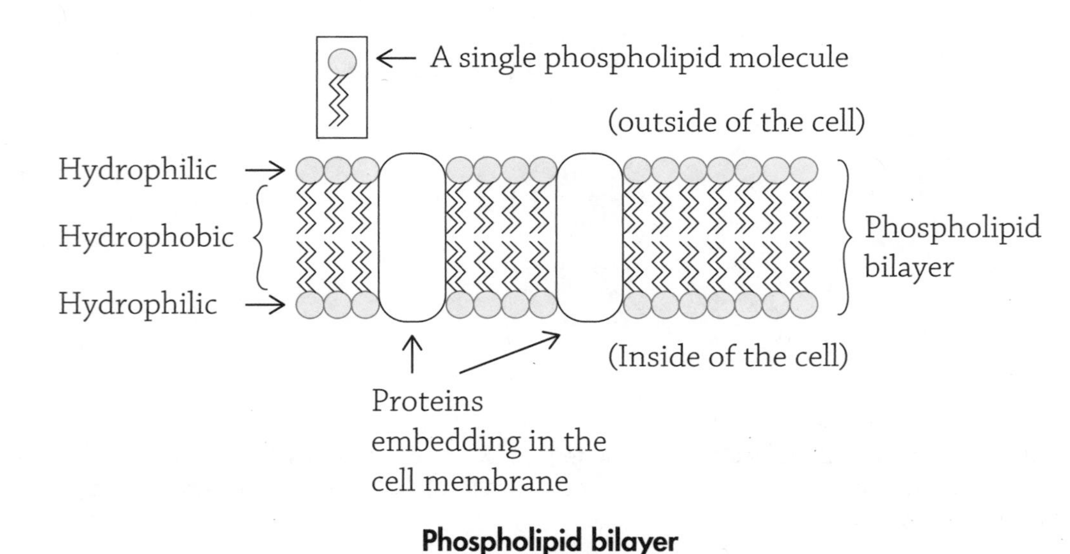

Phospholipid bilayer

Consider the adage "like dissolves like," and you will see why phospholipids behave as they do in aqueous (water) solutions. As we learned in Chapter 1 (Chemical Bonds and Reactions), phospholipids are amphipathic in nature: the phosphate head groups are hydrophilic (love water), and the two fatty-acid tails are hydrophobic (hate water). That is why a cell membrane is also called a phospholipid bilayer: it is composed of two layers of phospholipids arranged so the polar head groups face the water and the tails are tucked in the middle of the membrane, safely away from water.

Generally speaking, what form of matter (solid, liquid, or gas) usually comes to mind if you think of really cold temperatures? Solid! In cold conditions, molecules tend to slow down and get closer together (generally speaking). When the temperature increases, so does the speed and distance between molecules, meaning a substance becomes a liquid. If you are a cell, and your very existence is dependent upon that cell membrane, neither of those options (solid as a rock or melting into liquid) sounds particularly pleasant.

Cells need to maintain their cell membrane at a nice, even, not-too-solid-not-too-loose level. The temperature at which a membrane solidifies is in part dependent on what types of phospholipids make up the membrane. Do you recall from the chemistry unit the differences between saturated and unsaturated fats? Saturated fats are solid at room temperature (butter), and unsaturated fats are liquid at room temperature (oils). Even though the type of lipid we're currently talking about isn't a fat, the saturated-versus-unsaturated thing still matters in phospholipids.

If a cell membrane is made up of phospholipids that are composed of a fair amount of unsaturated fatty acids, that will increase the membrane's fluidity. As before, any double bonds in the unsaturated fatty acid tails will create a kink in the phospholipid, and they won't be able to pack very tightly together, making it more fluid. On the flip side, if the fatty acids in the phospholipids are mostly saturated (without any weird bends and kinks), they will be able to pack very tightly together and the membrane will be more rigid. This relates directly to our **must know** concept that a cell membrane's structure has a huge impact on its function (maintaining a nice, even fluidity).

Cell membrane with only saturated phospholipids (rigid)

Cell membrane with some unsaturated phospholipids (fluid)

Author: MDougM. https://commons.wikimedia.org/wiki/File:Lipid_unsaturation_effect.svg

This is an important concept to consider if you are a cell that lives in very hot or very cold conditions. Archaebacteria are prokaryotic cells that inhabit some of Earth's most extreme habitats. If a species of archaebacteria was living in a hot spring, it would endure temperatures near the boiling point. Clearly, this high of temperature would make the cell membrane very loose and flexible. To counteract this increase in fluidity, the cell membrane would be composed of mostly saturated phospholipids (to help stiffen it up a bit).

On the contrary, there are some species of fish that live in icy marine waters. Such brutally cold freezing temperatures make the fish's cell membranes too stiff; they have evolved to have a higher concentration of unsaturated phospholipids to counteract the stiffness and help increase flexibility.

Finally, there are some clever species of fungi and bacteria that can actively increase the percentage of unsaturated fatty acids in their phospholipid bilayer when there is a decrease in environmental temperature.

Membranes Are Also Composed of Proteins

Proteins play an important role in cells, and different types of cells have different types of membrane proteins. Generally speaking, a membrane protein can either span the phospholipid bilayer or sit on its surface. A **transmembrane protein** (also called an **integral protein**) extends through the bilayer and is often used in cell transport. Some proteins—**peripheral proteins**—instead sit on the surface of the membrane and play roles in things other than transport. Here are three examples, all of which we will talk about in more detail!

- **Transport** A transmembrane protein that extends across the phospholipid bilayer can act like a tunnel. It creates a hydrophilic channel for things to pass through that would otherwise be blocked (or slowed down) by the phospholipid bilayer. For example, aquaporin proteins are membrane channels specifically to help water pass. Each aquaporin channel allows *3 billion* water molecules to pass through, single file, *every second*! Some proteins take a more active approach and change shape to forcibly shuttle substances from one side to another. A protein pump that uses ATP in order to actively transport materials is a perfect example.

- **Enzymes** Membrane-associated enzymes can either span the entire membrane or reside on only one side (either inside the cell or outside the cell). Enzymes that are stuck onto the membrane can form a cooperative little group that catalyzes a series of reactions in a metabolic pathway.

- **Signal Transduction** When a chemical signal has a message for a particular cell, the cell "hears" the message because there are membrane proteins (receptors) that have a binding site for that particular signal. The receptor protein changes shape (while still remaining in the membrane), and essentially moves the message into the cell. The entire process of signal transduction will be covered in Chapter 8 (Signal Transduction and Cell Communication).

Cell Types

Cells can be divided into two large categories: **prokaryotic cells** and **eukaryotic cells**. Prokaryotic cells (bacteria) were the first cell type to evolve on Earth 3.5 billion years ago. They are simpler in structure and smaller in size. Prokaryotic cells lack **organelles**, which are specialized internal structures that perform specific functions (we will learn a lot more about organelles a bit later). Though all cells must contain DNA, prokaryotic cells do not house their DNA within a membrane-bound nucleus. Instead, bacterial DNA is condensed within a specific region called a **nucleoid**. Prokaryotic cells must undergo metabolic processes and create energy, so they have many enzymes needed to "run" their cellular reactions. Enzymes are made of proteins, and proteins are made by the bacterial **ribosomes** floating around inside. Some bacteria are motile (meaning they can move) and do so by using whip-like structures called **flagella** or smaller, hair-like projections called **cilia**. And, of course, a bacterial cell is surrounded by a cell membrane, plus an extra outside layer called a **cell wall**.

EXTRA HELP

Even though the prokaryotic cell has its DNA concentrated in a specific nucleoid region, it is still not considered an organelle because it is not a membrane-bound enclosure. It also can be confusing to see that organelle-free prokaryotic cells contain ribosomes. Ribosomes, however, are not classified as "organelles" because they are themselves not made of membranes.

Though prokaryotic cells are simple cells, that does not mean they are at a disadvantage. On the contrary, their simplicity allows them to quickly reproduce. Furthermore, quick reproduction can introduce genetic mistakes (mutations), which create variation in a population. If you combine those two qualities—a quickly growing population of cells, many of which have newly introduced genetic mutations—you create perfect conditions for evolution to occur! This has enabled bacteria to colonize an amazing array of different environments on Earth. Some prokaryotes called **archaea** (for "archaic," because they are closely related to the first cells to evolve on Earth 3.5 billion years ago) prefer what we consider "extreme" environments, such as deep-sea hydrothermal vents (where temperatures can exceed the boiling point!) and the Dead Sea (where nothing else can grow because of the high levels of salt). Healthy soil is rife with bacteria, and a single teaspoon can contain anywhere from 100 million to *one billion* bacteria! Bacteria are key in the food industry, such as fermentation of yogurt and cheese, and most importantly, bacteria keep us healthy. Bacteria provide us with antibiotics, and by colonizing our skin and digestive tract, good bacteria fill the ecological niches that would otherwise be taken over by pathogenic (disease-causing) microbes.

One major downside to their super-fast evolution, however, is their penchant to become resistant to the antibiotics that are supposed to stop their growth. You have most likely heard about the evolution of antibiotic-resistant bacteria. This occurs because a random mutation will enable a bacterial cell to survive in the presence of an antibiotic chemical that would otherwise kill it; this lone, lucky survivor quickly reproduces and creates offspring, all with that same, lucky, antibiotic-busting mutation. You can read more about the process of evolution in Chapter 15.

If prokaryotic cells are the small and simple types, eukaryotic cells are their larger and more complex cousins. The animals, fungi, protists, and plants are all composed of eukaryotic cells, and we'll learn a ton more about eukaryotic cell structure in the next chapter.

Though it may seem straightforward to use the presence of cells as the defining characteristic of "life," there is more to it. A living organism must also be able to reproduce and create progeny, and it must have a heritable genetic code (DNA). A living organism uses materials and undergoes metabolic processes, and it requires energy to survive. A living entity strives to maintain constant internal conditions (referred to as **homeostasis**) and responds to its environment. And finally, all living things evolve. To be alive is not only to be made of cells; "life" is an entire process.

What, then, about viruses? These little guys are composed of only nucleic acid (sometimes DNA, sometimes RNA) surrounded by a protein coat. There is no self-derived phospholipid bilayer, no cytoplasm, no organelles. Yet a virus appears to be an exception to the rule that to be considered alive, you must be composed of at least one cell. If you have ever come down with a bad cold, the seasonal flu, or something more exotic such as measles, you have definitely felt the wrath of those tiny little viral particles populating your body. A rhinovirus (the common cold) is inhaled, invades specific cells lining the respiratory tract, and once inside the cell, they replicate themselves to such high numbers the cell essentially explodes. The newly released viruses then continue on commandeering adjacent cells until the immune system catches wind of the assault and shuts down the invasion. The virus does, indeed, reproduce and create more progeny viruses, but it cannot do so without first invading and taking over a cell—that important building block of all life.

REVIEW QUESTIONS

1. The main structural component of the cell membrane is the [name the molecule] ______________.

2. How does archaea's preference to live in extreme environments relate to its ancestry?

3. Label the following diagram (A–E) of a cell membrane with the following terms: **hydrophilic phosphate group, phospholipid bilayer, protein, phospholipid, hydrophobic tails**.

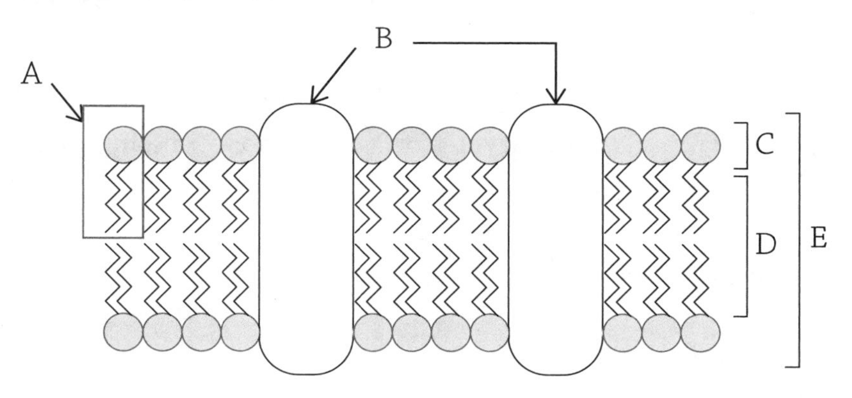

4. Why would transmembrane proteins (also called integral proteins) be useful in transport of things into and out of the cell?

5. Some cells have the ability to change the percentage of saturated versus unsaturated phospholipids in their cell membrane. Assume there is a bacterium whose membrane was composed of 50% saturated and 50% unsaturated phospholipids. If this cell suddenly was exposed to much colder temperatures, what might happen to the percentage composition of its cell membrane in order to try and maintain a "normal" fluidity?

6. Which cell type is smaller and lacks organelles?

7. Life is defined by a number of characteristics: list the eight characteristics of life covered in the chapter.

8. Choose the right term from each of the following pairs: A single phospholipid is *amphipathic*, meaning the phosphate head group portion of the molecule is **polar/nonpolar** and **loves/hates** water, whereas the two fatty acid chains are **polar/nonpolar** and **love/hate** water.

9. Choose the correct term: If a cell lived in very hot conditions (such as a volcanic deep-sea vent), its phospholipids would be composed of more **unsaturated/saturated** fatty acids in order to prevent the cell membrane from becoming too fluid.

10. List three possible roles for cell membrane proteins, and provide a brief description of each.

11. For each of the following, indicate whether it is found in a bacterial cell. Write Y if it is found in prokaryotes or N if it is not found in prokaryotes:
    a. Cell membrane
    b. Nucleus
    c. Flagella
    d. DNA
    e. Cell wall
    f. Ribosomes
    g. Mitochondrion (an organelle)
    h. Nucleoid

12. What is an advantage to prokaryotic cells' simplicity?

13. For the following multiple-choice question, choose the best answer. Which of the following is the smallest thing that could be considered alive?
    a. An atom
    b. A molecule
    c. A virus
    d. An amoeba
    e. A tree

14. How is a cell able to maintain internal conditions different from its surroundings?

Eukaryotic Cells and Their Organelles

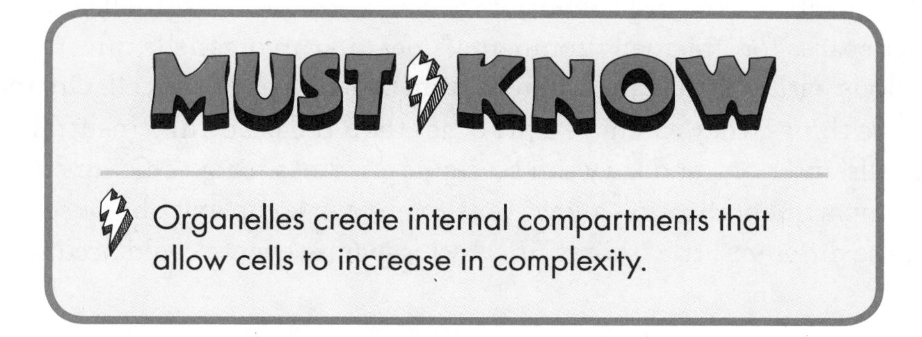

When a cell grows too big, it runs into a problem—its surface-area-to-volume ratio decreases. This means that there isn't enough membrane (the surface area) to deal with the large amount of cytoplasm (the volume). The cell membrane is the interface between the inside of the cell and the surrounding environment, so it's very important that there's enough membrane to deal with all the goings-on inside the cell. Keeping this in mind, it makes sense that by compartmentalizing the cell (our **must know** for this chapter), there is an increase in "surface" to deal with all that the cell needs to do. These compartments are the **organelles** of the eukaryotic cell. Organelles create smaller spaces for "microenvironments," meaning an organelle can create conditions within it that are different from other areas of the cell. Organelles can store things, they can have a pH other than the peaceful 7 (neutral) of most cells' interiors, and they can stockpile enzymes for specific functions. This compartmentalization allows cells to grow in complexity because it now has these different little "rooms" in which separate events could occur.

Organelles

Each organelle performs a very specific function in the eukaryotic cell and is only able to do so because it is a smaller compartment that can create special conditions within it. The **nucleus**, for example, is fondly called the brain of the cell because it houses the DNA (and thus controls the cell's functions). The **nuclear membrane** surrounds the nucleus and, unlike the outer cell membrane, it is riddled with holes called **nuclear pores**. These pores allow movement of "messages" (mRNA, which you will learn about in Chapter 11) between the DNA and the protein-producing **ribosomes** waiting in the cytoplasm. The number of ribosomes in a cell correlate with the amount of proteins the cell needs to make. Ribosomes can be floating around freely in the cytoplasm, waiting for instructions to make proteins that the cell intends to keep and use. Ribosomes that are stuck onto the endoplasmic reticulum, however, make proteins that are destined for export out of the cell.

The **endoplasmic reticulum (ER)** is an extensive network of membranes surrounding the nucleus. There is a ton of ER; almost half of a eukaryotic cell's membranes compose the ER! Two types of ER differ in both structure and function: the smooth ER and the rough ER. The **rough ER (rER)** looks rough because it has a constant supply of ribosomes stuck onto it. These ribosomes are constantly grabbing on to the rough ER and letting go; the ribosomes that latch onto the rough ER do so because they have instructions to make a protein destined for export out of the cell. The rough ER has the special role to help create proteins for export; the ribosomes thread their protein into the inner cavity (the lumen) of the rough ER.

The rough ER will then bleb off a tiny vesicle that contains this protein and send it on the beginning of its journey. A cell that specializes in exporting massive amounts of proteins tends to have a large amount of rER to help it do its job. The pancreas is an organ that makes many important types of hormones and digestive enzymes, including insulin. The pancreas's beta cells make and export insulin; beta cells, therefore, have a high amount of rough ER to help expedite the process.

BTW

The term lumen refers to a cavity of some sort. We will use the term again when we talk about the lumen of the small intestine and capillaries!

The rough ER has an organelle cousin called the smooth ER. The **smooth ER (sER)** lacks ribosomes, so it does not play a role in protein production. The sER, however, still has a number of important purposes, depending on the cell type in which it resides. Primarily, the sER makes more membranes for the cell (a worthy task, considering that organelles are made of membranes). The sER also plays a role in detoxification. The sER creates a space in which many catabolic enzymes can be concentrated. Alcohol and other drugs can stimulate the synthesis of more sER, and thus leads to an increased tolerance to these drugs, since they are broken down quicker. This highly active sER can also, unfortunately, wipe out and remove some types of antibiotics and other helpful medicines.

Think about this. If the smooth ER is used in detoxification, can you hypothesize what cell types would tend to contain a larger amount of smooth ER? To figure this out, think of the human body as a whole. What organs specialize in removing nasty things from circulation? The skin, the lungs, the kidneys . . . and the liver! The liver, specifically, carefully checks the blood and looks for poisons and other toxins. Liver cells, therefore, have a high amount of smooth ER to help them do their jobs. This is also why there is no need to go on a "detox diet." Your body is the result of millions of years of evolution; it can detox your blood better than any lemon-and-honey smoothie.

Remember those proteins created by ribosomes on the rough ER? These are the proteins destined for places other than the cell in which they were created. Considering that this protein has some specific target and purpose, it needs directions; these directions are provided by the **Golgi apparatus** (or **Golgi body**). The Golgi apparatus is responsible for taking freshly made proteins, modifying them (such as adding tiny sugar molecules or phospholipids), and then sending them to their final destination. Cells that specialize in secretion (such as our pancreatic beta cells) have a high number of Golgi bodies hanging around.

Lysosomes are organelles that act as the recycling centers of the cell. A lysosome can also be considered the "stomach" of the cell, since it's filled with digestive enzymes that break up large macromolecules into their smaller monomers. The lysosome is a perfect example of the importance of compartmentalization in a eukaryotic cell. A lysosome is a membranous bag full of digestive enzymes. When an old, nonfunctioning organelle needs to be recycled, a lysosome will engulf it and the enzymes will chew up the organelle into its macromolecule components. It would be disastrous if these digestive chemicals leaked out of the lysosome and into the cytoplasm of the cell . . . imagine the havoc! The cell would digest itself! Cleverly enough, the microenvironment inside the lysosome creates a type of fail-safe system. These digestive enzymes only work in acidic conditions, and the inside of the lysosome is, indeed, acidic. If the lysosome broke open and released a flood of enzymes into the cytosol, they would not work—the pH of the cytosol is neutral!

Endosymbiotic theory, the mysterious history behind mitochondria and chloroplasts

As mentioned earlier, the first cells to evolve were prokaryotic cells. There is a theory, **the endosymbiotic theory**, that hundreds of millions of years ago, a smaller prokaryotic cell was engulfed by a larger prokaryotic cell. Instead of the smaller cell being digested and destroyed, it stayed put and served a very specific (and helpful) purpose: it provided energy for the larger cell. The larger cell, in turn, gave the smaller cell a nice, safe place to live. This mutualistic relationship, the endosymbiotic theory suggests, is how the energy-producing organelles first came into being.

If you can guess, these two organelles are the **mitochondria** (found in all eukaryotic cells) and the **chloroplasts** (found in plant cells and photosynthetic protists). Both of these organelles have some unique and special traits: they can reproduce autonomously within the cell, they contain their own ribosomes, and, strangely enough, they have their own DNA! How do these characteristics support the endosymbiotic theory? Well, if chloroplasts and mitochondria originated from free-living bacterial ancestors, it makes sense that they maintain some remnants of their independent lives.

How cool is it that those old bacterial ancestors reside in every one of your own cells? You may have also heard of mitochondrial DNA (mtDNA) ancestry. Your mitochondria were a gift from your mom, because all of the organelles in a newly formed zygote are provided by the fertilized egg. Therefore, mitochondrial DNA is passed down through the mother's lineage.

IRL **There has been recent research suggesting we actually inherit mitochondrial DNA from both our parents . . . keep an eye out for further findings about your father's mtDNA.**

The mitochondrion is, yes, the powerhouse of the cell. What does that mean, specifically? The mitochondria supply the cell with a chemical energy source called **ATP**. It is able to create ATP because it creates a compartment within itself that serves as a reservoir for hydrogen ions (H^+), and this high concentration of H^+ is used to make ATP. The mitochondria also house many enzymes specialized in the methodical breakdown of the sugar glucose; these are in the innermost space called the **matrix**. We will cover all of this is more detail when we discuss cellular respiration in a bit. Similarly, the chloroplast also has a compartment within itself that holds a high concentration of H^+ for the same purpose (to create ATP). The **thylakoids** are the hollow sacs that hold H^+; the surface of the thylakoids is embedded with special pigment molecules that grab the sun's energy. The thylakoids—a stack of which is called a **granum**—are floating in a liquidy goo called the **stroma**. The stroma is the location of a bunch of enzymes used to create glucose from carbon dioxide. We will also discuss the process of photosynthesis later.

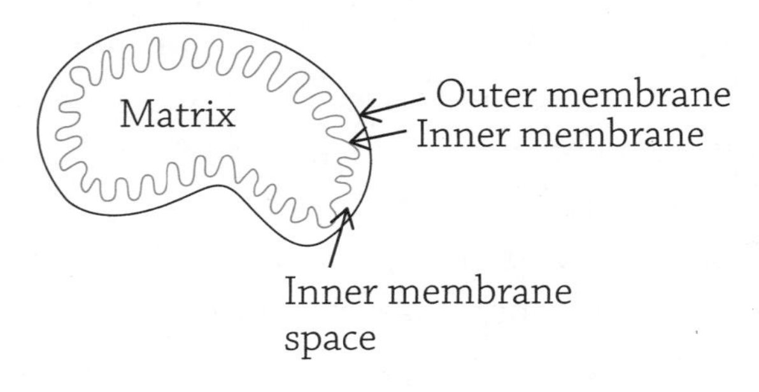

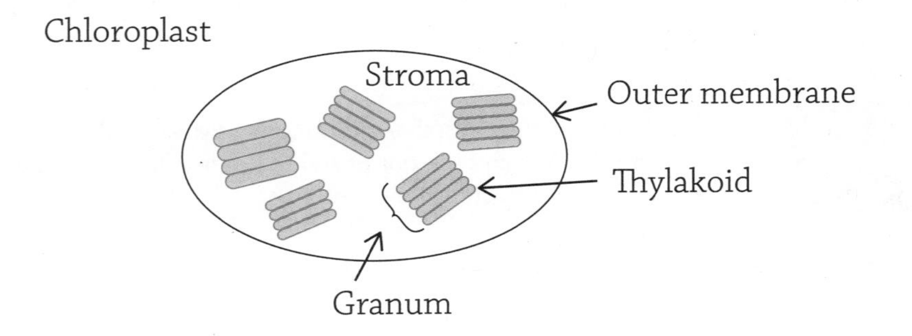

Plant versus Animal Cells

Both plant and animal cells are eukaryotic, meaning they are filled with organelles and are larger than their prokaryotic cousins. The differences between the plant and animal cells, however, are significant: plant cells have chloroplasts and are photosynthetic, whereas animals are defined as *not* having the ability to photosynthesize. Plants do not have skeletons, so they rely on the structure provided for them by their individual cells, a phenomenon called **turgor pressure**. A plant cell has an organelle called a **central vacuole** that fills with water and applies supportive pressure on the cell wall. Animal cells have neither a cell wall nor a central vacuole.

EXTRA HELP

When describing differences between plant and animal cells, it's important to remember a few things. It's true that plant cells most certainly do have mitochondria and undergo cellular respiration, just like animal cells do. The difference is, the glucose needed for cellular respiration in a plant cell is created from sun energy in the chloroplasts, whereas animals acquire their glucose by eating it (heterotrophs).

REVIEW QUESTIONS

1. How are organelles helpful to a eukaryotic cell?

2. A cell with a high amount of rough endoplasmic reticulum would be specialized to do what?

3. Match the organelle with its correct function:

Organelle	Function
Chloroplast	Cell's "stomach"
Golgi apparatus	Controls all the cell's functions
Lysosome	Allows mRNA to pass out of the nucleus
Mitochondrion	Makes proteins for export
Nucleus	Important in drug detoxification
Nuclear pores	Modifies and ships proteins to their final destination
Rough endoplasmic reticulum	Converts sun energy into glucose
Smooth endoplasmic reticulum	Converts glucose into ATP

4. What are two interesting traits of chloroplasts and mitochondria that suggest they were once autonomous (free-living) prokaryotic cells?

5. Label the mitochondrion with the following terms: **inner membrane space, inner membrane, matrix, outer membrane.**

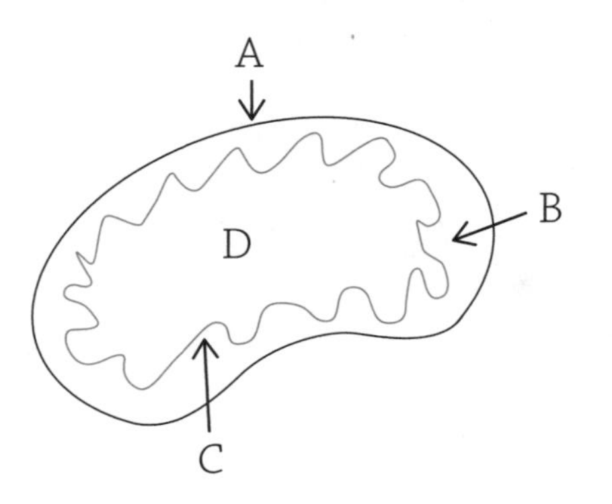

6. The glucose needed for cellular respiration in an animal cell comes from ________________, whereas the glucose needed for cellular respiration in a plant cell comes from ________________.

7. What three things does a plant cell have that an animal cell does not?

8. How is the lysosome a perfect example of creating a microenvironment within a cell?

9. Fill in the blanks with the correct organelles: When a protein is made for export, it is created by ____________________ stuck onto the surface of the ____________________.

10. True or False: Only plant cells have chloroplasts, and only animal cells have mitochondria.

11. Label the chloroplast with the following terms: **granum, outer membrane, thylakoid, stroma.**

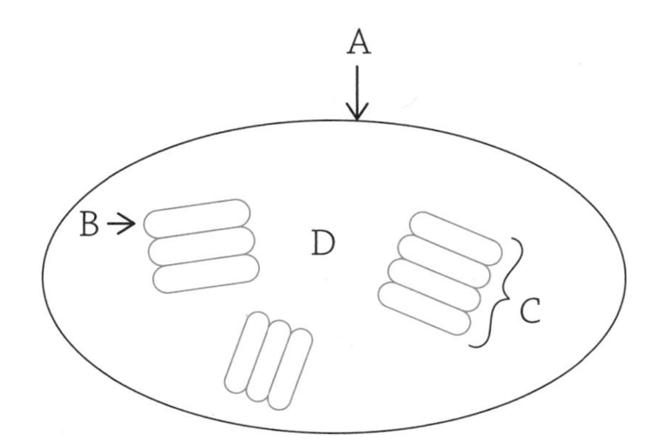

12. How is turgor pressure formed in a plant cell?

13. Which of the following cell type would most likely have a high number of mitochondria?
    a. Skin cell
    b. Lens cell
    c. Muscle cell
    d. Pancreatic cell

6 Cells and Energy Transformation

- Energy is neither created nor destroyed; a cell can only convert it from one form to another.

- The chemical ATP is the form of energy used by all cells.

- Cellular respiration is a catabolic process that releases energy.

- Photosynthesis is an anabolic process that stores energy.

- Both cellular respiration and photosynthesis utilize chemiosmosis (the diffusion of hydrogen ions) to create ATP.

I love this **must know**. It's a simple statement (welcome back, first law of thermodynamics), yet it perfectly shapes how we learn about energy transformation in cells. This entire chapter—whether we're learning about cellular respiration or photosynthesis—is based on the cell's need to convert energy from one form to another.

ATP

This book is about life, and one of the characteristics of all living things is the need for energy. The idea of energy is an amorphous understanding that it has something to do with your ability to move, to do things, to live. But what, exactly, *is* energy? This question is a hefty one, but for our sakes, I will address it from a cell's point of view: energy is a chemical called **adenosine triphosphate (ATP)**. More accurately, ATP is the chemical form of energy all cells use to power their cellular work. ATP is an organic molecule composed of an adenine, a ribose, and three phosphate groups.

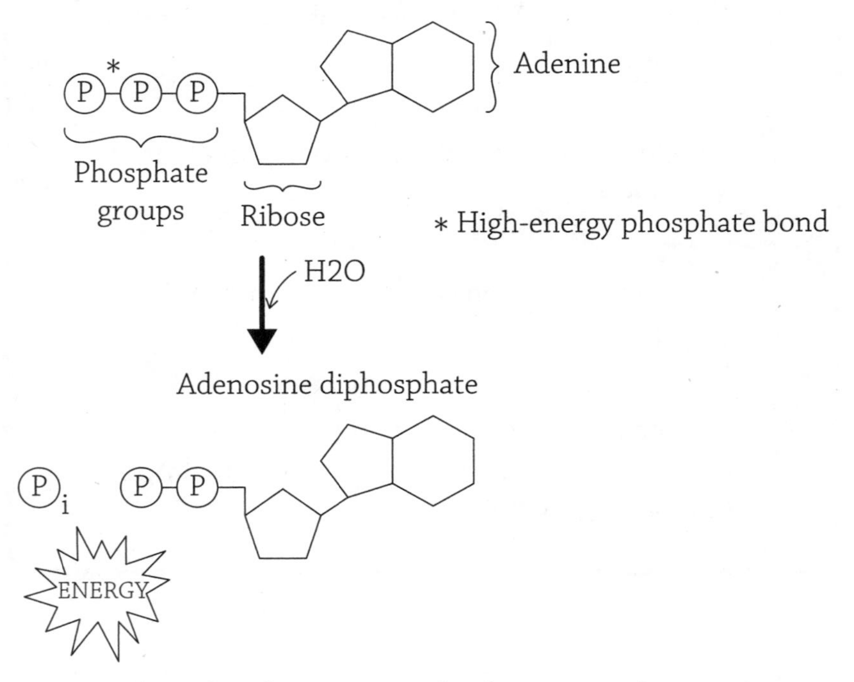

Adenosine triphosphate (ATP) and its low-energy-form, ADP

This is a cell's chemical energy currency. The covalent bond between the second and third (outermost) phosphate groups hold a large amount of potential energy. When a cell needs to power an activity (such as building molecules via anabolic reactions) or when a multicellular organism needs to do something (such as move muscles), then the third phosphate group in ATP is cleaved off, releasing the stored energy for the cell to use. This leaves adenosine diphosphate (ADP) and a free inorganic phosphate group.

BTW

The most common question from my students at this point is: what is that small subscript "i" next to the free phosphate group supposed to mean? It stands for "inorganic," which means the phosphate group is no longer part of a larger, carbon-based (organic) molecule. No big deal.

In order to recharge the battery (and reform ATP), there needs to be an input of energy in order to re-create ATP . . . keep this in mind as we discuss the process of cellular respiration in a little bit. The thing to focus on now is the idea that this chemical form of ATP needed to be "created" by the cell. Yet remember from Chapter 1 that there are these two ironclad rules of thermodynamics:

1. Energy cannot be created, nor can it be destroyed. It can, however, be converted from one form to another (our **must know**).
2. Any time energy is converted from one form to another (meaning it is being used by the critters of Earth), some of the potential energy is lost as heat. It is a thermodynamic rule that when energy is transformed, the process is never, ever 100% efficient (and most of the time, the lost energy escapes as heat).

Therefore, how can the cell create this chemical energy out of thin air? The answer is: it doesn't (Rule 1: energy can't be created from nothing!). Instead, the cell is following the second part of Rule 1: it is *converting* energy from one chemical form (glucose) into another (ATP). This glorious process is called **cellular respiration.**

Cellular Respiration

The process by which cells convert the chemical glucose into the chemical ATP is called cellular respiration. That is not to be confused with respiration, or breathing. The two are linked, however, because cellular respiration requires oxygen and generates carbon dioxide, and these gases are transported to (and from) the body's cells by breathing.

As we focused on earlier, energy can be converted from one form to another. Cellular respiration takes the sugar glucose ($C_6H_{12}O_6$) and, in the presence of oxygen, breaks up the sugar until it is nothing but carbon dioxide and water. By doing so, the energy stored in the sugar glucose is converted into another form of energy, ATP:

$$C_6H_{12}O_6 + 6O_2 \rightarrow 6CO_2 + 6H_2O + ATP$$

This equation is a neat, concise overview of the entire process which is *much more complicated*... we will delve into the details later. Right now, let's look at the idea of **catabolic** versus **anabolic** reactions, because it may help you understand.

Catabolic and anabolic reactions play a big role in this chapter. It will also be useful to understand the terms *reduction* and *oxidation*. In short, *catabolism* breaks things down, and *anabolism* builds things up. Now, when you think of breaking down and building up molecules, you must consider that you are either breaking or building covalent bonds. You need energy to create new bonds; you release energy when you break them. Therefore:

> Catabolic (breaking) reactions *release* energy.
>
> Anabolic (building) reactions *require* energy.
>
> Keep this in mind.

Oxidation refers to chemical reactions that involve an input of oxygen and a removal of hydrogens and electrons; *reduction* means that hydrogens and electrons are added. As we will see, when glucose is broken down in the catabolic (energy-releasing) reactions of cellular respiration, the glucose molecule is *oxidized* as many hydrogens and electrons are stripped away throughout the process. Furthermore, the "burning" of glucose requires an input of oxygen (once again, the definition of oxidation). On the flip side, photosynthesis is a *reduction* reaction because a bunch of hydrogens and electrons are added to CO_2 in order to create $C_6H_{12}O_6$; oxygen is also produced in the process. Therefore:

> Cellular respiration is a *catabolic* pathway that *oxidizes* glucose.
>
> Photosynthesis is an *anabolic* pathway that *reduces* carbon dioxide.
>
> Keep this in mind, too.

So, once again consider the overall equation for cellular respiration, but this time, look at the structural formulas for the carbon compounds glucose (a reactant) and carbon dioxide (a product):

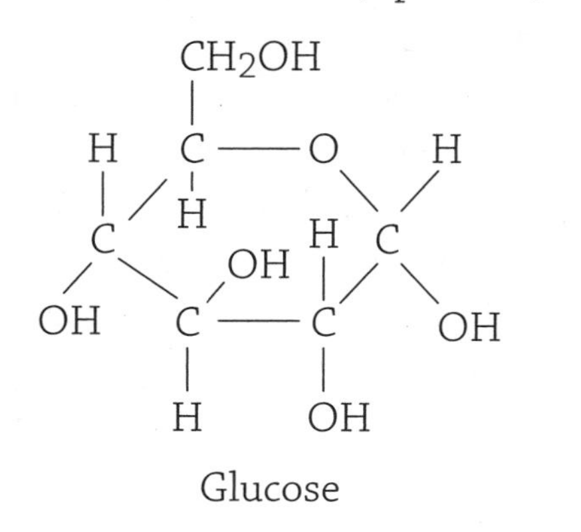

Glucose

O=C=O

Carbon dioxide

Glucose is a single molecule with six carbons, six oxygens, and twelve hydrogens. After the reaction commences, all the carbons locked into a molecule of glucose have been cleaved loose as six single carbon dioxide molecules (each with a single carbon). Was this an anabolic or a catabolic reaction? It was a catabolic reaction because one large molecule (glucose) was broken up into six small molecules (carbon dioxides). Was energy used or released in this reaction? It was released (which is the point of cellular respiration!).

ATP is indeed the energy currency of all cells. There's a problem, however—ATP is very volatile and cannot be stored for long amounts of time. Once a cell makes some ATP, it needs to be used right away. That is why glucose is so important: it is how the body stores long-term energy. And to take that one step further, how does our body store glucose for the long haul? It links all the glucose together into one lovely polymer called glycogen and stores it in our muscles and liver (think back to Part One!). So, when your cells need energy, it first releases individual glucose monomers from the stored glycogen, and then burns up the glucose to create ATP!

Glycolysis

The truth is more complex (and interesting) than the simple equation for cellular respiration: $C_6H_{12}O_6 + 6O_2 \rightarrow 6CO_2 + 6H_2O + ATP$. The entire process of cellular respiration takes place in three big steps: glycolysis, Krebs cycle, and the electron transport chain. Glycolysis ("glyco" = sugar; "lysis" = split) is a 10-step process where glucose is split into two molecules of pyruvate. The breaking and rearrangement of covalent bonds releases energy that is immediately transformed into two molecules of ATP per glucose. Furthermore, along this pathway hydrogens and electrons are pulled from the carbon molecules (which used to be glucose) and are carried over to the electron transport chain for later use (we'll get to that in a bit). The hydrogens and electrons are "held" by a molecule called NAD^+; once NAD^+ picks up its passengers (H^+ and electrons), it becomes NADH. Once the passengers are dropped off at the electron transport chain, the NAD^+ travels back to glycolysis to pick up more hydrogen ions and electrons.

EXTRA HELP

Much of the challenge of learning biochemistry and organic chemistry lies in the brand-new vocabulary. For example, in organic chemistry, the ending of some molecules' names can end with either "-ate" or "-ic acid."

This is pyruvic acid: $CH_3COCOOH$ ← It's a weak acid because of this hydrogen.

This is pyruvate: CH_3COCOO^- ← It lost the hydrogen and lost its "acid" designation.

Even though it's better for us to say pyruvate, it's forgivable if you refer to it as pyruvic acid.

Here is the simplified overview for the process of glycolysis:

Glycolysis - Simplified overview

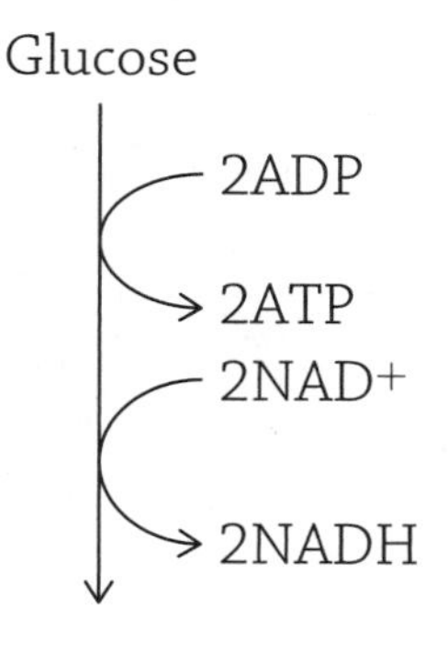

Glucose is broken up into two pyruvate molecules, and along the way, some ATP is made (using the energy released by breaking covalent bonds) and some of those NAD^+ molecules pick up their passengers (electrons and hydrogens). I am including what I refer to in class as The Truth (see below): the full 10 steps of the glycolytic pathway.

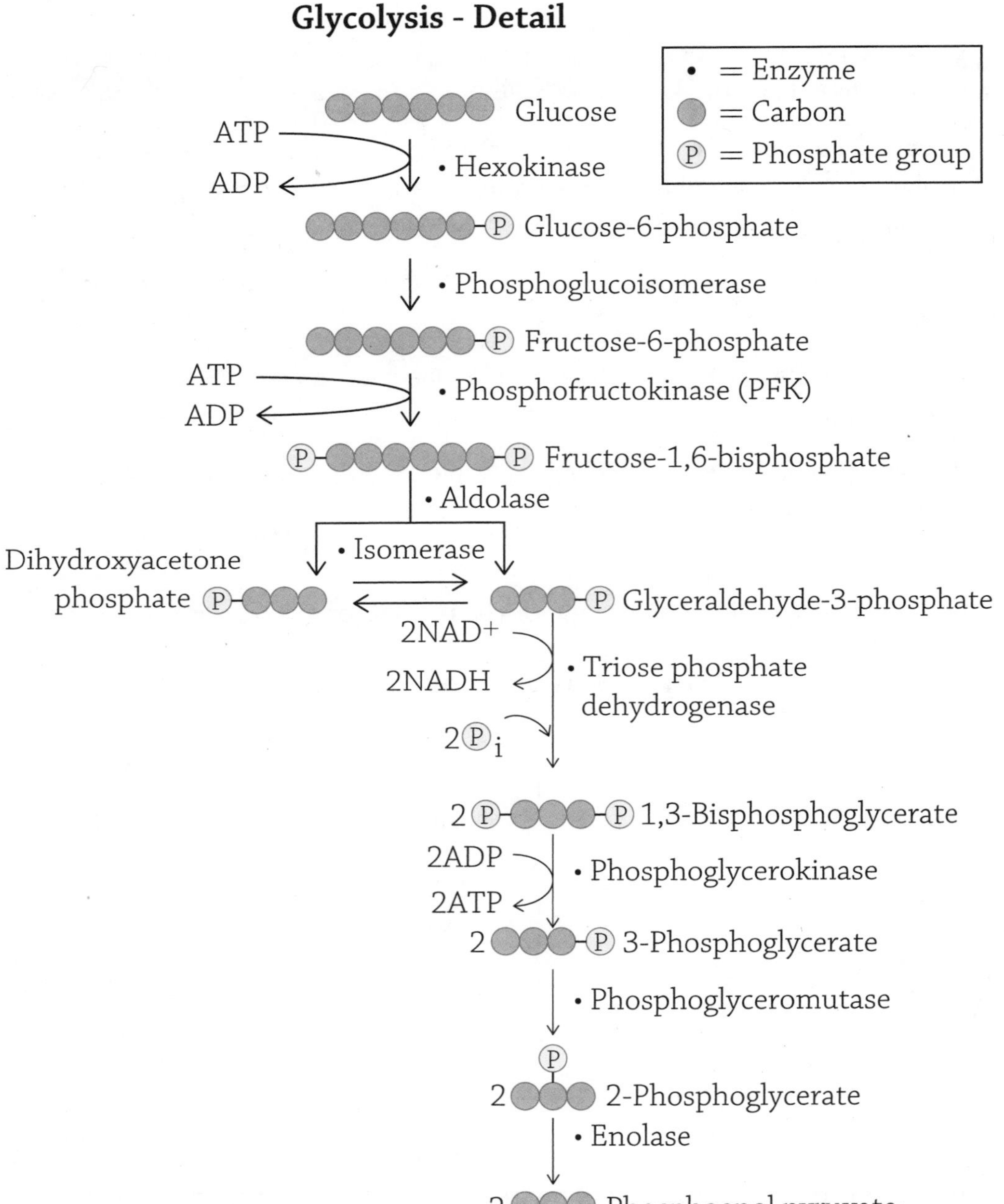

Glycolysis - Detail
• = Enzyme
= Carbon
P = Phosphate group
Glucose
ATP
ADP
• Hexokinase
Glucose-6-phosphate
• Phosphoglucoisomerase
Fructose-6-phosphate
ATP
ADP
• Phosphofructokinase (PFK)
Fructose-1,6-bisphosphate
• Aldolase
• Isomerase
Dihydroxyacetone phosphate
Glyceraldehyde-3-phosphate
2NAD+
2NADH
2Pi
• Triose phosphate dehydrogenase
2 1,3-Bisphosphoglycerate
2ADP
2ATP
• Phosphoglycerokinase
2 3-Phosphoglycerate
• Phosphoglyceromutase
2 2-Phosphoglycerate
• Enolase
2 Phosphoenol pyruvate
2ADP
2ATP
• Pyruvate kinase
Pyruvate

Each step shows you the carbon-skeleton structure of the molecule, and what enzyme is responsible for the reaction. There is a very good chance you do not need to know this much detail; if you do not, feel free to skip that part. For some people, however, it actually helps to see the detail.

BTW

Please notice that no carbon dioxide is released from glycolysis. All the glucose-derived carbon dioxide is exhausted during the Krebs cycle (which you will learn about in a bit!).

EXTRA HELP

The process of glycolysis doesn't occur in any organelle. Instead, the enzymes that catalyze the ten steps reside in the cytoplasm. This is significant because prokaryotic cells do not have organelles. Since glycolysis doesn't require organelles, it is found in all cell types! Now, consider this: what is the first cell type to evolve on Earth 3.5 billion years ago? Yup, a prokaryotic cell. Therefore, a logical deduction is that glycolysis was the first energy-producing biochemical pathway to evolve because it is found in simple, organelle-free prokaryotic cells.

Furthermore, glycolysis is an anaerobic process (it doesn't require oxygen to function). If this is the first biochemical pathway to evolve on Earth, back when cells first appeared on the scene, is it logical that it would not need oxygen to function? Considering that atmospheric oxygen didn't begin to accumulate until 2.7 billion years ago, it makes sense that glycolysis is an anaerobic process.

Control of Glycolysis

It is important for a cell to regulate the speed at which metabolic pathways produce products; it's wasteful to create a substance when it's not needed! Glycolysis has one important regulatory step that serves as a good review of an earlier concept: allosteric enzymes.

Recall that an allosteric enzyme has two "forms": a very-good-at-catalyzing-a-reaction (active) form, and a *not*-very-good-at-catalyzing-a-reaction (inactive) form. The enzyme can be "locked" into either of the forms, depending on the needs of the cell. Allosteric enzymes have the usual active site (into which a substrate binds), plus a *second* location called the allosteric site. If an *activator* binds at the allosteric site, the enzyme will take the active form; if an *inhibitor* binds at the active site, the enzyme will take on the inactive form.

Look at the detailed figure of glycolysis: the third step is under control of an enzyme called **phosphofructokinase (PFK)**. If PFK is functioning properly, the glycolytic pathway chugs along and produces ATP by oxidizing (breaking apart) glucose. But imagine that you're relaxing on your couch, playing a video game while snacking on cheesy puffs. Do you really need to produce a lot of energy at that moment? Probably not. Your cells respond properly by switching PFK to the inactive form, and glycolysis (and by association, ATP formation) slows down. Now here's a question: what would be a logical inhibitor molecule to bind to the allosteric site to tell PFK to power down for a bit? Hint: Think negative feedback! If you want to stop a metabolic pathway that does this,

Starting reactant A —Enzyme ① (Active form)→ B —Enzyme ②→ C —Enzyme ③→ D Final product

then the logical thing to bind to Enzyme 1 and tell it to stop making product D is a molecule of product D itself!

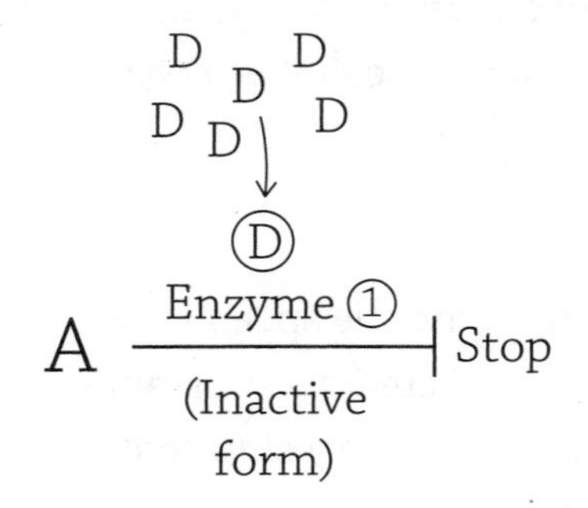

A high concentration of the end product ensures that a molecule of D (the inhibitor) will bind to the allosteric site and stabilize the inactive form of Enzyme 1. Therefore, the perfect choice for an inhibitor molecule to bind to PFK is a molecule of ATP! If you aren't actively burning up your ATP (and there's plenty floating around), an ATP molecule will bind to PFK's allosteric site and stabilize the inactive form. Now, imagine that while you are relaxing on the couch, a giant radioactive ant breaks through your front door, inspiring you to quickly get off the couch and run out your back door. This uses up all your ATP, including that one molecule that glommed onto PFK, thus removing the inhibition (which is good, because you need a lot of energy to run away from a giant radioactive ant). Furthermore, a *different* allosteric molecule (an activator) will take its place and lock it in the active conformation. Here is another thought question: what is a good choice of activator? It once again has to do with the whole point of glycolysis: to make ATP. You need to think a bit deeper, though, about what happens when you use up ATP. Does it just . . . *poof* . . . go away? No. When you use ATP (adenosine triphosphate), you release its energy by breaking free the last phosphate group:

$$\text{ATP} \rightarrow \text{ADP} + \text{P}$$

Once ATP is used, you're left with ADP (adenosine diphosphate). Therefore, a good signal to PFK that it needs to speed up and make more ATP is . . . ADP (the "dead battery" form of ATP)! ADP is indeed one (of a few) activator for PFK. I just love the logic.

Krebs Cycle

At the end of glycolysis, the glucose molecule has been broken up and stripped of some electrons and hydrogens. What you have left are two molecules of pyruvate. These two molecules contain three carbons each and still have a significant amount of energy stored in their carbon-carbon bonds. Before the two molecules move on to the second stage of cellular respiration (the Krebs cycle), each must first be "tweaked" a bit: one carbon is removed, a hydrogen and electron is stripped from it, and the remaining two-carbon molecule (called **acetate**) is energized by a molecule called "coenzyme-A" (see figure on next page):

Intermediate step (After glycolysis, before Krebs cycle)

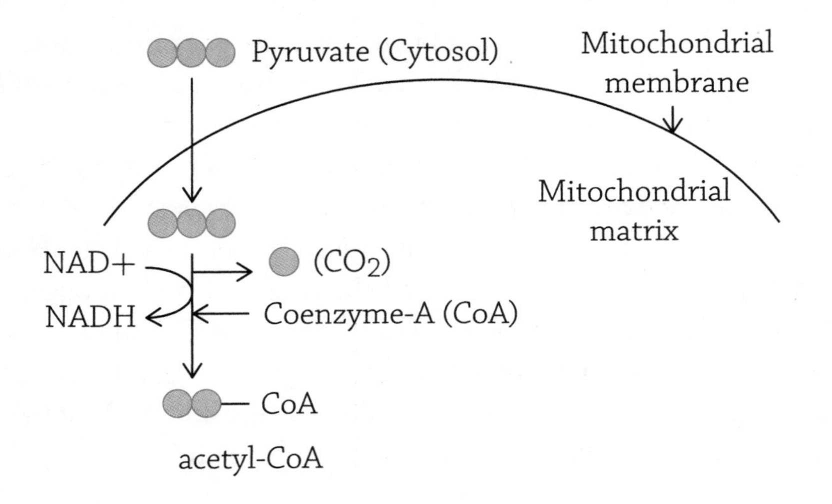

This intermediate step occurs in the mitochondrion and is necessary in order to get the carbon fragment ready for the next series of reactions in the Krebs cycle. The Krebs cycle (also called the Citric Acid cycle) is an eight-step cyclical series of reactions that occurs in the matrix of the mitochondrion. We refer to it as a cycle, because in these eight enzymatically driven steps, you cleverly end up back where you started.

Now the Krebs cycle will take over oxidation of the carbon molecule. After the tweaking step, you no longer have the three-carbon molecule called pyruvate, but instead, you have a two-carbon fragment called acetate. In order to be ferried over to the first step of Krebs, it must be held onto by a big moose of a molecule called coenzyme-A (the entire thing is called acetyl-CoA). The coenzyme-A (CoA) will drop off the acetate and head back to pick up another. (Why another? Recall at the end of glycolysis, we had split glucose into TWO molecules of pyruvate; therefore, there will be TWO molecules of acetate to drag into the Krebs cycle.)

There is energy still stored in the covalent bonds of acetate, and to release the energy and convert it into ATP, your mitochondria need to break those remaining carbon-carbon bonds. The two-carbon fragment is first "grabbed" by a molecule waiting in the wings; it's called **oxaloacetate** (or **OAA**). Oxaloacetate is a four-carbon molecule. You don't need to stress out about its actual structural formula, but it is important to focus on the numbers of carbons. If OAA has four carbons, and it picks up (and covalently bonds with) an incoming acetate molecule, a new molecule called **citric acid** is produced. How many carbons must be in citric acid? Yes! Six! (The four carbons of OAA + the two carbons of acetate = a six-carbon compound called citric acid.)

Since this is a cycle, we need to regenerate what we started with: a molecule of OAA (which will then happily snag another incoming acetate). We are, therefore, allowed to cleave off *two* carbons (the amount added by an incoming acetate), both of which are released as molecules of carbon dioxide. By doing so, we are breaking covalent bonds and converting that energy into molecules of ATP! In truth, there is a fair amount of rearrangement that needs to accompany this ATP production and carbon-dioxide release, but don't get lost in the details. Instead, keep in mind that the energy released from breaking and rearranging the covalent bonds in this remaining carbon compound (that *used* to be glucose) is being converted into the chemical form ATP (our **must know** concept!). In fact, each turn of the Krebs cycle yields one molecule of ATP. Here's a question: How many total molecules of ATP are made in the Krebs cycle, per glucose? The answer is two (the Krebs will "spin" twice per glucose molecule, because there were two molecules of acetate to deal with).

Now, as in glycolysis, these eight steps are accompanied by molecules of NAD+ (and one molecule of its chemical cousin FAD+) swooping in, grabbing some hydrogens and electrons, and shuttling them over to the electron transport chain.

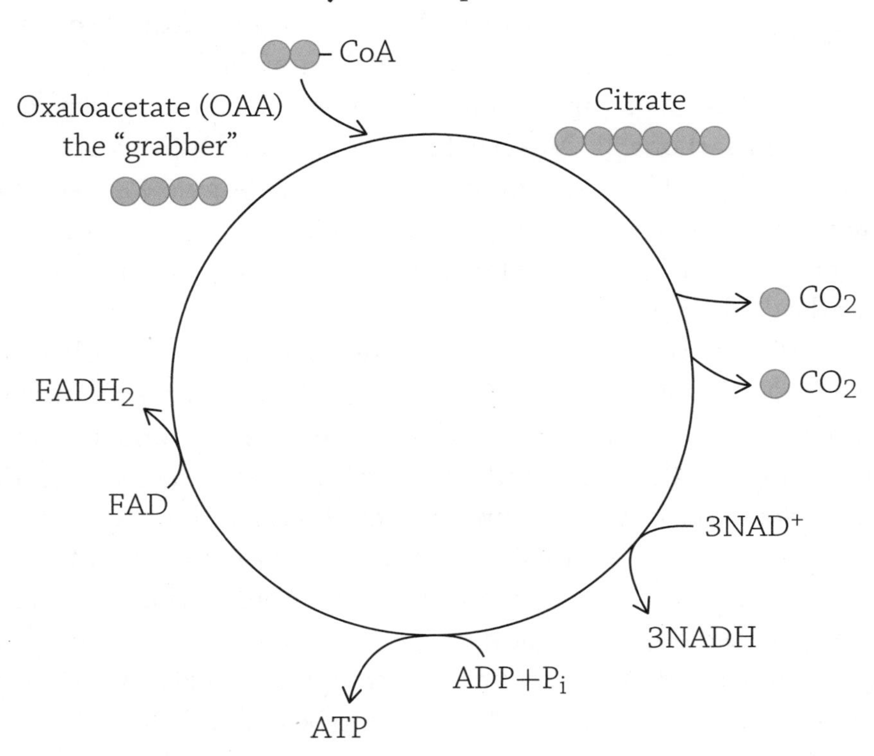

At this point, we have seen the last of our molecule of glucose. All the carbons have been cut free and "exhausted off" as carbon dioxide. And it is actual exhaust, by the way. When you exhale, carbon dioxide is released... all that carbon dioxide came from the lowly CO_2 generated by the Krebs cycle (and the intermediate step). The helpful hydrogens and electrons have been stripped free and shuttled over to the electron transport chain.

Electron Transport Chain

Up to this point, we have made a paltry amount of ATP. For a single glucose molecule, we produced two molecules of ATP during glycolysis and a total of two after the Krebs cycle (one per spin). That is not a lot, considering the amount of potential energy locked up in sugar. Luckily, the spirit of glucose lives on in the electrons and hydrogen ions that have been plucked from its carbon skeleton throughout glycolysis and the Krebs cycle. The NADH and $FADH_2$ molecules have traveled vast distances (on a cellular level) in order to arrive at their destination: the folded inner membrane of the mitochondrion.

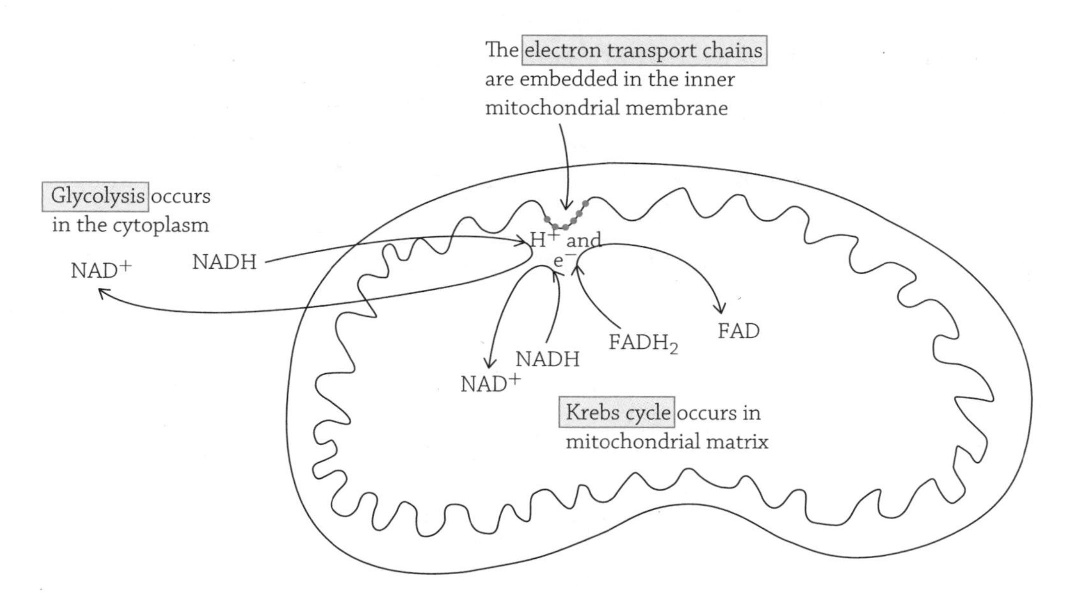

The locations of the three stages of cellular respiration. Notice that both glycolysis and Krebs send "shuttle buses" (NADH and $FADH_2$) over to the electron transport chain in order to drop off their passengers (electrons and hydrogen ions).

Now is the time you get a huge payoff from those H^+ and electrons, and it is through a process called **chemiosmosis**:

chemi—chemical

osmosis—referring to diffusion

Chemiosmosis is the diffusion of some sort of chemical, in this case, H^+. And in order for something to *want* to diffuse, it needs to first be at a high concentration somewhere. And in order to concentrate something in the first place, you need to use energy . . . that's where the electrons come in. These high-energy electrons use their energy to create a concentration gradient of hydrogen ions (they are stored in the inner membrane space of the mitochondrion). When the hydrogen ions diffuse back into the matrix, the only path they can take is through an enzyme called ATP synthase. This flow of hydrogen ions powers up the enzyme so it can make ATP. Consider this analogy: chemiosmosis is like a hydroelectric dam. In a hydroelectric dam, water is stored behind the dam wall (A). It can only pass through a small opening (B) before passing through a turbine (C), a machine that spins when the water passes through. The spinning of this machine creates energy. Eventually, the water reaches the region of lower water potential (D).

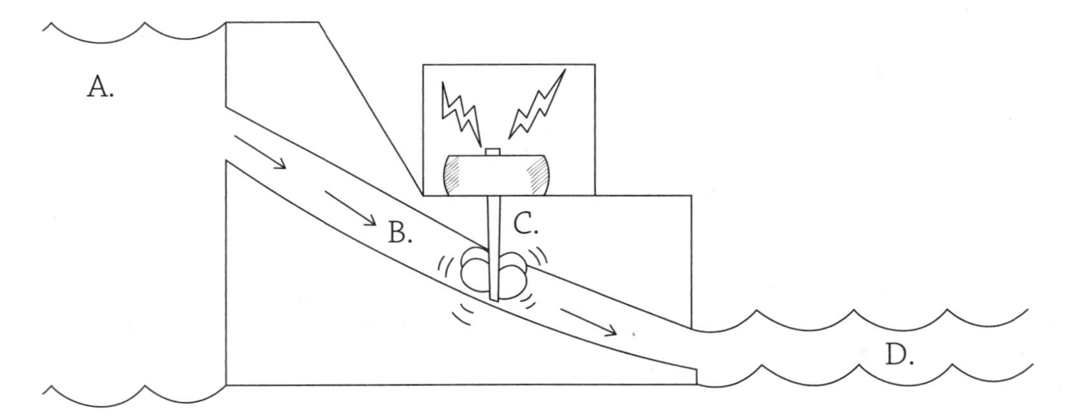

Chemiosmosis in a mitochondrion works the same way:

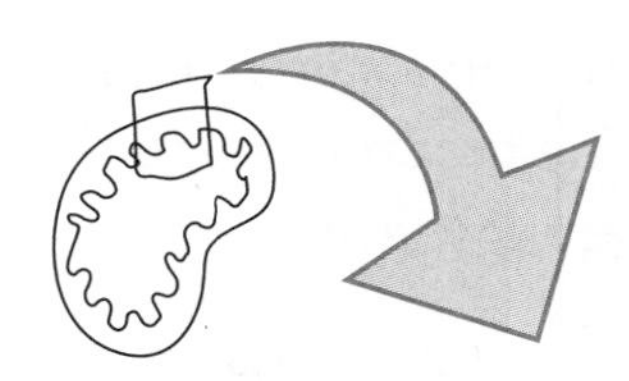

The "water" in a mitochondrion is the flow of hydrogen ions. A high concentration of hydrogen ions is stored behind the inner membrane "wall," and they can only diffuse through a small tunnel: the enzyme called ATP synthase. As the hydrogens flow through, ATP synthase spins (like a turbine) and generates energy (just like the turbine)!

Here is the complete story of how the electron transport chain makes a ton of ATP:

- The electron and H^+ carriers (NADH and $FADH_2$) bring over their passengers from glycolysis (in the cytoplasm) and the Krebs cycle (right there in the matrix).

- The electrons are dropped off at the first protein in a series of proteins, all of which want to grab on to an electron. The first protein in the chain holds on to the electron at its highest energy level.
- There's an oxygen molecule at the end of the electron transport chain, and because of its high affinity for electrons, it's "drawing" the electron toward it, down the chain of proteins.
- The next electron transport protein in line is willing to take the electron from the first protein, but only if the e– "cools off" a bit and loses some energy. This process continues, allowing this electron to slowly, controllably, lose its energy.
- This energy is used to power tiny hydrogen pumps (often called "proton pumps") embedded in the inner membrane. These pumps actively transport hydrogens that are hanging out in the matrix (remember: they were also dropped off by NADH and $FADH_2$) and move them into the inner membrane space.
- This creates a high concentration of hydrogen ions in a small space. These H^+ want to diffuse back into the matrix, but the only pathway available is through an enzyme called ATP synthase.
- The flow of hydrogen ions through ATP synthase causes the enzyme to spin and catalyze the reaction ADP + P → ATP.

Thus, the energy in electrons is converted into the chemical energy of ATP (our **must know** concept in action).

- Meanwhile, the electron gets to the bottom of the chain where it is grabbed by an awaiting O_2 and combines with a couple of hydrogens, thus creating water.

Each day, you produce about 230 milliliters (8 ounces) of water generated from cellular respiration. That is a LOT of molecules of H_2O produced at the end of many electron transport chains.

- This glorious, oxygen-driven process creates anywhere between 26–28 ATP per glucose!

IRL

As you know, cyanide is bad. What you may not know, cyanide kills because the chemical grabs hold of a certain protein of the electron transport chain and serves as a road block, stopping the flow of electrons down the chain. Without electrons to actively create a hydrogen ion gradient, there can be no ATP production . . . and a cell without ATP is dead (as are you, if you ingest enough cyanide). Interestingly enough, a major source of cyanide is the cassava root from South America. If you eat raw cassava, you will most likely die due to cyanide poisoning. If, however, you dry and boil the cassava root long enough, you will inactivate the poison from the plant tissue and turn the root into . . . tapioca!

NADH versus $FADH_2$—Why One Is Better Than the Other

The oxidation of glucose is a slow, controlled process. The energy is released slowly so it can be harnessed to do cellular work; otherwise, it would literally be tiny little explosions of burning glucose! Heat and light and *poof*! That wouldn't do anyone any good.

At key steps in glycolysis and the Krebs cycle, electrons and hydrogens are stripped off of the glucose molecule to be used in the electron transport chain. The electrons and hydrogens, however, need a "shuttle bus" to drive them over to the inner mitochondrial membrane. The molecular shuttle buses are the molecules NAD+ and FAD. Once these buses pick up their passengers, they are reduced into NADH and $FADH_2$. When electrons are transferred from glucose to NAD+, they retain almost all of their stored energy (remember: glucose is a *sugar* made with the energy of the *sun*, so any electrons stripped from this molecule are super energized and chock full o' energy).

The electron transport chain consists of a series of proteins embedded in the inner mitochondrial membrane. The chain starts at a high-energy end and finishes at a low energy end; with each step "down" the chain, the electron releases a bit of its stored energy. Oxygen is impatiently waiting at the very end of the chain, and it has such a love for electrons it "pulls" the electrons toward it. Once oxygen gets a hold of the electrons, it snags a couple of hydrogen ions and turns into H_2O.

The total amount of energy released in this drop—starting at the highest end of the electron transport chain and ending at the oxygen—is used to make ATP. NADH is very good at its job and drops off its passengers at the top of the chain, yielding 2.5 ATP per molecule of NADH. Now, if NADH is a bus and the electron transport chain is a hill, the bus drives to the top of the hill to drop off the electrons. $FADH_2$ is also a bus, but it's not quite able to make it to the top of the hill. Instead, it drops off its passengers a bit lower. Because of its failure to make it to the top, the electrons don't have as far to fall, and they don't release as much energy as their NADH-riding brethren. Therefore, each $FADH_2$ will yield only 1.5 ATP.

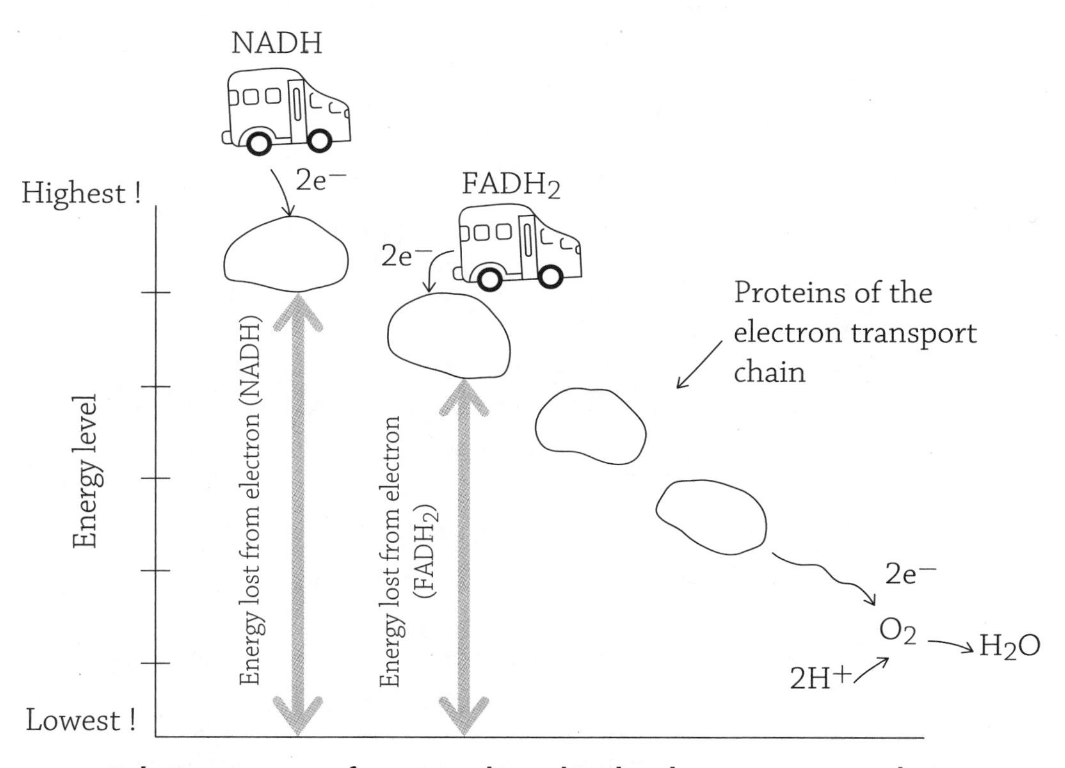

Relative amounts of energy released in the electron transport chain (NADH versus $FADH_2$)

My students always ask this excellent question: if the electron carrier NADH yields more ATP, then why did natural selection foster the formation of $FADH_2$? Why isn't there only the better shuttle bus, NADH? It has to do with the molecules from which the electrons and hydrogens are torn. For some reason, NAD+ couldn't do the job, but FAD was able to. It's better to get in there and grab any unused electrons, even if they are not as full of energy as others.

The Role of Oxygen (and what to do if there isn't any)

Let's refer back to the overall equation of cellular respiration:

$$C_6H_{12}O_6 + 6O_2 \rightarrow 6CO_2 + 6H_2O + ATP$$

Now we can see where each of these molecules played a role:

Molecule	What's its deal?
$C_6H_{12}O_6$	This glucose molecule was seen, briefly, right at the start of glycolysis.
O_2	This molecule had a duty of "pulling" electrons down the electron transport chain and then grabbing them when they were spent of their energy.
CO_2	The only time carbon dioxide is produced is during the Krebs cycle (and right before, in the intermediate step).
H_2O	This is what the oxygen formed once it grabbed the electrons at the end of the electron transport chain, plus a couple of hydrogen ions that were floating nearby.

The process of cellular respiration is an **aerobic** process, meaning it requires oxygen to run. This seems strange, considering that oxygen's role is seemingly small, right at the end of the end of the electron transport chain. Yet if there is no oxygen to grab the electrons as they hit the bottom of the chain, a subatomic particle traffic jam will form. A backup of the electron transport chain will also cause the Krebs cycle to come to a screeching halt; all that's left running is glycolysis. This process—glycolysis providing ATP in the absence of oxygen—is **fermentation**.

Fermentation

Fermentation is an **anaerobic** process, meaning it occurs in the absence of oxygen. When there's no oxygen, two major pathways of cellular respiration come to a screeching halt (Krebs cycle and electron transport chain); that leaves only glycolysis to run. Some cells, such as yeast, some bacteria, and our muscle cells, can create energy using only glycolysis when there's no oxygen around. Considering that glycolysis made two molecules of ATP per glucose (compared to the 34–36 ATP when the entire process of cellular respiration is running), it isn't as efficient. Yet, some ATP is better than none.

In order for glycolysis to occur, each of the 10 steps needs to happen. A couple of the steps requires that the carbon compound is oxidized—electrons and hydrogens are removed and transferred to NAD+, creating NADH. If NAD+ is not available to swoop in and grab its passengers (electrons and hydrogens), then glycolysis cannot proceed.

Consider our little passenger bus NADH. It's carrying its passengers: electrons and H^+. Where does NADH drop off its passengers? Yup, the electron transport chain. Now, if the chain is clogged because there's no oxygen sweeping away the low-energy electrons, then NADH can't swing by to unload its cargo. If NADH can't do that, it isn't able to turn back into its oxidized form of NAD+ and play its important role for glycolysis. If glycolysis stops . . . well, that's bad. Luckily, many cells can switch to a fermentation pathway in order to keep glycolysis running, even in the absence of oxygen.

Fermentation is the energy-converting process that relies solely on glycolysis. The covalent bonds of glucose will be broken and rearranged, yielding a net gain of two precious molecules of ATP. The big difference, however, is an "add-on" is necessary to oxidize NADH back into NAD+. Instead of NADH dumping its e− and H+ onto the electron transport chain, it uses a carbon molecule that's readily available (considering that it cannot continue to be broken up in the Krebs cycle): pyruvate! Once pyruvate is reduced—and depending on the cell type in which fermentation occurs—it will turn into either **ethanol** or **lactic acid**.

EASY MISTAKE!

It's easy to make the mistake of thinking the purpose of fermentation is the production of lactic acid or ethanol. The creation of those products is, actually, inconsequential. The purpose of the fermentation pathways is to regenerate NAD+ in order to *keep* glycolysis running in the absence of oxygen!

Ethanol Fermentation

Ethanol fermentation occurs in yeast cells and some bacteria. This is a two-step process that requires that pyruvate is first turned into a two-carbon molecule called acetaldehyde. The acetaldehyde then picks up the e− and H+ from NADH, turning into ethanol. Keep in mind, the point of this process is not to produce ethanol; the purpose is to regenerate NAD^+ in order to keep glycolysis running. The alcohol and carbon dioxide are both by-products, and not particularly wanted by the cells creating it. Humans, however, have harnessed yeast cells to do their bidding as long ago as 7000 BC; fermenting foods and brewing alcoholic beverages have been in our culture for a long time.

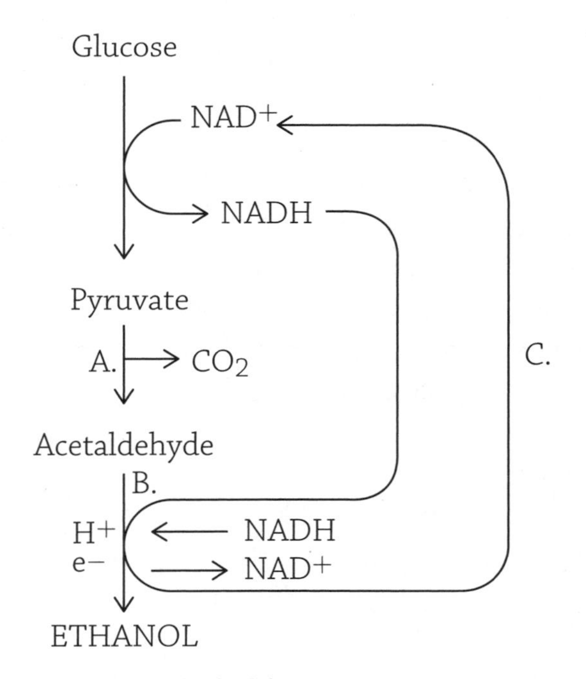

Alcohol fermentation

Let's walk through the above figure. First, pyruvate is changed into acetaldehyde (A), releasing a carbon dioxide in the process. NADH deposits its hydrogens and electrons onto acetaldehyde (B), which is reduced to ethanol. Meanwhile, the oxidized NAD+ returns to glycolysis (C) to keep the pathway going.

The carbon dioxide that is released from pyruvate conveniently provides the carbonation found in beer and other fermented drinks. In the brewing process, we stick yeast in an airtight vat with plenty of sugar for them to use for cellular respiration. Eventually, however, the oxygen runs out and the little yeasty-beasties switch to fermentation. The alcohol and carbon dioxide build up in this sealed system and, eventually, the poor yeast cells die in their own toxic pool of alcohol. Interestingly, this is the exact same fermentation that occurs when baking bread, but unlike beer, you don't need to be 21 years old to eat a dinner roll. When you allow bread to "rise" (ferment), the alcohol immediately evaporates as the carbon dioxide causes the dough to puff up.

Lactic Acid Fermentation

Lactic acid fermentation occurs in certain fungi, and you are now familiar with how we harness bacteria for the production of cheese and yogurt (the lactic acid produced is key to the production of these foods). In animals, muscle cells can switch from aerobic respiration to lactic acid fermentation when the need for ATP outpaces the supply of oxygen brought in by the blood. As in ethanol fermentation, this process serves as a means simply to regenerate NAD+. Once pyruvate is reduced by e− and H+, it turns into lactate, a weak acid. The lactic acid isn't wasted; it is carried over to the liver where it is converted back into pyruvate. The liver has plenty of oxygen, so the liver cells can send the pyruvate into their mitochondria to complete the process of cellular respiration (Krebs and electron transport chain)!

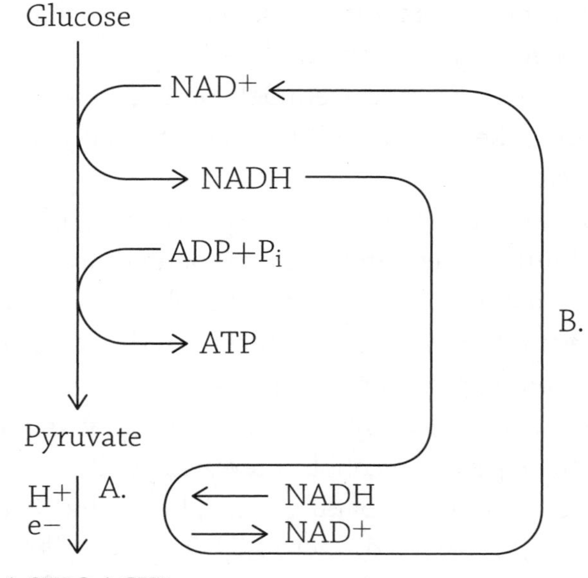

Lactic acid fermentation

In the figure above, we can see that NADH deposits its hydrogens and electrons onto pyruvate (A), which is reduced to lactic acid. Meanwhile, the oxidized NAD+ returns to glycolysis (B) to keep the pathway going.

You often hear people say their legs are burning because of the lactic acid generated after a hearty sprint or extended exercise session. Actually, this is not true. Yeah, acid burns, but the pain you feel is from tiny tears in the muscle tissue, not any fermentation by-product. Furthermore, lactate is not a bad thing; once oxygen is again available, it will be reconverted back into pyruvate and used in cellular respiration.

Photosynthesis: Light Reactions and the Calvin Cycle (AKA the dark reactions)

All eukaryotic cells undergo cellular respiration in order to convert the energy stored in glucose to the useable form of ATP. Where that glucose initially comes from, however, depends on the cell. If you are an animal cell, you must obtain the glucose from eating. If you are a photosynthetic cell, you make the glucose yourself through the process of photosynthesis.

Just like cellular respiration, the overall equation for photosynthesis sums up the process quite nicely: $6CO_2 + 6H_2O + \text{sunlight} \rightarrow C_6H_{12}O_6 + 6O_2$.

Notice that the equation for photosynthesis is the reverse of the equation for cellular respiration! This is significant, because if cellular respiration *releases* stored energy by *breaking* up glucose (a *catabolic* reaction), then photosynthesis needs an *input* of energy in order to *create* glucose (an *anabolic* reaction).

Photosynthesis can be broken up into two big halves: the **light reactions** and the **Calvin cycle**. The light reactions make ATP and NADPH; the Calvin cycle uses ATP and NADPH in order to create glucose from carbon dioxide. Think about it . . . if a cell needs to make glucose ($C_6H_{12}O_6$) from carbon dioxide (CO_2), it obviously needs some hydrogen atoms (there aren't any on carbon dioxide!). Furthermore, in order to glue together six carbon dioxides into a single molecule of glucose—including all those new hydrogens—there needs to be an input of energy (energy is required to create covalent bonds). These two things—a source of hydrogens and a source of energy—are provided by the light reactions!

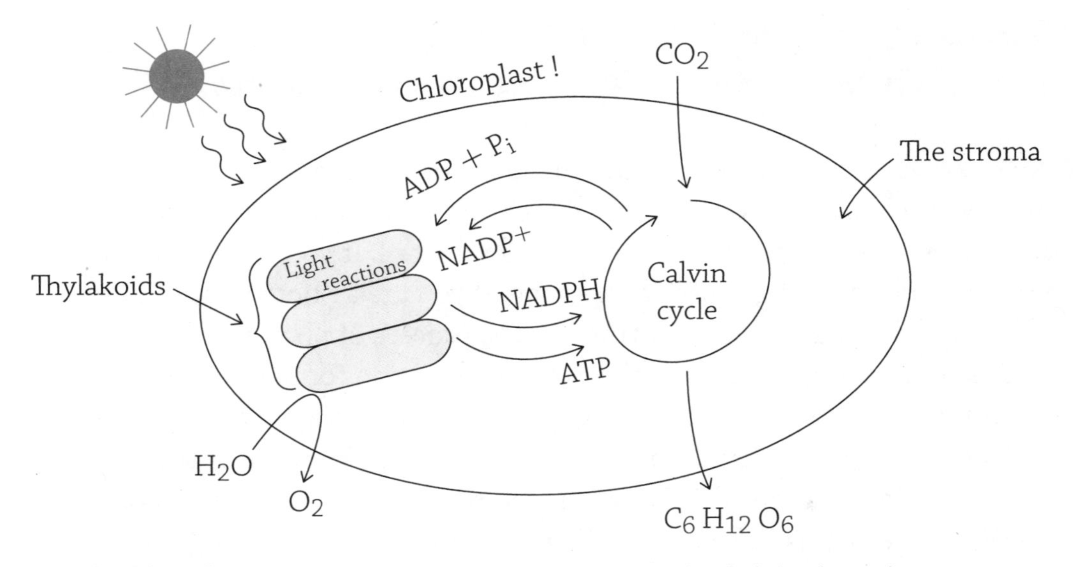

A single chloroplast. The light reactions are occurring on the thylakoid membrane, whereas the enzymes of the Calvin cycle reside in the stroma.

As you can see in the above figure, the light reactions provide the Calvin cycle with ATP (a source of energy). In turn, after the Calvin cycle uses the energy of ATP, it sends the "dead battery" (ADP and the lone phosphate) back to the light reactions to get recharged. What would be a good source of energy to use when recharging the dead battery? Yes, the sun! Our **must know** concept in all its glory: the chloroplast is transforming the energy of the sun into the chemical form of ATP.

What about that source of hydrogens? You have probably deduced that NADPH, like the NADH of cellular respiration, is a shuttle bus for electrons and hydrogens. Sure enough, if you refer to the previous figure, you can see that the "empty bus" ($NADP^+$), after having dropped off its passengers, goes back to the light reaction to pick up more.

You want an easy way to remember that NADPH is used in photosynthesis and NADH is used in cellular respiration? Imagine the "P" in NADPH stands for "photosynthesis"!

You probably know that chloroplasts are green. The green comes from pigment molecules called **chlorophyll** that are embedded in the thylakoid membrane. Pigments like to absorb light, which is why chlorophyll molecules are part of the light reactions. The pigments absorb so much light, in fact, an electron is energized to the point that it jumps out of the chlorophyll molecule. As in cellular respiration, this high-energy electron is passed down a series of proteins (an electron transport chain) that is embedded in the thylakoid membrane. And just as we learned in cellular respiration, the energy released by the electron is used to power tiny hydrogen pumps. These pumps move hydrogen ions from the stroma into the hollow thylakoid (called the **thylakoid space**). There is now a high concentration of H^+ stuffed into the thylakoid space, and they want to diffuse out back into the stroma. The only pathway through which they can diffuse, however, is through our old friend, the enzyme ATP synthase. The flow of hydrogens (chemiosmosis) powers the enzyme, and it catalyzes the reaction of adding P_i to ADP, creating ATP!

The electron that jumped down the chain isn't done, however. It is reenergized by another group of chlorophyll molecules and more sunlight. Once energized, it combines with $NADP^+$ and a lone H^+, creating NADPH.

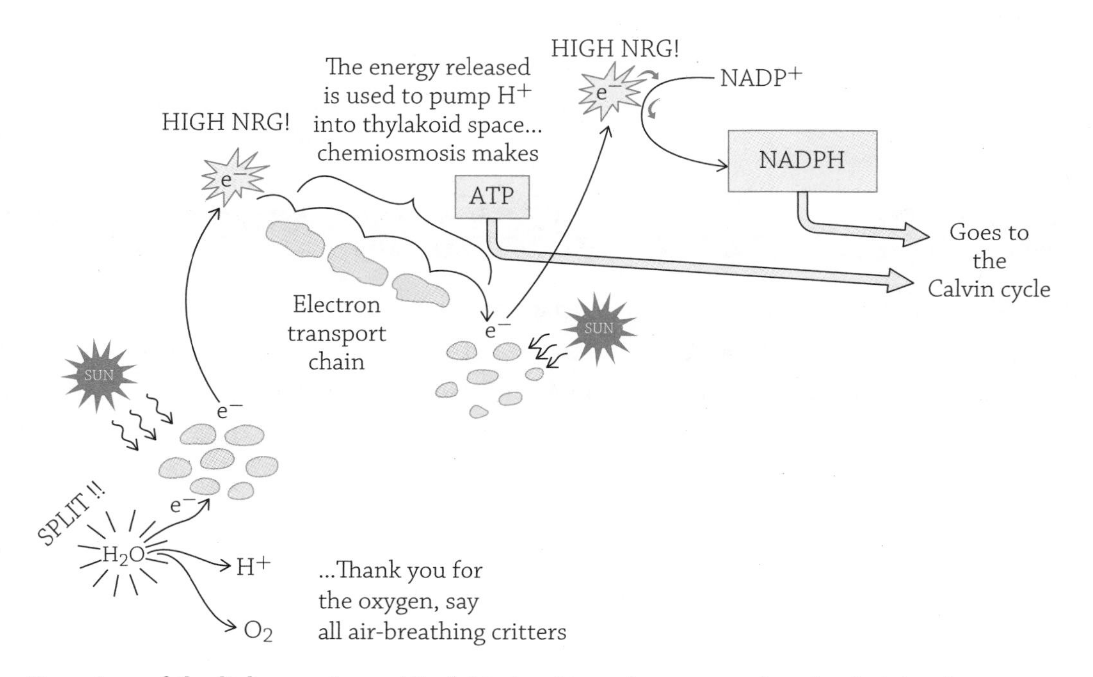

Overview of the light reactions. All of this is taking place on and in the thylakoid.

That little electron began its journey as a part of a chlorophyll molecule. It was energized by the sun, provided the power to create a hydrogen gradient inside the thylakoid, was reenergized by the sun a second time, and eventually ended up part of NADPH.

Do you see that molecule of water in the previous figure? It's important. We are missing an electron (the same one that ended up on NADPH). We need to split apart water to get a new electron to fill in the hole. When you rip apart water, you also release some H^+ (useful for chemiosmosis) and oxygen (useful for us!).

BTW

Want to blow your mind? What is the final-final resting place of that energized electron? It takes a ride on NADPH and goes over to the Calvin cycle, where it will become part of a brand-new molecule of sugar. When cellular respiration tears apart that glucose molecule and uses the electrons to power the mitochondrion's electron transport chain, the electron doesn't need sun to boost it up to a high energy level . . . why? Because it's already energized because it just absorbed all this sun energy in the light reactions of a chloroplast when the glucose was made! Our mitochondria are harnessing the power of the sun, though it has been converted into different forms along the way!

EXTRA HELP

All the oxygen found in the final glucose molecule ($C_6H_{12}O_6$) originated from the carbon dioxide molecules. The oxygen from the water molecules (H_2O) is instead released into the atmosphere after the water is split.

REVIEW QUESTIONS

1. Choose the best answer: The energy in ATP is released by cleaving the bond between:
   a. The first phosphate and the ribose
   b. The adenine and the ribose
   c. The third phosphate group and the second phosphate group

2. Cellular respiration is the process where cells convert the stored energy of ______________ into the useable form, ______________.

3. Cellular respiration is a(n) [choose one: **anabolic/catabolic**] reaction because energy is released as a molecule of glucose is broken down into smaller molecules of carbon dioxide.

4. What is the overall purpose of fermentation?

5. In photosynthesis, the light reactions create ________ (a source of energy) and ________ (a source of hydrogen).

6. Cellular respiration and photosynthesis both use chemiosmosis (the diffusion of hydrogen ions) in order to power ATP synthase and make ATP. Where is the high concentration of hydrogen ions stored in the mitochondrion? Where is it stored in the chloroplast?

7. If adenosine triphosphate is analogous to a battery, the fully charged form is [choose one: **ATP/ADP**] and the dead battery form is [choose one: **ATP/ADP**].

8. In glycolysis, two ADP are converted into two __________, two NAD^+ are reduced to form two ______________, and glucose is oxidized into two molecules of ______________.

9. What is the role of oxygen in the mitochondrion's electron transport chain? What molecule does oxygen end up in?

10. Choose the correct term from the pair: Photosynthesis is a(n) **anabolic/ catabolic** reaction because energy is required in order to build a molecule of glucose from smaller molecules of carbon dioxide.

11. The Calvin cycle creates glucose from carbon dioxide. In order to do so, it needs a source of ________________ (provided by NADPH) and ________________ in order to create new covalent bonds (provided by ATP). Both NADPH and ATP are provided by the __________________ occurring on the thylakoid membranes.

Cell Transport

MUST KNOW

- Gradients are necessary for a cell to live.
- Water diffuses from a region of high water potential to low water potential.

Gradients are life for a cell. Our **must know** is based on this idea: a cell without a chemical gradient (meaning it is at equilibrium) is dead. A chemical gradient is a form of potential energy, and without energy . . . well, that's no good. Think back to the previous chapter. A cell was able to generate a whole bunch of ATP by using the process of chemiosmosis, or the diffusion of hydrogen ions down their concentration gradient. That concentration gradient of H^+ was not easy to maintain! It required a constant flow of electrons to power those little hydrogen ion pumps. Without that gradient, ATP synthase wouldn't work. These gradients are everything to a cell. What we will discuss now is how, exactly, a cell maintains those gradients (active transport), and what processes are working to undo gradients (diffusion). When we learn about the diffusion of water, keep our second **must know** in mind: water will diffuse from a region of high water potential to low water potential.

Passive versus Active Transport

If you have a passive personality, it means you just go with the flow and don't try and force things to happen. Passive transport is similar—it's letting something just flow without the expenditure of energy. Passive transport is the movement of a substance from a region of high concentration to a region of low concentration. For example:

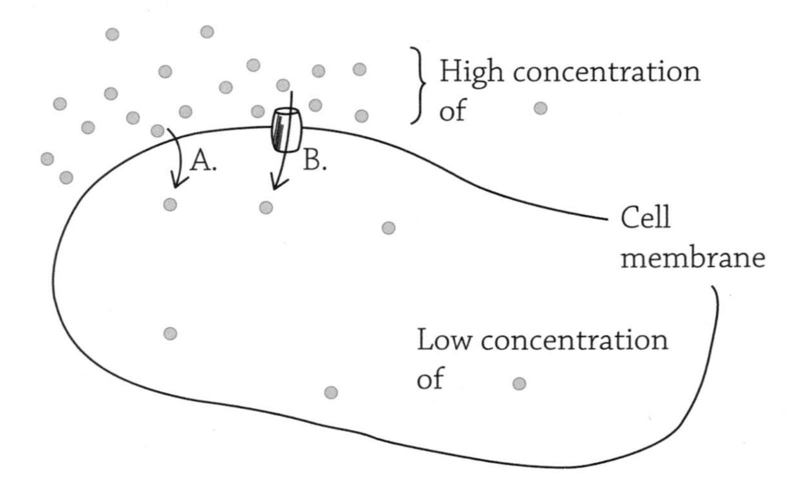

Diffusion. (A) Simple diffusion; (B) facilitated diffusion.

This happens spontaneously without an input of energy. For example, if you stand at one end of the room and spray perfume, your friend at the other end of the room would eventually notice. The scent particles were initially concentrated around you, yet after a relatively short amount of time, the particles diffused to areas of low concentration (the other end of the room). **Diffusion** (part A in the figure above) is the passive movement of molecules through a gas (like the air) or a fluid. In biology, diffusion is usually from the perspective of a **solute** moving through water across a semipermeable cell membrane.

Osmosis is the term to use if you're talking specifically about the diffusion of water. Finally, if a molecule wants to diffuse but is otherwise blocked by the cell membrane, it may need a bit of help with a protein tunnel of some sort. The term **facilitated diffusion** (part B in the figure above) refers to diffusion of molecules through a transport protein (the tunnel). All three of these processes work to wipe out concentration gradients, and do so without the slightest input of energy.

BTW

*The term **solute** refers to something dissolved in a liquid (salt, for example). The **solvent** is the liquid part, usually water.*

Active transport, simply speaking, is the opposite of diffusion: active transport moves things against their concentration gradient (from low concentration to high), and it requires energy, along with a transmembrane protein pump:

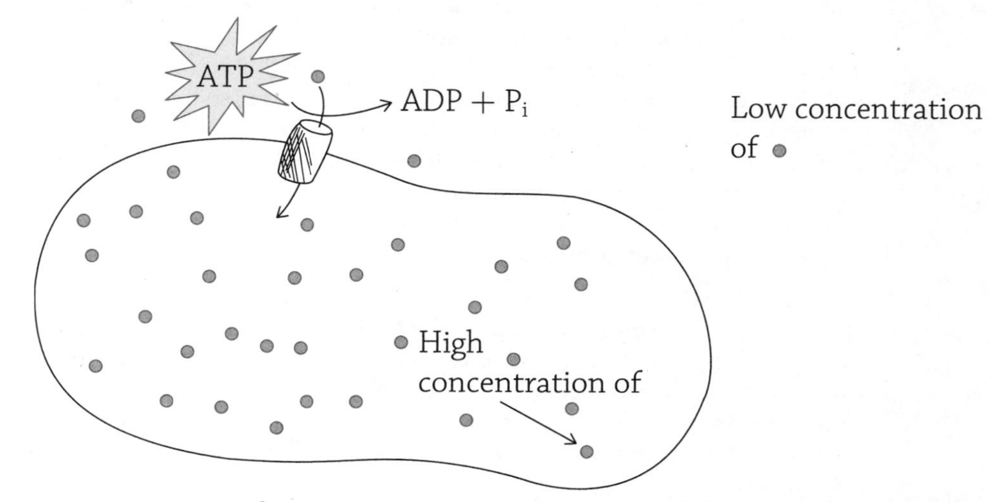

Active transport, involving a transmembrane pump and the expenditure of energy

Active transport is critical to a cell because, as we mentioned before, a cell at equilibrium is unable to generate energy; active transport battles against equilibrium! In a mitochondrion, there needs to be a high concentration of hydrogen ions in the inner membrane space in order to create ATP (chemiosmosis). Active transport created that concentration gradient in the first place. Recall that the energy lost as the electron dropped down the electron transport chain powered the proton pumps embedded in the inner membrane. If not for this concentration of H^+, the cell would die.

A common mistake is assuming if there is a transmembrane protein involved, it must be active transport. This is not the case! Facilitated diffusion also requires a protein. The difference is it is only acting as a tunnel—no energy needed—as opposed to the energy-powered protein pump of active transport.

IRL

Once upon a time (early 1900s), there was a new "miracle" diet drug that promised to make people lose weight fast. And sure enough it did . . . it also caused their body temperatures to spike as high as 110°F, followed quickly by coma and death.

This drug—2,4-dinitrophenol—worked by making the mitochondrial inner membrane leaky, allowing hydrogen ions to sneak back into the matrix without passing through the enzyme ATP synthase. This destroyed the mitochondrion's ability to create ATP via chemiosmosis. In turn, the body would desperately try harder to create the hydrogen gradient by sending even more electrons down the electron transport chain, as quickly as possible; the cells would burn through all the available glucose in a futile effort to ramp up ATP production. Because the mitochondria are working so hard, body temperature goes through the roof. Furthermore, for every electron that hits the end of the electron transport chain, there needs to be an O_2 to take it away (as water). If the rate of electrons flying down that chain increases, so does the need for O_2. Many deaths occurred due to excessive fever (111°F!) and unsustainable respiration rates.

Soon thereafter, this "miracle drug" was designated as dangerous and not safe for human consumption (duh). But here's the punch line: it's still being marketed today, even though the consequences of consumption have been studied and widely documented. The moral of these studies? If you hear of some "miracle drug" that is purported to do some magical deed, chances are it's at best an ineffective waste of money, at worst a dangerous and potentially deadly concoction.

Another type of active transport is on an even greater scale: endocytosis (also called phagocytosis) and exocytosis. In **endocytosis**, the cell uses energy to rearrange its internal skeleton so it can reach out pseudopods ("fake arms") in order to engulf a particle and package it within a **food vacuole**. The thing being engulfed could be a speck of food, an invading bacterium, or even an old and worn-out organelle that needs to be recycled. After the item stuck in the food vacuole fuses with a **lysosome** (another type of vacuole that contains hydrolytic enzymes), the item is digested into smaller bits. **Exocytosis** then moves the vacuole over to the cell membrane to expel any waste.

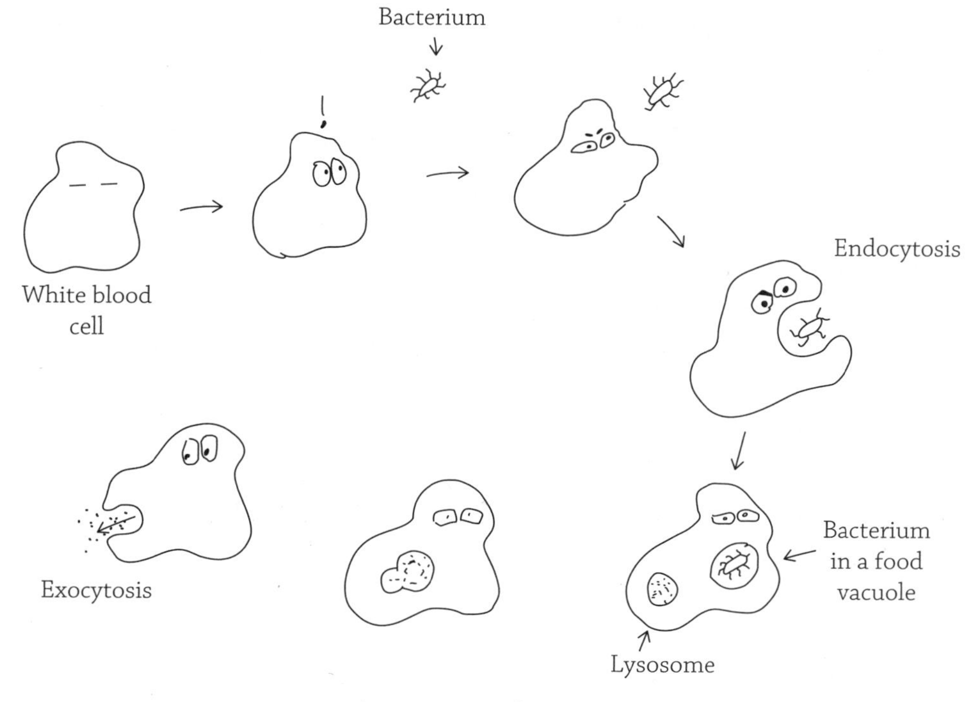

Endocytosis and exocytosis

The white blood cells of your immune system rely on endocytosis in order to phagocytose ("eat") invading bacteria and other foreign bodies.

Osmosis and a Cell's Surroundings

The diffusion of water can mean life or death for a cell. Imagine if you were a single-celled critter, like a paramecium, living in a freshwater pond. Your life is composed of happy days filled with feasting on bacteria and scooting about your habitat in search of pockets of the perfect temperature water (in other words, trying to maintain homeostatic conditions). You maneuver into a corner of the pond in which the water has an unusually high salt concentration due to road salt seeping into your habitat. All of a sudden, the water begins to leave your cell body, leached away by the saltier surroundings of the pond water. You respond instinctively (not like there's a lot of higher reasoning happening in a protist) and zip away from this salty corner and return to your previous perfect water conditions.

What happened here? First of all, a cell needs to maintain a constant amount of water inside itself. Losing too much water can lead to the cell shriveling up, and too much water can cause the cell to expand and explode. The ideal condition for a cell is to live in an aquatic environment where there is an equal rate of water loss and water uptake; this is called an **isotonic** environment. The paramecium strayed into water with a higher solute concentration than what is in the cell—this is called **hypertonic** conditions and it leads to water loss *from* the cell. If instead a cell is surrounded by pure water without a speck of solute (or at least less solute than what is in the cell), this is called a **hypotonic** solution and might lead to an influx of water *into* the cell. The following is a table to help you understand what these different solutions mean to a cell living within them:

Type of solution surrounding the cell	Amount of solute in the solution (compared to what's in the cell)	Direction of osmosis	Fate of an animal cell	Fate of a plant cell
Isotonic	The same	Water moves into and out of cell at the same rate	Happy!	Happy!
Hypotonic	Less	Water moves into the cell	A cell may burst if it doesn't have a way to pump out the excess water	Happy! A plant cell in hypotonic conditions enjoys turgor pressure
Hypertonic	More	Water moves out of the cell	A cell may shrivel	A cell may plasmolyze (and the plant will wilt)

When considering the movement of water across a cell membrane, you need to keep in mind that the selectively permeable cell membrane is preventing the solute from passing through. Therefore, only the diffusion of water (osmosis) can occur across the cell membrane. The diffusion of anything goes from high concentration to low; water will also diffuse from a region of higher concentration water to a lower concentration of water. When a cell is floating in a hypotonic solution, there is a lower concentration of solute surrounding the cell than what is inside the cell. That means there is a higher concentration of water surrounding the cell, and it will diffuse down its concentration gradient (into the cell). The opposite is true for a cell floating in a hypertonic solution. There is a higher concentration of solute surrounding the cell, and less water. The inside of the cell has more water, and so it will diffuse out.

In order to simplify things a bit, we can use our **must know** term **water potential** to help us describe the direction of osmosis. It is an actual numerical value that takes two things into consideration: the amount of solute added to the water (Ψ_s, or **solute potential**) and the amount of pressure exerted on the system (Ψ_p, or **pressure potential**). I know this sounds weird and confusing, but let me try to explain.

1. Water potential (symbolized by this → Ψ) is an actual number (measured in units of pressure called megapascals, or MPa) that can be calculated, and it can help us decide the direction water will flow:

$$\Psi = \Psi_s + \Psi_p$$

2. Water will move from a region of *high water potential* to *low water potential.*

As you can see, the overall value of a solution's water potential takes two things into consideration: the solute potential (Ψ_s) and the pressure potential (Ψ_p). Start with this fact: pure water in an open container has an overall Ψ of zero. "Pure water" means there is nothing dissolved in it ($\Psi_s = 0$) and if it's in an open container, there is no change from normal atmospheric pressure ($\Psi_p = 0$). Now let's consider what happens when you alter either the amount of solute or the pressure.

Pressure Potential Ψ_p

The pressure potential is the amount of pressure exerted on a solution. Look back at what is required to have an overall zero water potential—there cannot be any pressure other than just atmospheric pressure (saying it's in an "open container" means just that, only the atmosphere is pushing down on the solution). Once pressure is either added or removed from the solution, the pressure potential can become either negative or positive. If a cell has a positive pressure, that means the cell is very full of water and it's pressing against the cell membrane, like an overinflated balloon. If, instead, there is a vacuum applied and water is being pulled from the cell, the cell is under a negative pressure.

Solute Potential Ψ_s

Water potential defines the amount of water in a solution: a high water potential means that there is a lot of water in the mix (and not much stuff dissolved in it). Pure water has the highest water potential possible . . . because it's pure water. If you start shoveling solute into the water, you are lowering the water potential because, per volume, some of the water is being displaced by the stuff dissolved in it. Now, if *pure water* has a solute potential of zero, guess what it becomes if you add stuff to it? Negative! You can never have a solute potential larger than zero (can't make pure water any more watery). As solute concentration increases, the Ψ_s becomes more negative. In order to calculate the numerical value for the solute potential, you need to use this equation:

$$\Psi_s = -iCRT$$

i = ionization constant

C = molar concentration

R = pressure constant ($R = 0.00831$ liter MPa/mole K)

T = temperature in kelvins (K = °C + 273)

It makes sense that you need to know the molar concentration of the solute (C); the more concentrated the solute, the lower the solute potential. The "R" is a pressure constant that never changes, and the temperature must be in kelvin (add 273 to the Celsius temperature). That "i" variable, however, may not be as obvious. It's the ionization constant, and it takes into consideration how many "pieces" the solute breaks into upon dissolving. Glucose ($C_6H_6O_6$), for example, will dissolve when mixed with water, but that doesn't mean the actual molecule breaks up into pieces. If you mix

100 molecules of glucose in a small volume of water, you will still have 100 molecules of glucose. Molecules that ionize, however, are different. Table salt (NaCl) will ionize into Na^+ and Cl^- when dissolved in water. If 100 molecules of NaCl are dissolved in a small volume of water, you will end up with 200 separate bits once each molecule of NaCl splits into two separate ions. Therefore, the ionization constant for NaCl is *two*! Finally, notice the entire equation begins with a negative sign. Makes sense, because if you dissolve *anything* in water, it is no longer pure water and will have a solute potential below zero (i.e., negative).

EASY MISTAKE!

It's so easy to forget that little negative sign when calculating the solute potential. Just remember: pure water has a solute potential of zero, and if you dissolve anything in water, you lower the solute potential . . . making it negative! Your solute potential will never be a number greater than 0.

EXAMPLE

▶ What is the solute potential of a 0.15 M NaCl solution at 24°C?

▶ Below, I will walk you through how to solve the previous problem using the equation for solute potential, $\Psi_s = -iCRT$.

$i = 2$ (because NaCl will break up into Na^+ and Cl^- in solution)

$C = 0.15$ M

$R = 0.00831$ liter MPa/mole K

$T =$ temperature in kelvins ($24°C + 273 = 297$)

$$\Psi_s = -iCRT = -(2)(0.15M)(0.00831 \text{ liter MPa/mole K})(297)$$

$$\Psi_s = -0.74 \text{ MPa}$$

Living with Osmotic Challenges

As you read above, cells tend to be happiest when they are in isotonic conditions. But if cells only lived in such a narrow selection of conditions, imagine all the ecological niches that would remain open and unexplored! Therefore, natural selection has come up with some ways for cells to live in water conditions that would otherwise be dangerous. For a single-celled critter to live in hypotonic conditions (fresh water), it has to have a way to deal with excess water entering the cell. Luckily, a species of paramecium has a special organelle called a **contractile vacuole** that pumps excess water back out of the cell. In an earlier chapter, we learned that plant cells have a large organelle called a central vacuole that holds and stores excess water. It expands and inflates the outer cell membrane, but it won't pop; instead, the cell membrane presses up against the plant cell wall, providing the pressure needed to keep the cell wall stiff. This is called *turgor pressure*, and it is how plants maintain their structure. If, instead, a plant cell is surrounded by a hypertonic solution, the water will seep out and the cell membrane will shrink and pull away from the cell wall. This causes a loss of turgor pressure, and the plant will wilt.

IRL Think about what your fingertips look like after soaking in a bath for a long time—they get pruney and wrinkly, right? And if you're wrinkling up, it means your cells are losing water and your bath must be a hypertonic environment? But that doesn't make sense, considering that your bath is freshwater. The truth is, your bath is a hypotonic solution, and because water is moving into your cells, you're actually puffing up (not shriveling!). There is not a consensus on why this happens, but there's a couple really good hypotheses. One idea is that the strange, wrinkly look is because your outmost epidermal layer is swelling and increasing its surface area, so it has to become folded in order to fit on the area provided by your fingertip. Another hypothesis is that wrinkling of the surfaces of the fingers and toes provides an evolutionary advantage when hanging on to wet surfaces, because the higher surface area provides a better grip!

REVIEW QUESTIONS

1. What two things are needed in order for active transport across a cell membrane to occur?

2. Compare and contrast diffusion, facilitated diffusion, and osmosis.

3. Is the process of endocytosis passive or active transport?

4. A cell living in hypotonic conditions must deal with osmosis occurring [choose one: **out of the cell/into the cell**].

5. A paramecium is a single-celled organism that normally lives in a freshwater environment. What would happen if it was transferred to a saltwater habitat?

6. What is the solute potential (Ψ_s) of a 0.23 M NaCl solution at 20°C?

7. Glucose molecules do not easily pass across a cell membrane, even if there is a higher concentration of glucose molecules on the outside of the cell than on the inside. What type of cell transport would help the passive movement of glucose molecules into the interior of the cell?

8. Why is it bad to disrupt a cell's chemical gradients?

9. For each of the following, indicate if the solution surrounding the cell is hypertonic, isotonic, or hypotonic:
   a. A plant cell floating in pure water: ___________________.
   b. A protist cell normally living in a freshwater lake transferred to a marine environment: ___________________.
   c. A bacterium in a solution that has the same water potential as inside the bacterial cell: ___________________.

10. Some cells that live in hypotonic conditions have evolved to rely on an organelle called a ________________________ to pump out the extra water that diffuses inward.

11. If a cell is in an open container and has a solute potential of −2.0 MPa, what is its overall water potential?

12. Which of the following cells is *not* relying on active transport?
    a. A respiring cell releasing carbon dioxide into the environment
    b. A freshwater paramecium maintaining homeostasis via a contractile vacuole
    c. Uptake of glucose by the cells lining the small intestine
    d. Neurons creating a concentration gradient of sodium and potassium ions

Signal Transduction and Cell Communication

MUST KNOW

- In signal transduction, the original signal never enters the cell.
- The purpose of signal transduction is to move the message into the cell in order to elicit a response.
- The G protein-coupled receptor is found in all eukaryotes.

Cells are in constant communication with their surroundings, and the signals they receive are frequently in the form of chemicals. Considering that the cell membrane provides a barrier separating the inside of the cell from the surrounding environment, the signal must somehow pass through the membrane and into the cell's interior. If the chemical signal is lipid soluble, it is easy! The saying "like-dissolves-like" applies here, because if the signal is a lipid and the membrane is made of lipids, the signal molecule will just pass through. More often than not, however, the signal is made of protein (many hormones are proteins) and thus cannot pass through the cell membrane. For example, the hormone epinephrine (also called adrenaline) is responsible for telling cells in the liver and muscles to release glucose from the storage polymer glycogen. If epinephrine is a protein and cannot pass into the cell, how is it supposed to cause the hydrolysis of glycogen *within* the target cell? As our **must know** concept states, the original signal never enters the cell. The answer to this problem lies in the process of signal transduction.

Three stages are needed for a signal molecule to illicit a cellular response without actually entering the cell:

1. Reception
2. Transduction
3. Response

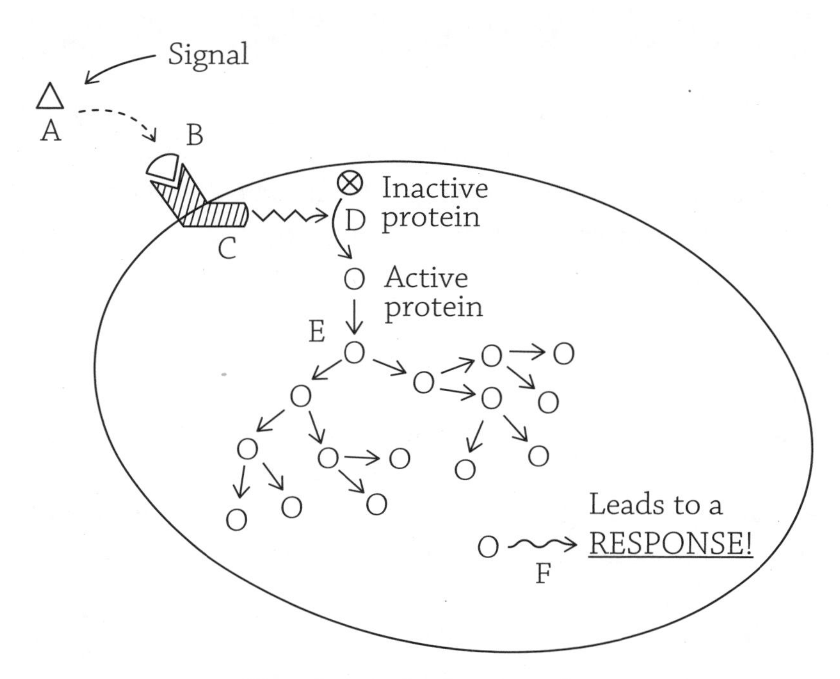

Signal transduction overview

Let's take a look at what's happening in the figure above. The initial signal (A) never enters the cell. Instead, it binds (B) with a transmembrane receptor protein. As soon as the receptor binds the signal, the inside portion changes shape (C) and activates another protein (D). The activated protein activates other proteins in a chain reaction (E). Eventually, the swarm of activated proteins lead to some sort of cellular response (F).

Step 1: Reception

Reception is what it sounds like: the signal is "received" by the cell. The cell hears the signal when the ligand binds with a membrane receptor. The receptor is a transmembrane protein that spans the width of the cell membrane. The part

EXTRA HELP

Here is another example of a protein's tertiary (3D) structure being very important to its function. Not only that, the ability of a protein to be flexible and change shape is also very important. An enzyme's function relies on these structural qualities, as does our signal transduction membrane receptors.

that extends to the outside of the cell has a specific shape that fits perfectly with the ligand. This is important because it ensures only the correct target cell (with the correct receptor) actually hears and responds to the signal. Something happens at this step that magically moves the message into the cell without the actual signal getting through. Once the transmembrane receptor binds the signal, it changes shape.

When the inside portion of the receptor protein changes shape, it essentially moves the signal into the cell. But before we get to the next step of transduction, let's look at two specific types of receptors: **G protein-coupled receptor** and the **ligand-gated ion channel.**

BTW

The phrase "conserved through evolutionary history" means an adaption was so helpful to the survival of the organism that the trait was maintained through countless rounds of natural selection. If a species has a characteristic that isn't very helpful (or, even worse, is disadvantageous), then those individuals would not survive and pass down those characteristics to the next generation. The G protein-coupled receptor, however, was so pivotal and critical to proper cellular functioning, it was maintained through many generations. Since the G protein-coupled receptor first arose in an early eukaryotic ancestor, it is now in almost all eukaryotes: fungi, protists, and animals (and maybe plants, though this is not yet known for sure).

G Protein-Coupled Receptor

This is probably the most important type of receptor protein to study. The G protein-coupled receptor is the main signaling pathway in a wide variety of eukaryotes, which suggests it evolved very early and has been conserved through evolutionary history.

When a signal molecule binds to the "outside-the-cell" portion of the G protein-coupled receptor, the receptor protein becomes activated and changes its shape on the cytoplasmic (inside) portion. There is a separate protein called (appropriately enough) a G protein that hangs out next to the cytoplasmic portion of the receptor. Once the receptor is activated, it also activates its buddy, the G protein. Finally, this activated

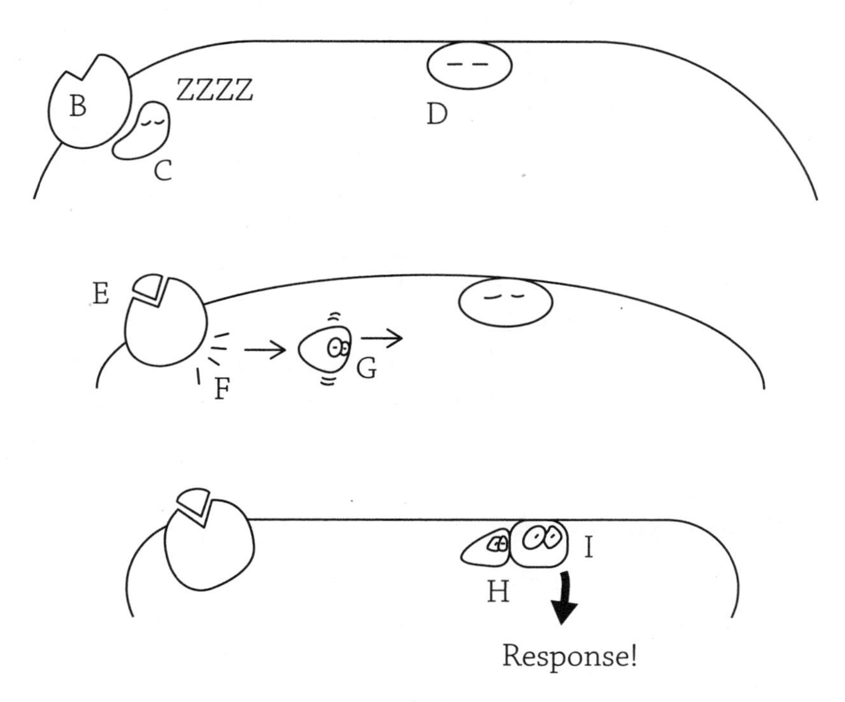

G protein-coupled receptor

G protein slides along the cell membrane over to an inactive enzyme and activates it! The purpose of this now-activated enzyme could be many different things (it depends on the specific cell type in which this signal is being sent).

As we can see in the figure above, if the signal (A) has not been yet grabbed by the receptor (B), then the partner G protein (C) remains quiet and inactive. There is an inactive enzyme (D) waiting for its own signal. Once the receptor binds the signal (E), the receptor activates the G protein (F), and the activated G protein slides along the inner membrane (G). The G protein binds to the awaiting enzyme (I), activating it and leading to a response.

Because it is the most abundant type of mammalian cell-surface receptor, there are numerous cool examples of G protein-linked receptors in action. For example, some bacterial toxins wreak havoc with G protein function and cause diseases such as cholera, whooping cough, and botulism. Another cool (and less deadly) example is some folks' ability to taste the chemical phenylthiocarbamide (PTC). The gene *TAS2R38* codes for a G protein-coupled taste receptor located in the cells of the taste buds. This receptor binds to the chemical PTC, resulting in a large number of proteins being made that stimulate neighboring neurons to send a "bitter" signal to the brain. For many people, the gene that codes for the PTC receptor is mutated. These people don't "hear" the bitter signal of PTC. A long time ago (during our hunting/gathering days), having the ability to easily detect bitterness could help reduce the chances of ingesting toxic plant materials (plant toxins often have a bitter taste). Nowadays, a robust ability to detect the PTC signal tends to influence our food preferences and make some items—such as broccoli and rhubarb—unpalatable and bitter.

Ligand-Gated Ion Channel

The ligand-gated ion channel acts as a doorway for ions such as Na^+ or Ca^{2+}. The key to unlocking this door is the signal molecule, which itself does not pass into the cell. The "message" is instead the influx of ions that just occurred because the signal molecule swung open the gate!

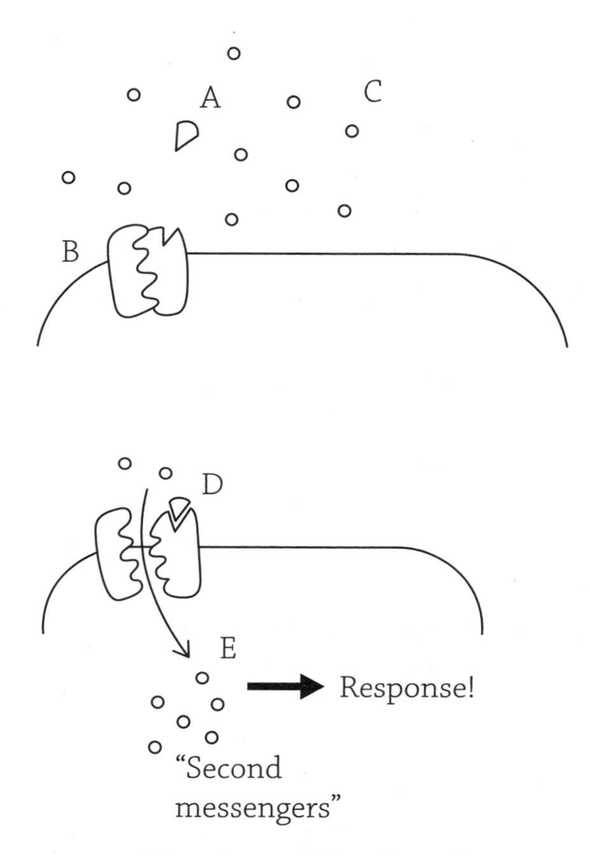

Ligand-gated ion channel

From the figure above, we can see that before the signal (also called the ligand, A) binds the receptor, the receptor remains closed (B), blocking the entry of extracellular ions (C). Once the signal binds (D), the receptor gate swings open and the ions (also called "second messengers") rush in (E), leading to a response.

Ligand-gated ion channels are key to nervous system function. When a signal must traverse the open space between two adjacent neurons (the space is called the synapse), the chemical messenger is a neurotransmitter. Neurotransmitters aren't allowed into the cell, and instead they latch onto ligand-gated ion channel receptors. Once this happens, it signals the second neuron that it needs to “fire” in the forms of a wave of positive charge that whooshes down the cell. This wave occurs because the ligand-gated ion channels open and Na^+ flows in (more about neuron function in Chapter 26).

Step 2: Transduction

Now that the cell has “heard” the signal, it must be amplified. This is a super helpful step because even if only one, single, signal molecule managed to find its way to the target cell, it is enough to create a large coordinated response. Once the message has been moved into the cell, it can now be amplified and made louder through transduction. The transduction step is based on a cascading series of molecular interactions that boosts the signal as it is moved through the cell. It’s like a cascade of dominoes!

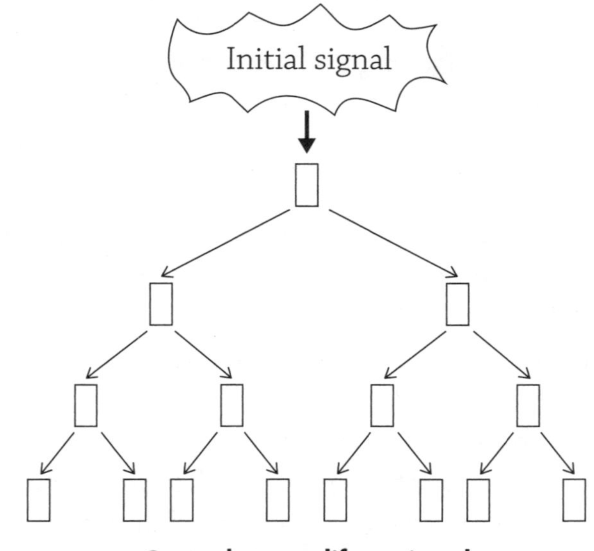

Cascades amplify a signal

Usually, each step involves activating a protein and making it change its shape. This shape change is caused by phosphorylation, or adding a phosphate group to a protein. Phosphorylation (and dephosphorylation) of proteins is a widespread means of regulating the activity of the proteins. The enzyme responsible for transferring a phosphate group from ATP and sticking it onto the target protein is called a kinase.

Protein kinases are enzymes that transfer phosphate groups from ATP to a protein. Protein kinases play a major role in signal transduction pathways. They often activate other protein kinases and create a cascade, just like the domino analogy. If you swap out the dominos with protein kinase enzymes, you have created a phosphorylation cascade.

Since adding a phosphate changes a protein from inactive to active, you also need a way to stop the activation. Protein phosphatases are enzymes that remove phosphate groups from proteins, inactivating the pathway. This is important because if the initial signal is no longer present, you don't need the response to keep going!

BTW

If you see an enzyme and it has "kinase" in its name, it's a hint to the reaction it catalyzes. Kinase enzymes specialize in transferring phosphate groups to (or from) ATP. The glycolytic pathway alone has many different kinase enzymes.

Step 3: Response

Finally, we get to the point of all this . . . the response. It can be any cellular activity that occurs in the nucleus or in the cytoplasm. Many signaling pathways result in a gene being expressed, because the initial signal told the cell "Make this specific protein! Now!" The pathway may instead result in the activity of a protein that's already waiting in the cytoplasm. Earlier, I mentioned epinephrine as an example signal, and the response was the activation of an enzyme that cleaved apart glycogen in order to release glucose molecules. Your liver cells shouldn't send out a flood of glucose molecules unless it hears the proper signal (epinephrine) to do so.

Yeast and Shmoo Love

Here is a romantic example of the entire process of signal transduction: yeast mating. And yes, yeast can "mate." Usually, they reproduce by asexually budding, but what's the fun (and genetic variation) in that? The simple baker's yeast—*Saccharomyces cerevisiae*—has the option to mix up its alleles (and thus increase variation) by fusing with another cell. Each yeast cell has two mating types: ***a*** and α. To ensure maximum variation, yeasty beasties prefer to mate with the *other* mating type: ***a*** mates with an α. In order for a yeast cell to "sense" another nearby yeast cell of the correct mating type, it relies on chemical pheromones that are released by the cells into the surroundings:

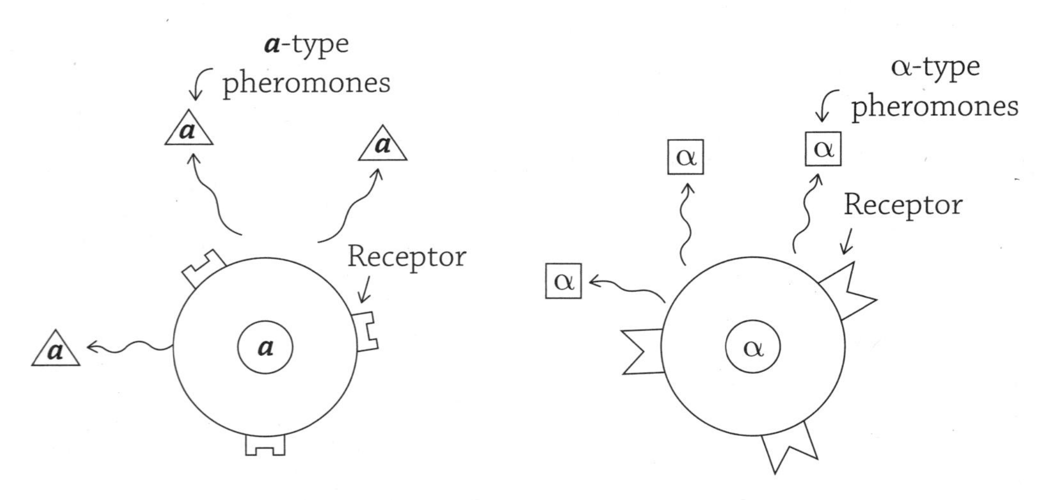

Yeast cells sending out mating signals

In order for an ***a*** cell to "hear" the signal from a nearby α cell, the α's signal must bind to the ***a*** cell's G protein-linked receptor (and vice versa).

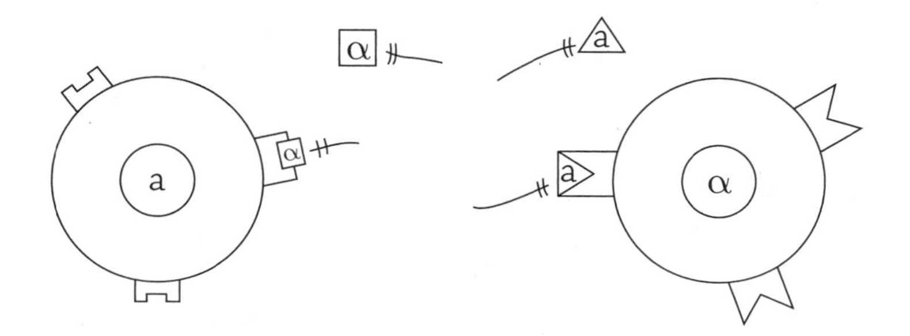

Yeast cells "hearing" the signal from the other mating type

The protein mating signal, once bound, activates a signal transduction pathway. This causes a signal cascade of a bunch of protein kinases activating other protein kinases, and so on and so forth. The kinases eventually descend upon the cell's nucleus and spark the activity of a certain transcription factor (more on that in Part Three: Genetics) that initiates transcription of genes necessary for the cells to elongate and grow toward their mating partner. Specifically, the shape change occurs in the direction toward the highest concentration of pheromone, and the spherical little yeast cell forms a pear-like shape called . . . wait for it . . . a *shmoo*!

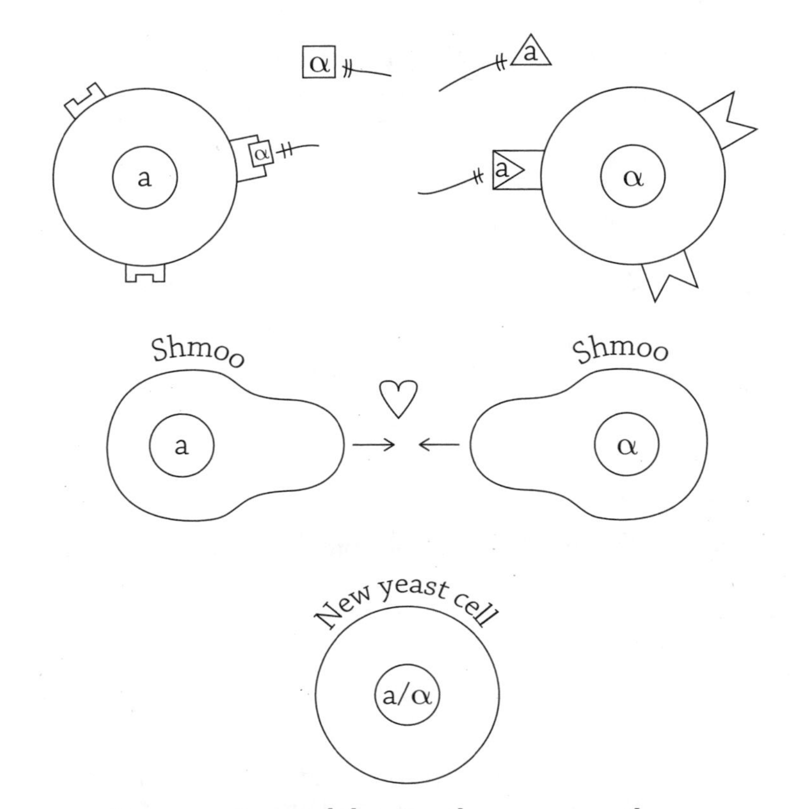

Yeast mating and the signal response pathway

Shmooing takes a lot of energy! The signal transduction pathway ensures that the cells only respond when a mating partner is close enough to make the attempt. Making a love connection is exhausting.

REVIEW QUESTIONS

1. Which category of chemical signal (lipid or protein) relies on the process of signal transduction to move the message into the cell?
2. Why does the receptor have to be a transmembrane protein, instead of a surface protein?
3. Put the following steps of G protein-coupled receptor function in order:
    a. Cytoplasmic side of the receptor changes shape
    b. G protein activates an enzyme on the inside of the cell
    c. Signal binds the receptor
    d. G protein is activated
4. What is the purpose of the transduction step of signal transduction?
5. List the three stages of signal transduction.
6. As soon as a signal binds the receptor, what occurs to move the message into the cell?
7. Put the following steps of ligand-gated ion channel function in order:
    a. Second messengers rush into the cell
    b. Receptor gate opens
    c. Receptor is closed
    d. Signal binds the receptor
8. Which receptor type is key for nervous system function?
9. Protein kinase enzymes play an important role in signal transduction because they ______________ the message by creating a ____________________ cascade.

10. Which of the following is a response elicited by the yeast mating signal pathway?
    a. A protein kinase cascade
    b. Activation of transcription factors
    c. Directional growth
    d. All of the above

PART THREE

Genetics

As we learned earlier, the building block of life is the cell. A cell must be able to perform its particular function, and it must be able to divide and create more cells. All the information a cell needs to do its job is contained within its deoxyribonucleic acid (DNA). Without DNA, there cannot be cells; without DNA, there can be no life.

DNA and RNA: Structure of Nucleic Acids

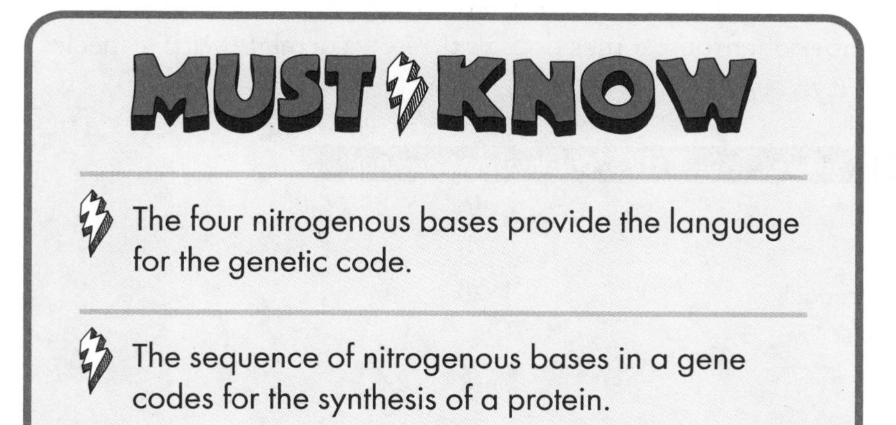

ll of the collective DNA held within a critter (bacterium, protist, fungus, plant, or animal) is called its **genome**. Prokaryotic cells have it easy; their genome is one single loop of DNA (along with tiny "bonus" loops of DNA called **plasmids**). Eukaryotic cells have way more DNA, and it's organized into separate linear segments called **chromosomes**. The number of individual chromosomes depends on the species; humans have 46 chromosomes, and each cell's nucleus (with the exception of the egg and sperm . . . more on that later) contains the exact same collection of 46 chromosomes. But don't assume that all other critters must have fewer than 46 chromosomes because, you know, humans are so special and complex. The number of chromosomes doesn't directly correlate with a species' complexity:

Species	Chromosome number
Human	46
Chicken	78
African hedgehog	90
Fruit fly	8
Radish	18
Pitcher plant	78

The number of chromosomes doesn't tell you how many individual genes there are. For example, even though chickens have more chromosomes than humans, they still have about the same number of genes as us (20,000–25,000). A **gene** is a segment of DNA within a chromosome that codes for a certain protein. And no matter the life-form, we all share a similar DNA language; all DNA consists of the same selection of individual monomers, called nucleotides.

BTW

Your genome is like a library, and each individual book in that library is a chromosome. Each book (chromosome) is filled with letters (nucleotides), and when the letters are grouped, they spell individual words (genes). The Must Know concept is to understand that the four different kinds of DNA nucleotides create the language for our genetic code. The DNA alphabet only needs four different "letters" to create everything from humans to chickens, bacteria to toadstools.

Nucleotides

Even though you are quite different from a chicken, your DNA is composed of the same selection of nucleotides as your friend the fowl. As we learned in a previous chapter, there are four different nucleotides, each defined by the type of nitrogenous base it carries: adenine, thymine, guanine, or cytosine. These four nitrogenous bases provide the language for all life's genetic code. The language of ribonucleic acid (RNA) is also composed of four nitrogenous bases: three of which are shared by DNA (adenine, guanine, and cytosine) and one of which is unique to RNA (uracil).

The nucleic acids (both DNA and RNA) are large carbon-based polymers. Recall that a polymer is a large molecule that is composed of smaller, repeating subunits called monomers. The monomer of DNA and RNA is a nucleotide:

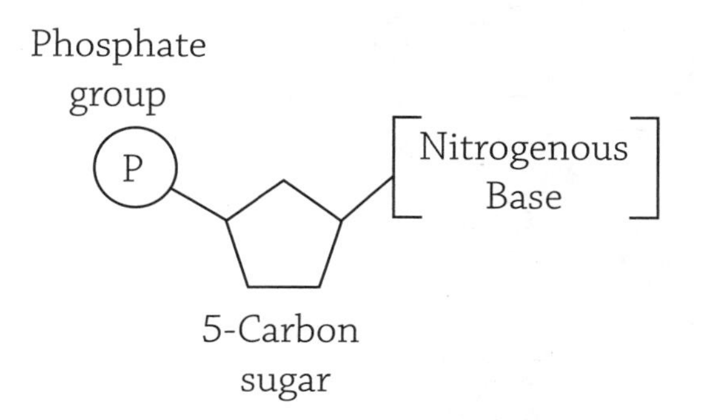

Basic structure of a nucleotide (either DNA or RNA)

Each nucleotide has three components to it: a 5-carbon sugar (ribose in RNA, deoxyribose in DNA), a phosphate group, and a nitrogen-containing base. Recall that the **must know** concept focuses on the four different options for nucleotides, and the difference is found in the nitrogenous base. The sugar and phosphate groups of the nucleotide never change. This is a good thing, because when monomers link together to create the larger polymer, the sugar and phosphate groups hold on to each other by forming strong covalent bonds. This forms the sugar-phosphate "backbone" of the DNA molecule (as seen in the following figure).

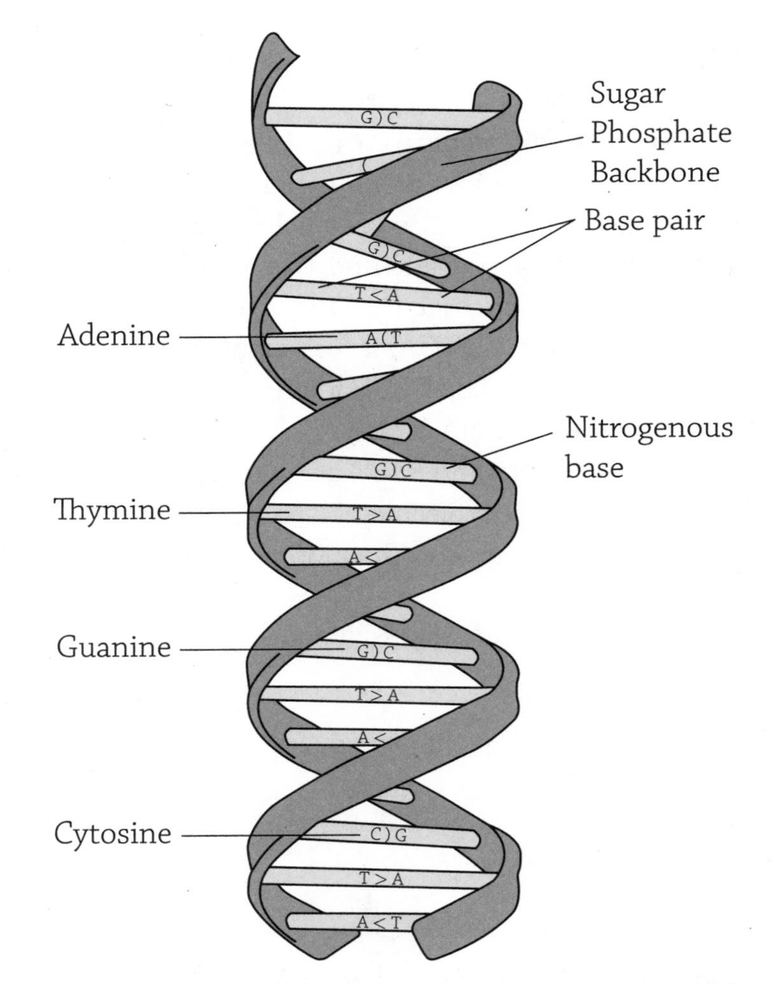

Structure of double-stranded DNA. Notice the location of strong covalent bonds along the sugar-phosphate backbone, and the weaker hydrogen bonds between the adenine-thymine and guanine-cytosine pairing.

Author: MesserWoland. https://commons.wikimedia.org/wiki/File:DNA-structure-and-bases.png.

The nitrogenous base is hanging off the sugar part of each nucleotide and is not part of the alternating sugar-phosphate backbone. If you are talking about the double-stranded structure of DNA, the base is pointing "inward" toward the center of the double helix, and it is gently holding on to the second strand of the double helix. This bond is not the strong covalent bond of the backbone, but instead a special intermolecular bond called a hydrogen bond. Hydrogen bonds occur along the length of the entire DNA molecule, holding the two strands together in what is called a **base pairing**. The hydrogen bonds occur between specific pairings of nitrogenous bases of the two strands:

DNA Base-Pairing Rules

Base	Its "partner"	Number of H-bonds shared
Adenine (purine) 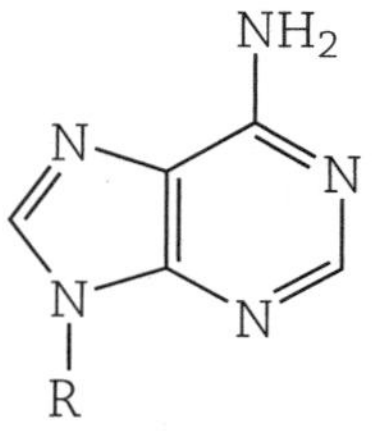	Thymine 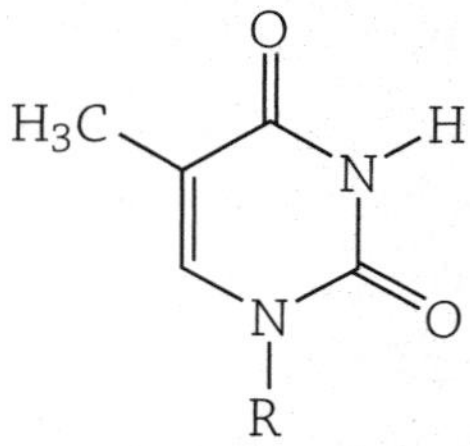	Two
Thymine (pyrimidine) 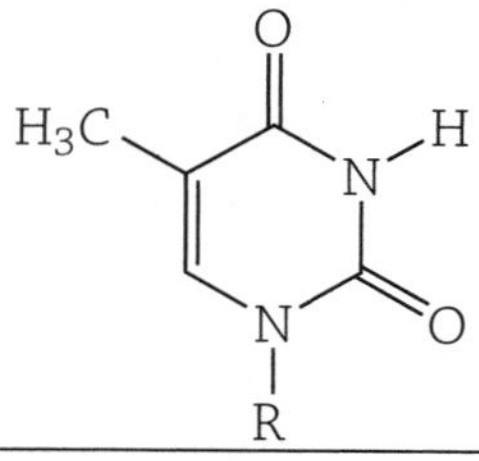	Adenine 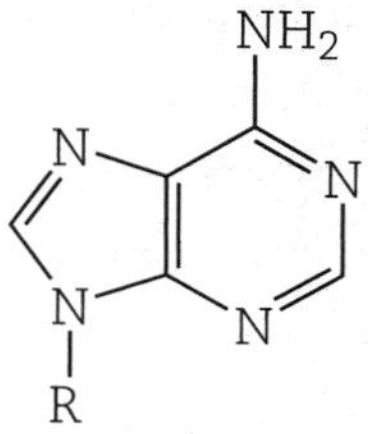	Two
Guanine (purine) 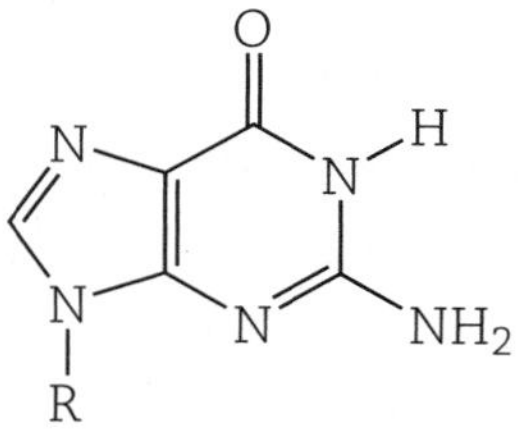	Cytosine 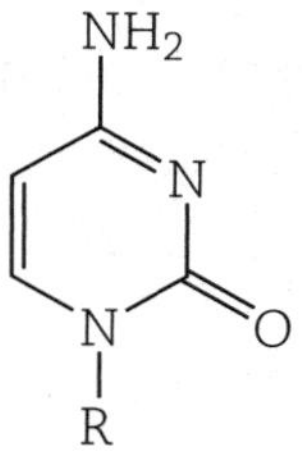	Three

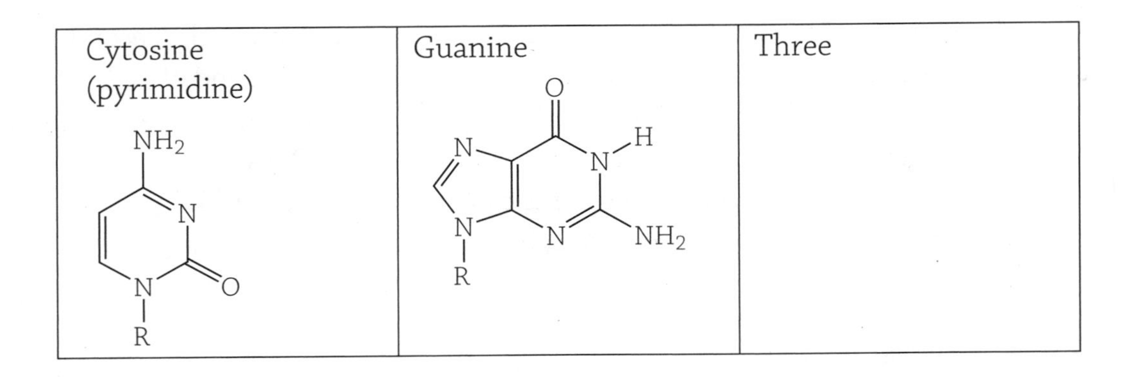

Cytosine (pyrimidine)	Guanine	Three

In DNA, adenine always pairs up with thymine, and guanine always pairs up with cytosine. There are, in fact, two reasons for this specific pairing. One has to do with the different shapes of the two types of nucleotides: purines versus pyrimidines. Think of it this way: structurally speaking, why wouldn't it be a good idea to pair up [adenine + guanine] and [cytosine + thymine]? Look at the adenine and the guanine . . . those are both the big, double-ringed purine bases. That leaves the two smaller pyrimidines (cytosine and thymine) to base-pair. Depending on the pairing, your DNA molecule would not have a consistent width!

The second reason the base-pairing rules exist has to do with the number of hydrogen bonds formed between each pair (look at the third column of the previous table). Adenine and thymine make a great pair because it consists of one large base and one small base, and they both like to form *two* hydrogen bonds. The guanine and cytosine pairing also consists of one large base and one small base, plus they both want to form *three* hydrogen bonds.

Note that, in order to maintain a consistent width, there must always be a single-ringed nucleotide paired with a double-ringed nucleotide. The second reason it must be A-T and G-C pairing is due to the numbers of hydrogen bonds each base wants to make.

Structure of RNA

Even though DNA tends to get the most attention, RNA is just as important. It is very similar in structure, and our **must know** concept applies here, too: the four different bases of RNA provide the language for its genetic code. There are differences, however, between the structures of RNA and DNA:

1. Instead of the sugar deoxyribose, RNA nucleotides contain the sugar *ribose*.
2. RNA is *single stranded* instead of double stranded, like DNA.
3. RNA has the nitrogenous base *uracil* instead of thymine. If base pairing occurs between DNA and RNA, the uracil will partner up with adenine (see below):

EASY MISTAKE! If you solve a question about the base sequence of RNA, don't forget to use U instead of T!

Base	**Its "partner"**	**Number of H-bonds shared**
Adenine (purine) 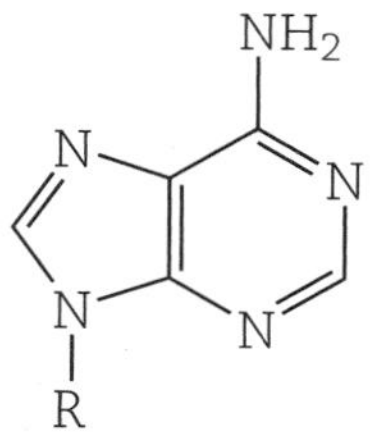	Uracil (pyrimidine) 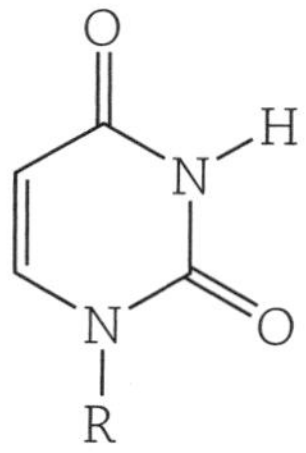	Two

There are two very important types of RNA that will play a pivotal role in gene expression: **messenger RNA (mRNA)** and **transfer RNA (tRNA)**. We'll definitely talk about them later. But for now, the most important thing to focus on are these nitrogenous bases. These four letters create the genetic language. Even though there are only four options (small, when compared to the 26 letters of the English alphabet), it is enough to create an unlimited number of different proteins! But first, let's talk about how the cell organizes your genetic library.

DNA Chromosome Structure

The chromosomes in your nuclei aren't just floating around like strewn-about strands of spaghetti. In order to enable 46 total chromosomes to fit into a single cell's nucleus, the DNA needs to be efficiently packed.

If you stretch out all the DNA from a single cell's chromosomes and line them up end-to-end, it would be six-feet long!! That's right . . . from a *single cell.* Six feet of DNA. Whoa.

The DNA of a chromosome is packaged into tight little bundles by wrapping it around special proteins called **histones**, sort of like a thread being wrapped around a spool. This mixture of DNA and histone proteins is called **chromatin**. This phenomenon of tightly wrapping the DNA around histones will be significant later, when we talk about transcription and gene expression.

REVIEW QUESTIONS

1. Rank the following terms from *smallest* to *largest*: chromosome, gene, genome, nucleotide.

2. The upright portion of the DNA backbone is composed of alternating ____________ and ____________ portions of the nucleotides. This backbone of the DNA is very strong because it is linked by ____________ bonds. The two strands of the double helix hold onto one another because of ____________ bonding between the ____________ of complementary nucleotides.

3. What would be the impact on DNA structure if base pairing occurred between cytosine-thymine and guanine-adenine?

4. List the three structural differences between DNA and RNA.

5. What would be the RNA sequence complementary to the DNA sequence ACTGACA?

6. What are the two reasons for the base-pairing rules (G-C and A-T)?

7. What is the name of the monomer of DNA? ____________ Which part of the monomer is the basis for the genetic language? ____________

8. Based on the base-pairing rules of DNA, write the complementary sequence to GGACACTT.

9. A strand of DNA is wound around special proteins called ____________. This mixture of DNA and protein is called ____________.

10. Which of the following is a correct pairing between DNA bases?

    a. Adenine paired with uracil
    b. A two-ringed base paired with a single-ringed base
    c. Pyrimidine paired with another pyrimidine
    d. Thymine and adenine sharing three hydrogen bonds

10 Replication of DNA

MUST KNOW

- The double-stranded nature of DNA allows for stability and heritability.

I love this **must know** concept. It blew my mind when I first thought about it. DNA is double stranded, right? But *why*? You're soon going to learn about the process of correctly copying both strands of DNA, and when expressing a gene, carefully selecting the one strand that actually carries the correct protein code. It seems that it would be so much simpler if there was only one strand of DNA to deal with. But the double strandedness of DNA is the reason it can be inherited. In order to fully explain this concept, we have to first learn some of the mechanics of DNA replication.

All cells come from preexisting cells. When a cell divides, it must first create a perfect copy of its DNA. When the cell then splits in half, both of the new "daughter cells" will have a full complement of DNA, identical to the original cell. The process of the cell copying all of its chromosomes is called **DNA replication.** The process is fairly simple and straightforward.

Helicase

An enzyme called **helicase** unwinds and unzips the double-stranded chromosome. When the two strands separate, the sugar-phosphate backbone remains completely intact; instead, the weaker hydrogen bonds between the base pairs are broken. The two strands now have exposed hydrogen bonding that wants to snap back together, so proteins called **single-stranded binding proteins** latch onto the unpaired nitrogenous bases to keep the two strands apart. The point at which the DNA is "unzipped" (meaning the hydrogen bonds between the base pairs are broken) is called the **replication fork**.

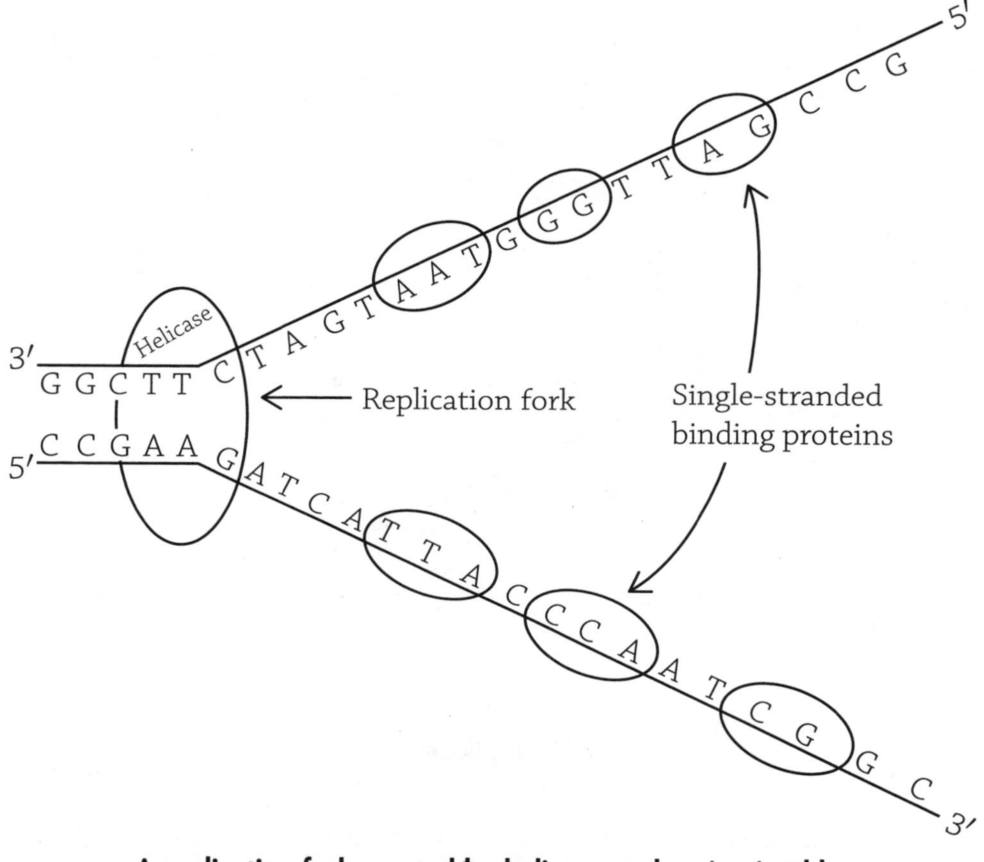

A replication fork created by helicase and maintained by single-stranded binding proteins

DNA Polymerase

Another enzyme called **DNA polymerase** (please refer to labeled A in the following figure) matches up free nucleotides (B) that are complementary to the unzipped (exposed) DNA template. These free nucleotides are always available, floating around inside the nucleus.

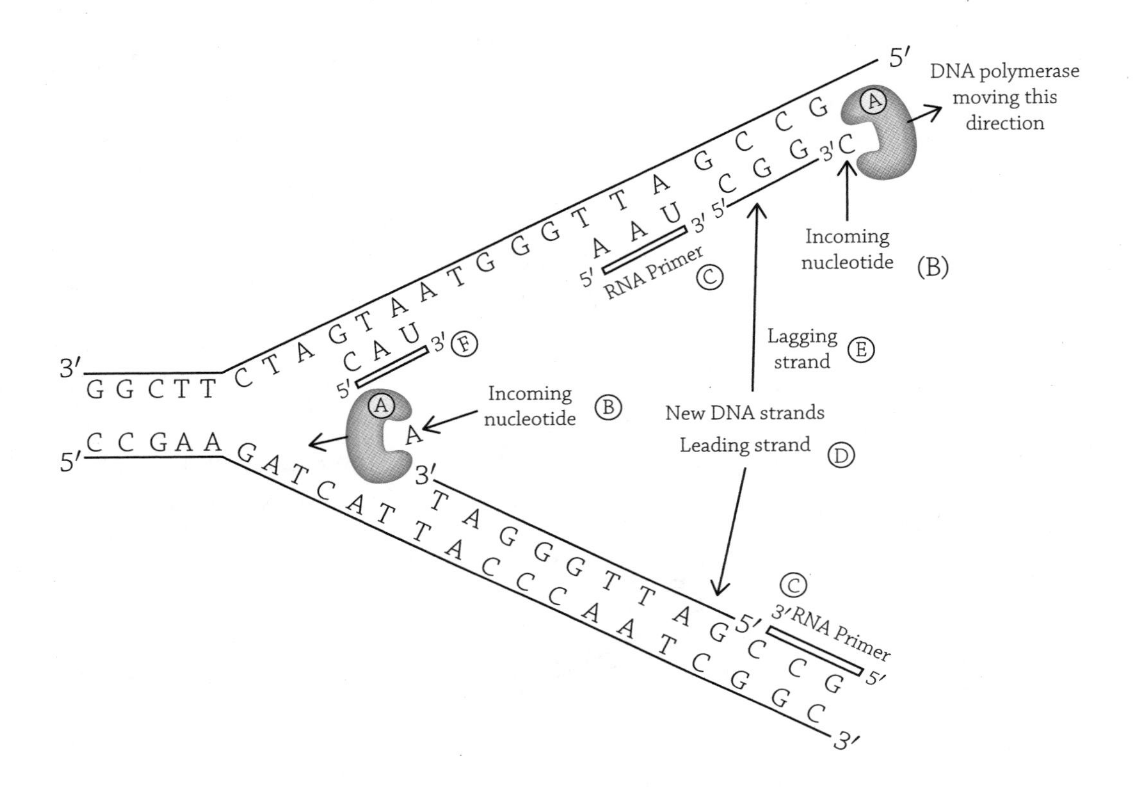

DNA replication

DNA polymerase can match about 50 nucleotides every second! Not only is it fast, DNA polymerase is very good at its job and rarely makes a matching error (if there is a mistake, it quickly fixes it). When all is said and done, there is only around 1 mistake per 1 billion nucleotides. DNA polymerase, however, needs a bit of help getting the process started. In order to add a new nucleotide, there must be a free 3′-hydroxyl (-OH) group to add onto. This sounds confusing, but just look closely at the ends of a piece of DNA:

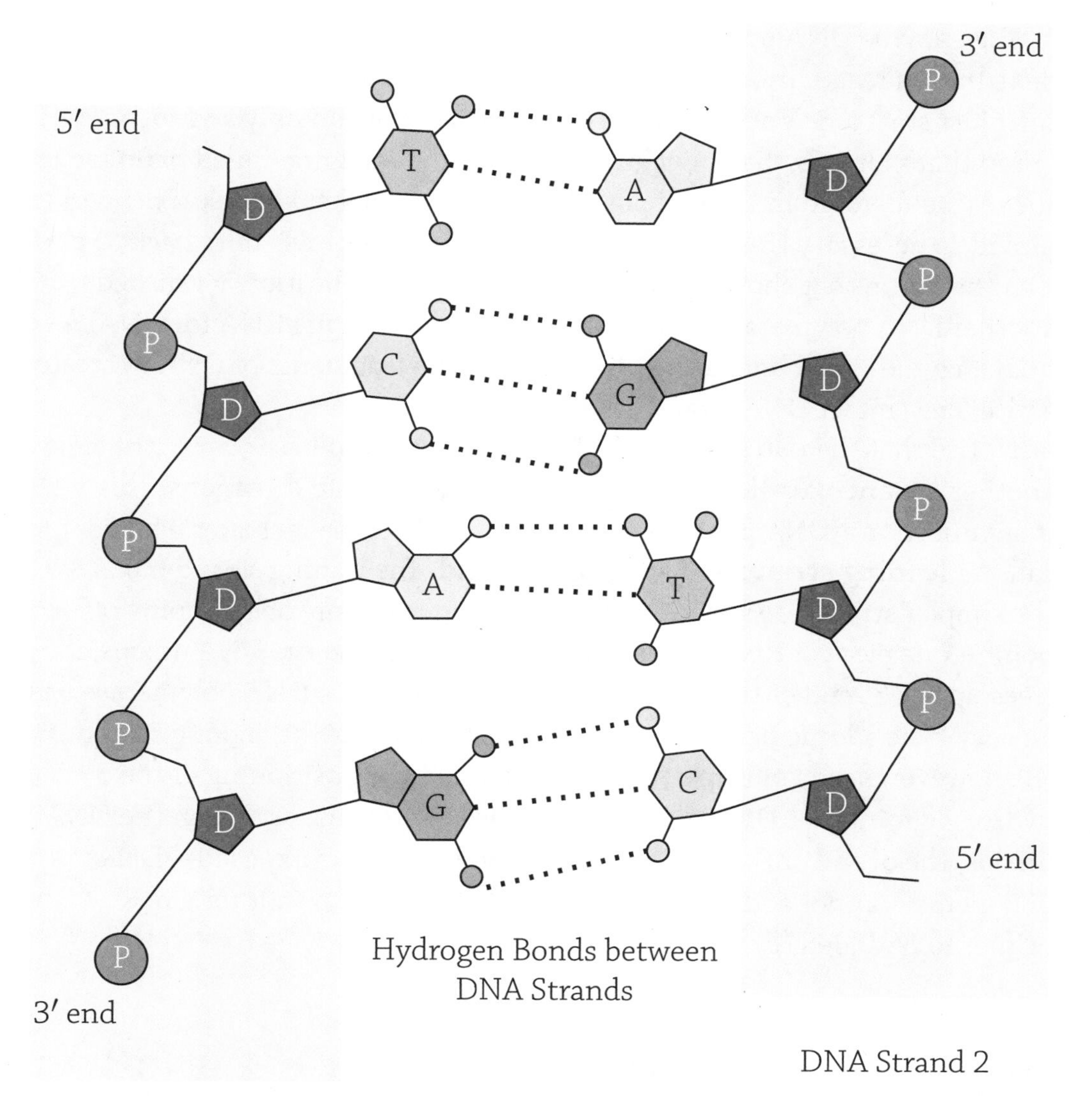

A strand of DNA showing basic structure. Notice the 5′ and 3′ ends of a strand of DNA.

Author: Boumphreyfr. This file is licensed under the Creative Commons Attribution-Share Alike 3.0 Unported license. https://commons.wikimedia.org/wiki/File:Dna_strand.png.

The enzyme DNA polymerase adds a new nucleotide through a dehydration reaction, which needs an –OH group to react with. That's why it can't just start a new strand of DNA from scratch; it can only add to an –OH group that's already there. Luckily there's another enzyme called **primase** to the rescue! Beforehand, it will put down a tiny bit of RNA (called a primer) to provide a necessary free-hanging 3′-hydroxyl group (please refer back to part C in the figure that shows DNA replication). Once replication is finished, another DNA polymerase will swoop in and swap out the RNA for DNA. A final enzyme called **ligase** will glue together the fragments of DNA to create a nice, new, intact DNA strand.

Notice that the two strands of a DNA molecule that base-pair with one another are **antiparallel**, meaning they run in opposite directions. This is significant for DNA polymerase to do its job, because it creates what is called a **leading strand** and a **lagging strand**. The leading strand (D) is the simpler strand to synthesize because it requires only one jumping off point—one piece of RNA to serve as a primer. Then, as the DNA unzips, it frees up the 3′-end of the new, growing strand of DNA; DNA polymerase has an easy time adding new nucleotides onto that 3′ end. The lagging strand (E), however, needs multiple primers because of the positioning of the 3′ hydroxyl group that DNA polymerase so desperately needs. When the DNA unzips, the 5′ end of the new growing strand of DNA is exposed—DNA polymerase cannot add onto that end! Instead, it has to wait for a new primer to be added (F).

Dehydration Reaction

As mentioned before, a covalent bond (specifically called a phosphodiester bond) forms between the newly added nucleotide and the growing (new) strand of DNA after a dehydration reaction occurs. This is commonly how monomers of a larger polymer are "glued" together. The word *dehydration* gives a clue to how this reaction proceeds . . . to dehydrate something means to remove water!

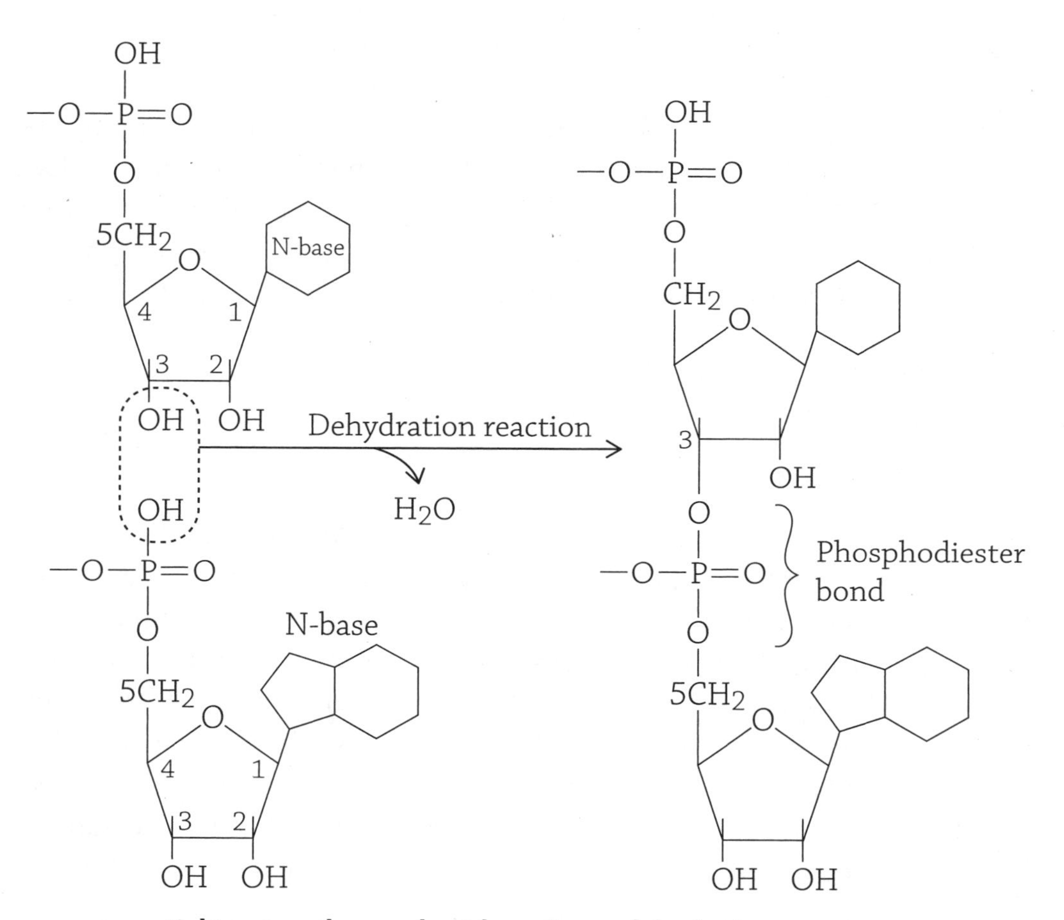

Linking together nucleotides using a dehydration reaction

Two Identical Chromosomes

Once the process is complete, there will be two identical double-stranded chromosomes, each composed of one of the original DNA strands and one strand that is composed entirely of new nucleotides attached by DNA polymerase. This is called the **semiconservative model of DNA replication**.

Okay, now we can relate this back to our **must know** concept. DNA replication seems so difficult, considering the 3′ and 5′ ends, and subsequent leading and lagging strands. It would be simpler if a cell only had to copy one strand of DNA, right? But then what would that replicated piece of DNA look like? Would it be genetically identical to the original, and would the resulting daughter cell get the exact same DNA as the original cell? Nope. Because of complementary base pairing, if the original strand's DNA sequence was this:

ACGTTACACAGGG

then the copied DNA would be this:

TGCAATGTGTCCC

Not identical at all! Complementary, yes, but not identical. And when a cell replicates its DNA, it has to be exactly the same. This is taken care of if DNA is a double-stranded molecule made of two complementary strands of DNA. When both strands are replicated, you get another double-stranded DNA molecule, genetically identical to the first. Another important consideration has to do with the stability of the molecule. A single-stranded molecule of nucleic acid (whether its DNA or RNA) isn't very stable; it will degrade quickly. You know what you don't want your DNA to do? Degrade. A double-stranded molecule is much more stable . . . and that's a good thing.

REVIEW QUESTIONS

1. What must occur before a cell divides?

2. The double-helical structure of DNA is antiparallel. This means that the 3′ end of one strand (which has a ______ group hanging off of it) must line up with the ______ end of the complementary strand (which has a phosphate group hanging off of it).

3. Which strand needs multiple RNA primers in order for DNA polymerase to create a complementary copy: leading strand or lagging strand?

4. Free nucleotides are added to a growing strand of DNA through a ______ reaction, which creates a new ______ bond.

5. Why, exactly, can DNA polymerase only attach free nucleotides to the 3′ end of a growing DNA molecule?

6. Fill in the blanks with the correct term: **DNA polymerase, helicase, ligase, primase, single-stranded binding proteins.**
   a. The enzyme ______ unwinds and unzips DNA in order for replication to occur.
   b. The enzyme ______ must first synthesize a small piece of RNA in order for another enzyme to begin adding DNA nucleotides to the growing strand.
   c. Once DNA is unzipped, ______ stick to the unpaired bases in order to keep the two strands apart.
   d. The enzyme ______ glues together fragments of newly synthesized DNA by creating new phosphodiester bonds.
   e. The bulk of the work of creating a new strand of DNA is done by ______, the enzyme responsible for pairing up complementary nucleotides.

7. What are the two reasons DNA must be double stranded?

8. Which of the following statements correctly describes the semi-conservative model of DNA replication? Once DNA is copied, it will have:
    a. Two strands, both from the original DNA molecule
    b. One strand from the original DNA molecule and one created from new nucleotides
    c. Two strands, both created from new nucleotides

9. True or False: Because of the need for primers during DNA replication, the daughter cell's chromosomes will contain some RNA fragments.

10. Which of the following statements about DNA replication is correct?
    a. If a cell undergoing DNA replication has 46 chromosomes, each of the resulting daughter cells will also have 46 chromosomes.
    b. After DNA replication, thymine bases will be replaced by uracil bases.
    c. DNA replication doesn't occur until immediately after a cell divides.
    d. The leading strand needs multiple primers when undergoing replication.

Gene Expression and Differentiation

MUST KNOW

- Differentiation occurs when a cell begins to select the genes it will express in order to become specialized in its function.

- Transcription is the most important step of differentiation.

- Correct grouping of codons in mRNA (keeping the correct reading frame) is necessary for proper translation of proteins.

- The proteins that are created by a cell depend not only on the genes but also on the control by which genes are expressed.

- An operon makes sure that genes are expressed only when they're needed.

Let me tell you a story: a long time ago, you didn't exist. Then, one cell fused with another, and you did exist. At least the *beginning* of you existed. Every organism that is the product of sexual reproduction was formed after a sperm cell fused with an egg cell, creating what is called a **zygote** (the cell formed after the egg and sperm merged). This single cell needed to do many rounds of DNA replication and division before it formed the glorious specimen that is YOU.

But along the way, other clever things happened. If that beginning zygote simply divided over and over and over again, you would be, what, a huge blob of cells without any specific shape. Think about what you look like now: you have skin, and bones, and muscles, and all sorts of interesting bits. In order for that to occur, the cells in your body—that all have the exact same DNA, by the way—had to start to get picky about what genes they were going to "express." The evolution of advanced multicellular organisms occurred because cells began to differentiate. **Differentiation** means a cell begins to express the proteins from specific genes that were otherwise ignored, and by doing so, the cell starts to specialize in form and function.

You may often see references to stem cell research and their use in medicine. Stem cells have amazing potential because they are *undifferentiated*, meaning they have not begun the process of picking and choosing which genes to read (and which cell type to turn into). If a person has a failing organ or needs a skin graft, it would be ideal to collect some of their own stem cells and guide them down the differentiation pathway for whatever organ or tissue type that is needed.

This first step—the cell choosing which genes it cares about and which proteins it wants to create—is the most important step of cell differentiation (one of our **must know** concepts). But I am getting ahead of myself. Let's first talk about what gene expression actually means, and how a cell goes about reading a gene and creating the protein that it codes for.

Transcription

Your DNA never leaves the nucleus. It's too precious and important to go wandering about the cell. It's the original blueprint for *you*, so it better stay locked in a safe location. Recall from Part Two, however, that the cell machinery responsible for making the proteins (ribosomes) reside in the cytoplasm, outside of the nucleus. There needs to be a temporary message that can take the instructions from the genes in the nucleus and shuttle them to the ribosomes waiting in the cytoplasm. This "go-between" instruction is the piece of messenger RNA (**mRNA**) that's created in the nucleus. It is a single piece of RNA complementary to only a single gene, and the process of creating it is called **transcription**. The procedure is similar to DNA replication, in that the double-stranded DNA molecule has to unzip by breaking the hydrogen bonds between the two strands, and a polymerase enzyme will then match complementary nucleotides to the exposed template strand. There are, however, significant differences:

1. Only the DNA double helix around the gene being expressed is unwound.
2. Only a single gene is transcribed (not the entire chromosome). There is never a need for a cell to make the proteins of all 25,000 different genes . . . that's crazy! Why would a cell in your eyeball need to make the protein for hair growth? Ew.
3. The polymerase that transcribes is called **RNA polymerase**, and as its name suggests, it matches up RNA nucleotides to the exposed gene template (makes sense, since it's making a piece of mRNA).
4. Only *one* side of the DNA double helix for any given gene is transcribed. There are two strands for each gene, but the two strands are not identical—they are complementary. Therefore, the proper recipe for a protein is contained in only one of the two strands; this is designated the **template strand.**

For example, here is a segment of a gene in DNA:

GGTACCTGTGTGTAAATAAGACTCAG ← template strand

CCATGGACACACATTTATTCTGAGTC

After this double-stranded portion of the DNA unzips, RNA polymerase will create a piece of mRNA complementary to that top template strand. Also, keep in mind that RNA has uracil nucleotides instead of thymine.

Here is the resulting mRNA strand created by RNA polymerase:

CCAUGGACACACAUUUAUUCUGAGUC

It is a **must know** to realize this moment of selecting and transcribing only certain genes is absolutely key to cell differentiation. And since RNA polymerase targets individual genes for transcription, it needs some sort of "landing strip" to help it position properly. These landing sites are called **promoters**, and it's simply a segment of DNA with a certain sequence of nucleotides. One promoter is situated before every gene. In eukaryotic cells, in order for the large, ungainly RNA polymerase to land on the promoter, it also needs the help of a collection of proteins called **transcription factors**. RNA polymerase can't position itself properly without the help of these special proteins. Once RNA polymerase reaches the end of the gene it is transcribing, it lets go of both the DNA and the piece of mRNA it just created.

You just learned about the *how* of transcription, but it's just as important to consider *how much*. A cell can regulate the amount of a given protein that is produced by controlling how many times RNA polymerase scoots down a gene and creates a piece of mRNA. If a cell wants a LOT of a certain protein, it needs to make a LOT of the appropriate mRNA. To do so, there are these special transcription factors (called **activators**) that bind way upstream from the gene's promoter at special sequences in the DNA called an **enhancer** region. Even though the activators bind far away, they somehow help RNA polymerase to land at the promoter. This seems weird, right? How can something hanging out on the DNA way before a gene have any influence on RNA polymerase landing at the promoter? Surprisingly, the DNA molecule has to bend in order for these activators to work:

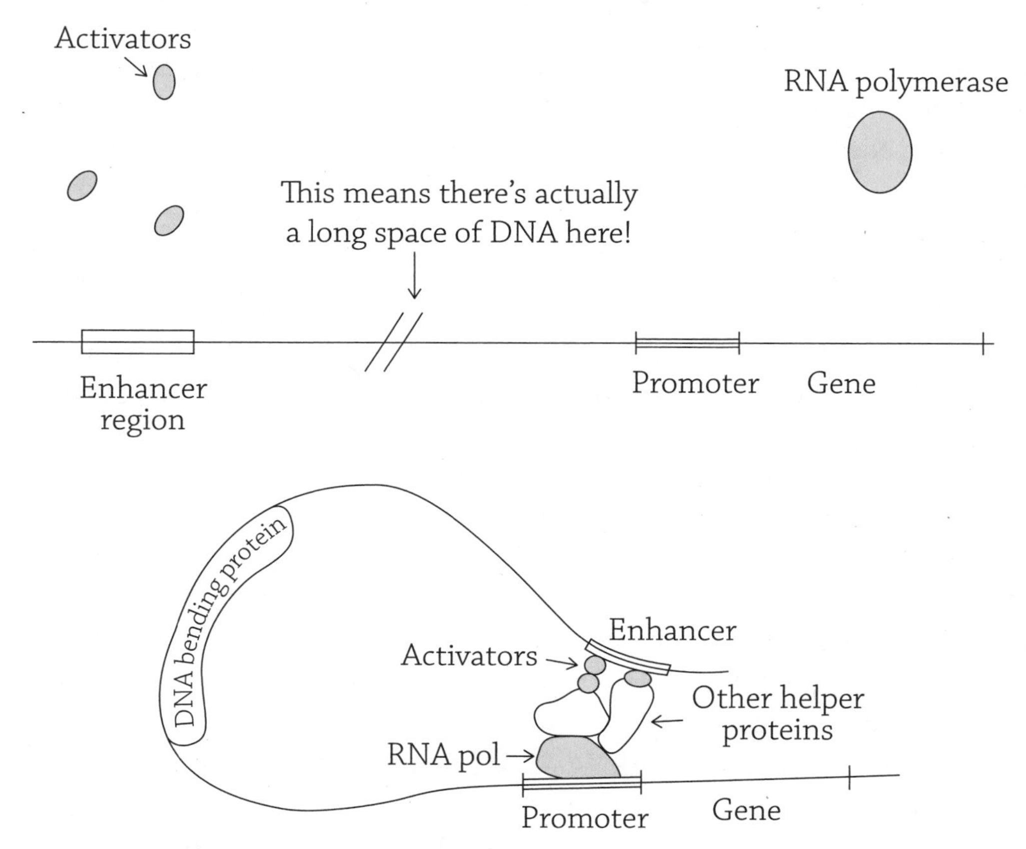

Specific transcription factors (activators) helping RNA polymerase to land

With the help of a DNA bending protein, the strand of DNA loops around to bring the activators close to the promoter. They, along with other helper proteins, help align RNA polymerase on its promoter, facilitating transcription. Recall that our **must know** concept states that the cell choosing which genes to transcribe is the key step to a cell differentiating into a specific cell type. We will talk about transcription factors in more detail when we learn about the specifics of cell differentiation later on.

Translation

The cell now has a brand-new piece of mRNA, fresh off the RNA polymerase assembly line. The instructions of the mRNA now will be "read" by the ribosomes, who will then assemble the proteins coded for by the mRNA (and originally coded for by the gene). Do you remember where the ribosomes are hanging out? Yes, they're waiting in the cytoplasm, either floating around freely or stuck onto the rough endoplasmic reticulum. The freshly transcribed piece of mRNA needs to now move from the nucleus into the cytoplasm (easy, because the nuclear membrane has a bunch of pores in it for this very purpose).

Once the mRNA gets to the cytoplasm, it's time for the ribosomes to get to work. A ribosome will grab onto the beginning (5′ end) of the mRNA molecule (we learned about the 3′ and 5′ ends of nucleic acids in Chapter 3). Each mRNA molecule carries the message for the final protein product, originally coded for by a particular gene safely stored away in the nucleus. Keep in mind that a protein is a polymer of individual amino acids. In order to decipher the correct sequence of amino acids, the ribosome needs to look at every three "letters" in the mRNA (each letter is a single nucleotide and its particular nitrogenous base). A grouping of three bases in mRNA is referred to as a **codon**. Each codon tells the ribosome what amino acid is next in the growing protein. It is a **must know** that the correct grouping of codons in mRNA—and keeping the correct reading frame—is necessary for proper translation of proteins. Unfortunately, there's an easy way for the ribosome to mess up. Here is an example of a catastrophic mistake:

The fat bat sat but had the flu

What if, instead, the ribosome thought it should start grouping into sets of three starting at the first "H"?

Hef atb ats atb uth adt hef lu

Oh, that's not good. The groupings of three are now totally different! Just as it doesn't make sense when we read it, it doesn't make sense when making a protein. The ribosome would piece together seven amino acids, but they would be the wrong amino acids. Furthermore, notice that the original protein should have eight amino acids. Because of the unfortunate shift, there's a last grouping of two letters, which is not enough to code for an amino acid (a codon contains three bases). When a ribosome is translating a protein, this type of mistake is called a **frameshift mutation**. It occurs because the ribosome didn't line up properly when it grabbed the mRNA. In order to make sure it aligns correctly, it needs a start point in order to set the correct **reading frame**. Luckily, there is such a start point! It is a special codon called the **start codon**, and it's always adenine-uracil-guanine (**AUG**) and codes for the amino acid methionine. This is where the ribosome situates itself when it lands at the beginning of the mRNA. By doing so, it sets the correct reading frame because it creates the proper grouping of three bases. Consider our example mRNA sequence from earlier:

mRNA:

CCAUGGACACACAUUUAUUCUGAGUC

If you look carefully, can you figure out where the ribosome would land and establish its reading frame?

CC [**AUG**] [GAC] [ACA] [CAU] [UUA] [UUC] [UGA] [GUC]

EXTRA HELP

When translating a mRNA sequence, take a moment to first find that start AUG (it's key to setting the correct reading frame). If there isn't an AUG in the sequence you're translating, don't worry . . . sometimes the question doesn't bother providing a start codon. Just set your reading frame starting from that very first RNA base!

The ribosome essentially ignores any nucleotides before the AUG start codon. Now each of those codons code for a single amino acid in the final protein. The ribosome relies on another type of RNA to bring over the correct amino acid for a given codon. These are called **transfer RNA (tRNA)** and they are molecules of RNA with two important "ends"—one end that attaches to a specific amino acid (of which there are 20 options), and the other end that wants to hydrogen bond with the specific codon it is destined to recognize. The part of the tRNA that hydrogen-bonds with a codon in mRNA is called the tRNA's **anticodon**.

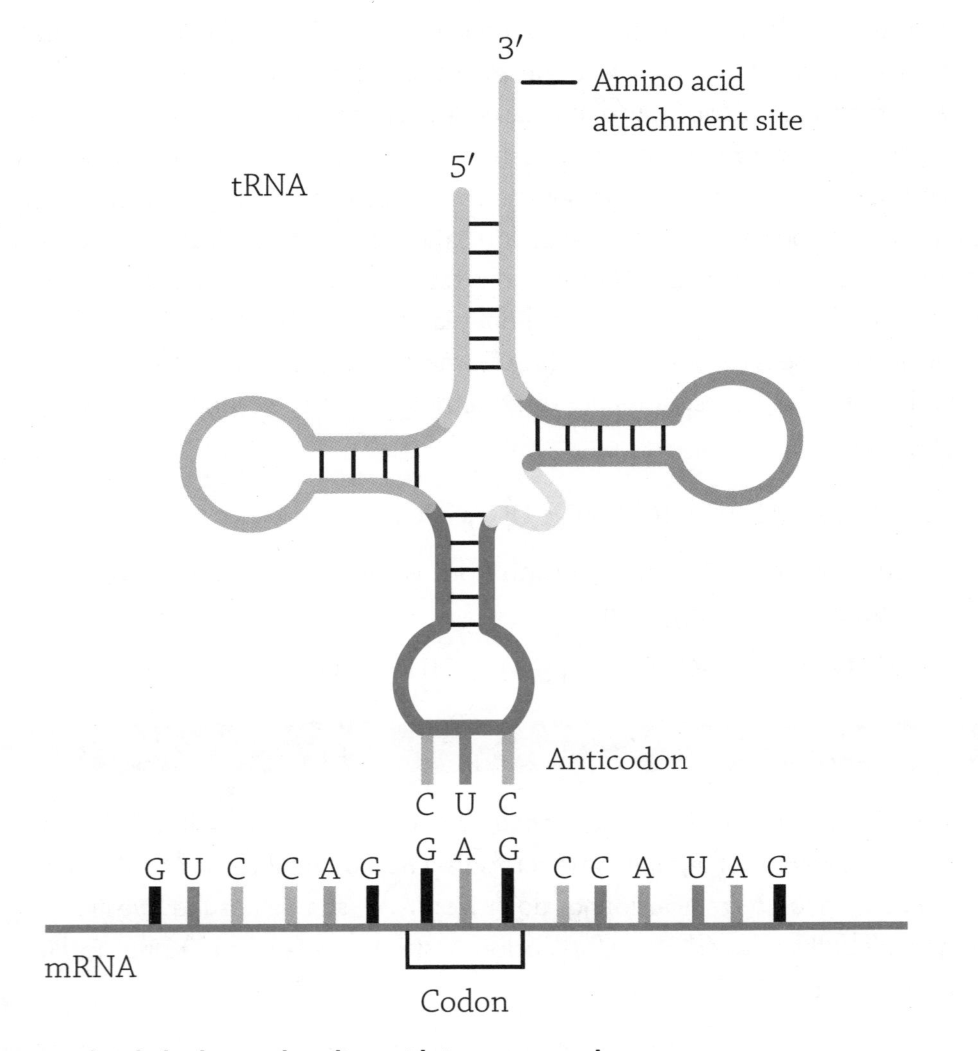

A tRNA molecule hydrogen-bonding with its correct codon

Author: Yikrazuul. https://commons.wikimedia.org/wiki/File:TRNA-Phe_yeast_en.svg

After the correct tRNA binds and brings in the starting methionine amino acid, the ribosome shifts down the mRNA to the next codon. A different tRNA will arrive, carrying with it the correct amino acid. A special covalent bond called a **peptide bond** will form between the two adjacent amino acids, and the first tRNA releases its amino acid. It can then leave the ribosome, float off into the cytoplasm, and find another methionine amino acid to grab. Meanwhile, the ribosome shifts down another codon, and the process begins anew.

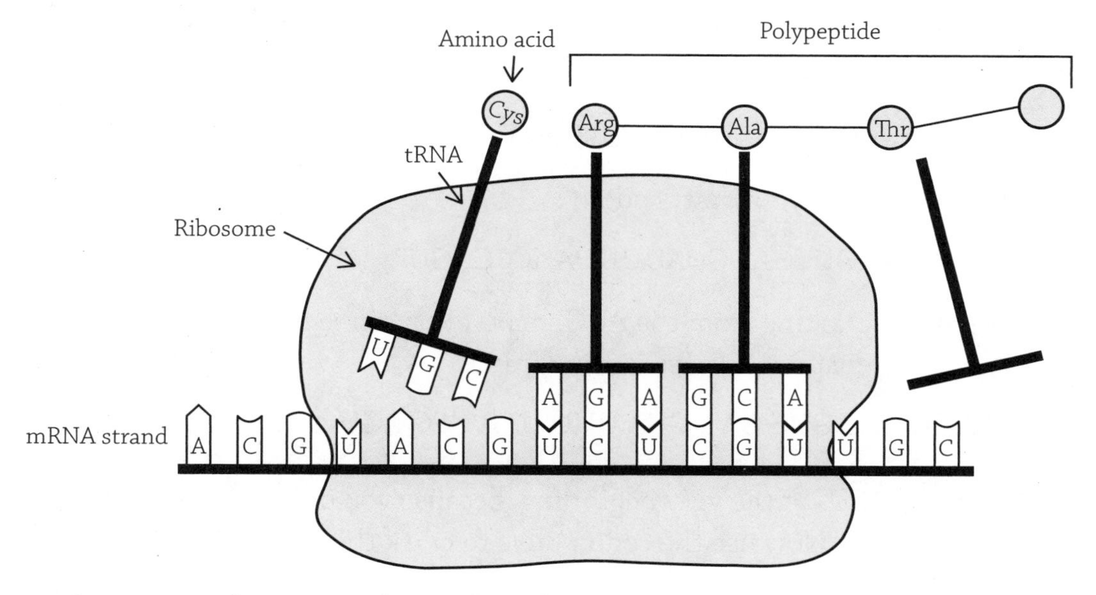

A ribosome coordinating translation of a mRNA into a protein

Author: Sarah Greenwood. https://commons.wikimedia.org/wiki/File:Protein_Synthesis-Translation.png

Similar to the ribosome knowing where to start making a protein by landing on the start codon, the ribosome knows when it's done once it hits the **stop codon**. Unlike the start codon (which actually coded for the specific amino acid methionine), the stop codon doesn't add an amino acid to the protein. Instead, a special protein called the **release factor** comes over and sets free the ribosome, the mRNA, and the protein. Process complete!

BTW

Once a "naked" tRNA leaves the ribosome (it's naked because it lost its amino acid), and scoots off to grab another one, it doesn't just grab any of the 20 possible amino acids. Remember the actual structure of a tRNA molecule: it's a piece of RNA with a specific sequence of bases for its anticodon. The anticodon of a tRNA never changes, so a tRNA molecule will always recognize the same codon and bring in the same amino acid.

You are able to crack the genetic code of a gene. All you need is a codon table and to remember the **must know** step of setting the correct reading frame. How do you do that, you ask? First, circle the start codon (AUG), and mark every codon (three bases) thereafter. Then you use a codon table to translate. For example, for this mRNA sequence, determine the sequence of amino acids of the protein:

CUAUGAACGUCGAGGUAUUUUAACCGAGUA

Step 1 Find your start codon:

CU**AUG**AACGUCGAGGUAUUUUAACCGAGUA

Step 2 Starting from the AUG, mark every three bases thereafter to ensure a correct reading frame:

CU **AUG** AAC GUC GAG GUA UUU UAA CCG AGU A

Ignore the "CU" at the very beginning, because you only start translating at the AUG start. Now use the codon table to crack the code:

Second letter

First letter	U	C	A	G	Third letter
U	UUU UUC } Phe UUA UUG } Leu	UCU UCC UCA UCG } Ser	UAU UAC } Tyr **UAA Stop** **UAG Stop**	UGU UGC } Cys **UGA Stop** UGG Trp	U C A G
C	CUU CUC CUA CUG } Leu	CCU CCC CCA CCG } Pro	CAU CAC } His CAA CAG } Gln	CGU CGC CGA CGG } Arg	U C A G
A	AUU AUC AUA } Ile **AUG Met**	ACU ACC ACA ACG } Thr	AAU AAC } Asn AAA AAG } Lys	AGU AGC } Ser AGA AGG } Arg	U C A G
G	GUU GUC GUA GUG } Val	GCU GCC GCA GCG } Ala	GAU GAC } Asp GAA GAG } Glu	GGU GGC GGA GGG } Gly	U C A G

Codon table

Okay, so when you use a codon table, simply consider the three bases in a single codon. Find the first letter along the left-side column, and notice that every codon along that row starts with the same letter. Next, find the correct second letter by choosing the correct column (look at the top of the table). Finally, choose your specific codon (decided by the final third letter).

AUG = methionine (MET . . . three-letter abbreviations are fine)

AAC = asparagine (ASN)

GUC = valine (VAL)

GAG = glutamic acid (GLU)

GUA = valine (VAL)

UUU = phenylalanine (PHE)

UAA = STOP!

Therefore, your final protein has a total of six amino acids, each held together by a total of five peptide bonds:

MET-ASN-VAL-GLU-VAL-PHE

Even though there are more bases at the end of your mRNA strand, if they occur after the STOP codon, just ignore them.

EXTRA HELP

Notice in our above protein that the amino acid valine was coded for by two different codons: GUC and GUA. There are a total of 20 different amino acids but a total of 64 different three-letter combinations of A, C, U, and G. That means that for all of the amino acids (besides methionine) there are at least two different codon options. Furthermore, for the most part, the change occurs in the third base. The third base is referred to as the "wobble position" because it can often be swapped out for another base and give you the same amino acid.

Cell Differentiation and Epigenetics

Let's go back to the fact that you used to be a single, undifferentiated blob of cells. They were, in fact, stem cells, because they had the potential to turn into any cell type. Eventually, the cells in your body changed into different tissues and formed different organs. The cells that compose these different body parts are very different, both in what they look like (form) and what they do (function). The cool thing is, they all originated from that early blob of stem cells, and they all have the same genes. But because of those special transcription factors called activators, each cell could begin to pick and choose which genes to focus on. For example, a pancreatic beta cell needs to produce a lot of insulin, so it has the activators specifically to increase transcription of the insulin gene. The beta cell, however, does not have the specific activators to "turn on" the gene for stomach acid production because, well, why would it want to do that? The pancreas cell has the gene for hydrochloric acid, but it never bothers to transcribe it. The gene isn't gone, it's just quiet. Some cells go one step further than just not bothering to transcribe a gene; a cell can take steps to turn OFF huge swaths of a chromosome by blocking RNA polymerase. So, essentially, there are two ways to tweak which genes are expressed: large-scale regulation and small-scale regulation.

Large-Scale Regulation

Recall that it is necessary to wrap DNA around histone proteins in order to pack all that DNA into a tiny nucleus. This mixture of DNA and histone proteins is called chromatin. Large-scale regulation of DNA expression has to do with modifying chromatin structure in large segments of a chromosome. Modification of the chromatin structure is necessary because RNA polymerase can't transcribe the genes on a segment of DNA that is wrapped around histones; the genes are essentially turned off. Luckily the DNA can unspool and unwrap whenever RNA polymerase needs access to certain genes.

There are clever ways for a cell to control whether a segment of a chromosome is being used or kept quiet. For example, if the cell adds these little chemicals called **methyl groups** to certain bases in the DNA, the genes in that region will not be transcribed (and the genes in that area will remain "off")!

H_3C —

A methyl group attached to the DNA

If, on the other hand, these other chemicals called **acetyl groups** are added to the histone proteins, then the DNA is forced to unwind a bit and transcription can occur.

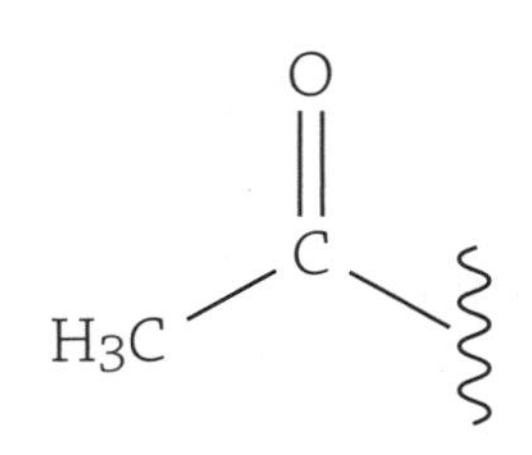

An acetyl group attached to the DNA

This is an interesting concept: a cell can control what genes are being expressed without actually changing the genetic sequence. It's not changing the message, it's just deciding whether the message is read. This ability to control the genome without changing the actual nucleotide sequence is called **epigenetics**. The word *epigenetic* literally means "in addition to changes in genetic sequence." Interestingly enough, these epigenetic changes to chemical changes to your DNA or histones are heritable!

IRL Epigenetic changes are a natural way for your cells to control gene expression. However, epigenetic changes have also been linked to a number of diseases, including some cancers.

Small-Scale Regulation

Even when there is a huge stretch of DNA exposed and ready to be expressed, the cell still picks particular genes to be transcribed. This is the small-scale (finely tuned) level of gene regulation, and it is all under the control of the specific transcription factors called activators we learned about earlier. In general, any transcription factor helps RNA polymerase to position itself correctly on the promoter. Some of these transcription factors aren't picky about what genes they help RNA polymerase target; they are, therefore, called general transcription factors (A). Before RNA polymerase can begin the process of transcribing the gene into mRNA, these general transcription factors first land at a sequence within the promoter called a TATA box (B). These little proteins help RNA polymerase to align itself properly on the promoter (C).

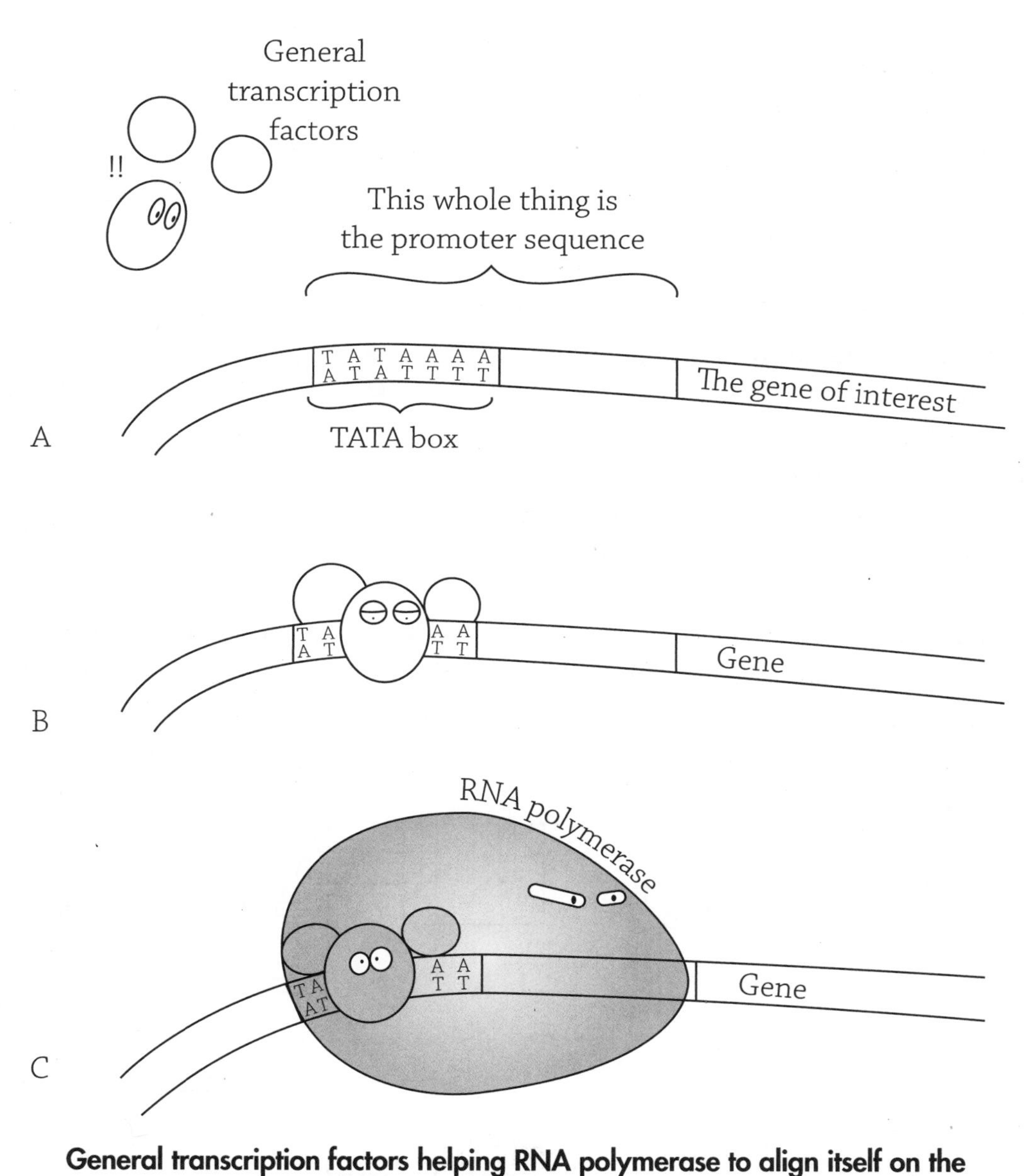

General transcription factors helping RNA polymerase to align itself on the promoter

If general transcription factors are involved in the low-level transcription of all genes, then specific transcription factors are involved in the substantial transcription of particular genes. This is a key step in the process of cell differentiation! In order for a cell to differentiate, it must begin to pick and choose which genes are important to its specific function. For example, a pancreatic cell needs to "switch on" the insulin gene, whereas a liver cell must switch on the gene for albumin. How does a particular cell know which genes it must express? This is possible with the help of the activators.

BTW

Unlike general transcription factors, activators bind somewhere other than the promoter. Seems weird, right? How can activators help RNA polymerase bind at the promoter if they themselves bind somewhere else? Strangely enough, when activators bind, they bind way upstream (meaning before) of the gene and its promoter. When they do, they cause the entire piece of DNA to bend, and the activators grab onto RNA polymerase and help it land appropriately. We talked about this earlier in the chapter, in case you want to review how this actually happens.

So, each cell has a selection of activators, which are needed to unlock the specific genes that the cell wants to express. These activators bind to a region called the **enhancer region**. The enhancer region is made from a mix-and-match of little DNA sequences called **distal control elements**. Are you confused yet? I'm so sorry. Let me try and give you an analogy.

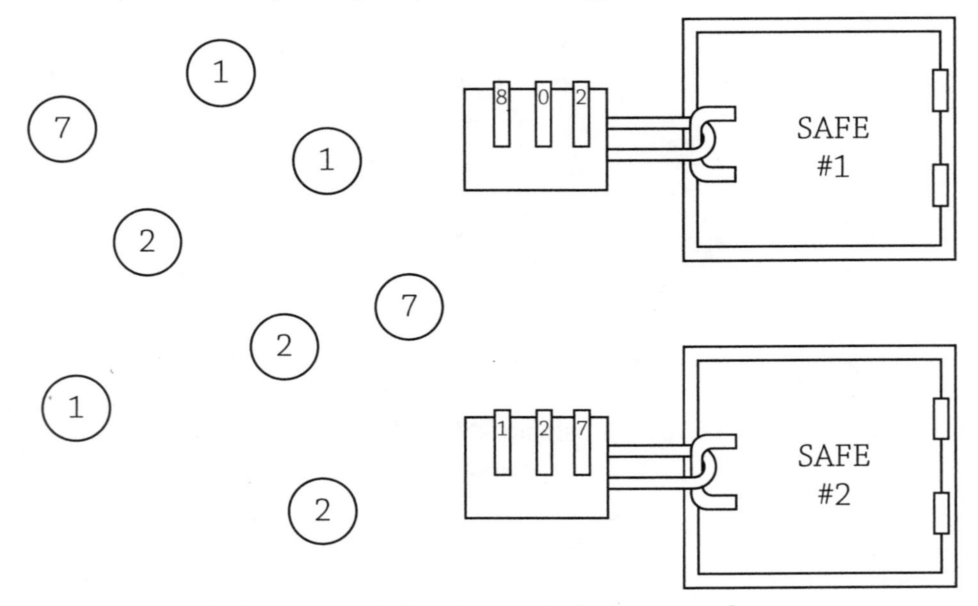

Activators and distal control elements analogy

Let's pretend there are two safes, each containing a different set of blueprints for something cool. You are only allowed to build the really cool thing stored within the safe for which you have the correct keys. In the above example, which safe are you allowed to open? Yup, safe #2, because you have available to you the "number keys" one, two, and seven. Specific transcription factors work in the same way: the entire lock is the enhancer region, the specific numbers in the lock are the distal control elements, and the floaty number keys are the activators (a cell's specific selection of transcription factors).

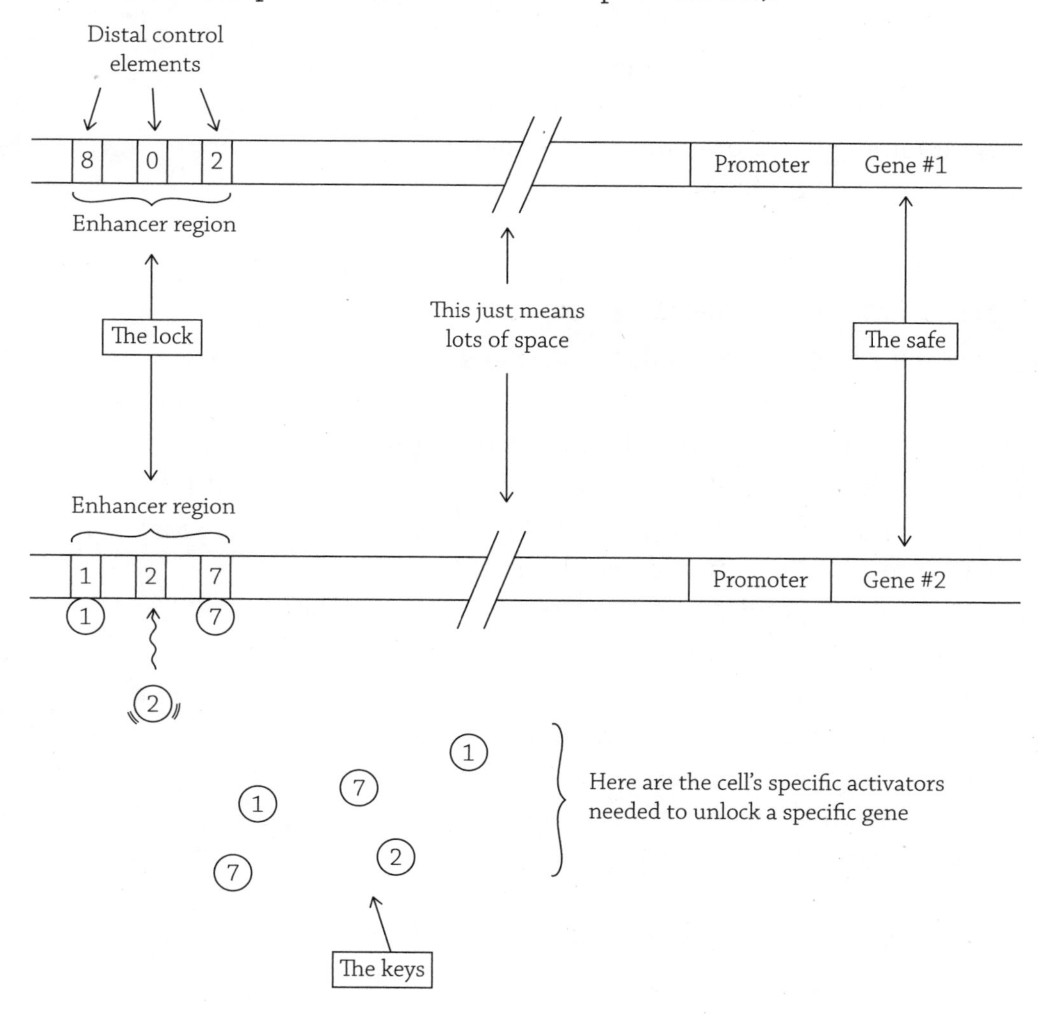

Activators and distal control elements

Please recall our **must know** concept, that the proteins created by a cell depend not only on what genes are present, but whether a certain gene is allowed to be expressed. A cell knows if it is supposed to express a gene if it has the correct activators that bind the combination of distal control elements in the enhancer region associated with that specific gene. A pancreatic cell will have the correct activators that match the enhancer region associated with the insulin gene; a liver cell will NOT have the correct combination of enhancers associated with the insulin gene (why would a liver cell want to produce insulin??). Instead, the liver cell will have a different combination of activators that will instead bind to the enhancer region linked to the albumin gene, a protein that liver does indeed want to make!

If a cell doesn't express a gene, that doesn't mean the gene is destroyed and removed—it's just ignored.

Operons: Regulation of Gene Expression in Prokaryotes

Prokaryotic cells (or more informally, our friends the bacteria) do not have nuclei nor organelles. But all cells must have DNA, and bacteria do indeed have DNA: a single, circular piece that is composed of approximately 1,500 genes (give or take, depending on the species). Even though the number of genes is relatively small compared to us huge, complex, and (arguably) advanced eukaryotes, bacteria still must finesse their way around their genome and only express those genes that are needed and necessary at any given moment. Prokaryotes have evolved a very clever system to control the expression of multiple genes simultaneously: the **operon**.

Keep in mind that the term "expression" here means RNA polymerase is landing at a gene's particular promoter and creating a complementary bit of mRNA that will then be translated by a ribosome into the final protein product. When a gene is expressed, the protein encoded for by that gene is going to be produced by the cell.

Oftentimes, it behooves a cell to express a handful of genes, or not bother with expressing any of these genes. This usually applies to biochemical pathways that have multiple steps that all work together toward a common goal:

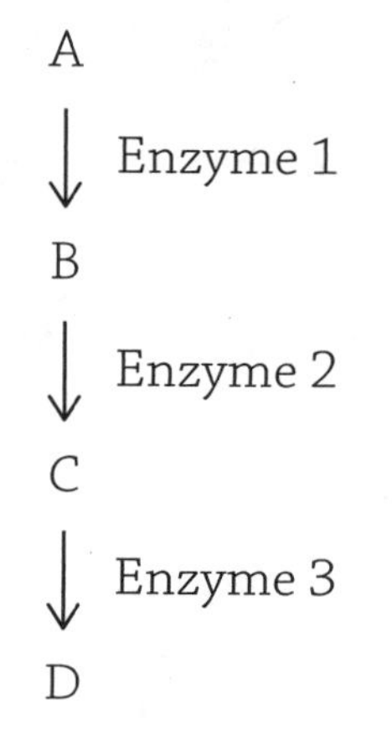

Biochemical pathway to create the compound D from the starting compound A

The figure above shows a multistep biochemical pathway, and each of the three steps are catalyzed by a specific enzyme. This example could model the *synthesis* of a compound (D) from a starting compound (A), or it could model the *digestion* of a compound (A) that needs to be broken down in three stages before the cell gets the final product needed (D). The point here is that both of these examples require multiple enzymes to get the job done, and stopping halfway through wouldn't necessarily do the cell any good. That is why operons are so helpful: it puts the expression of multiple genes under the control of a single promoter. In our example, the genes of the operon are coding for the enzymes (enzymes 1, 2, and 3) needed to complete the biochemical pathway. In order for the suite of genes to be expressed in appropriate conditions, the operon needs other components in order to control whether RNA polymerase will land at the promoter or not. An operon has a total of four components, each needed for this expression finessing.

Operon component	Purpose
Structural genes	These genes are the point, here. These genes code for the proteins that are under control of the single promoter.
Promoter	The sequence in the DNA where RNA polymerase will land and begin transcription.
Regulatory gene	A gene upstream (before) of the structural genes that produces a repressor protein.
Operator	A smaller region of DNA within the promoter. This is where the repressor protein will bind (and block RNA polymerase).

The figure below shows the general structure of a bacterial operon. In this diagram, the regulatory gene has been transcribed and translated, producing a repressor protein (which may or may not bind to the operator within the promoter). RNA polymerase is poised to transcribe the structural genes. If the repressor is unable to bind the operator, RNA polymerase can do its job; if the repressor protein binds the operator, RNA polymerase will be blocked (and the structural genes will not be expressed as proteins).

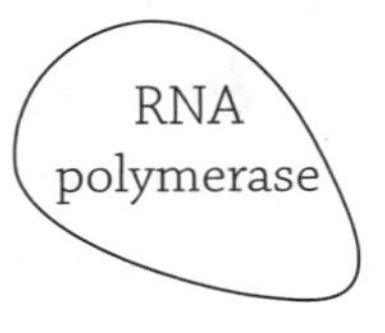

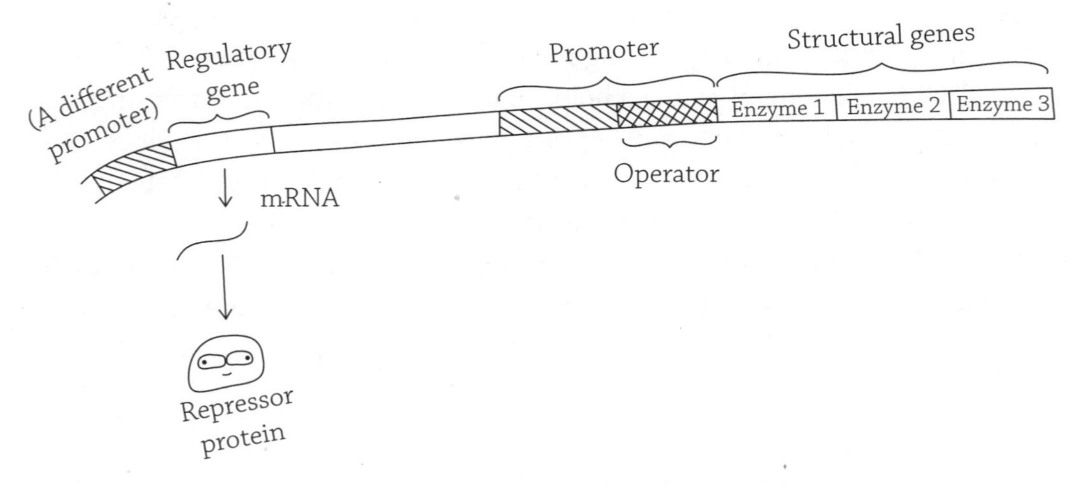

The general structure of a bacterial operon

We will talk about two different operons, both found in *E. coli* bacteria: the **lactose (*lac*) operon**, and the **tryptophan (*trp*) operon**. I'll start with the *lac* operon.

lac Operon

I mentioned earlier that the structural genes of an operon can create enzymes to either *digest or synthesize* a particular compound. The *lac* operon creates enzymes necessary for the digestion of the sugar lactose.

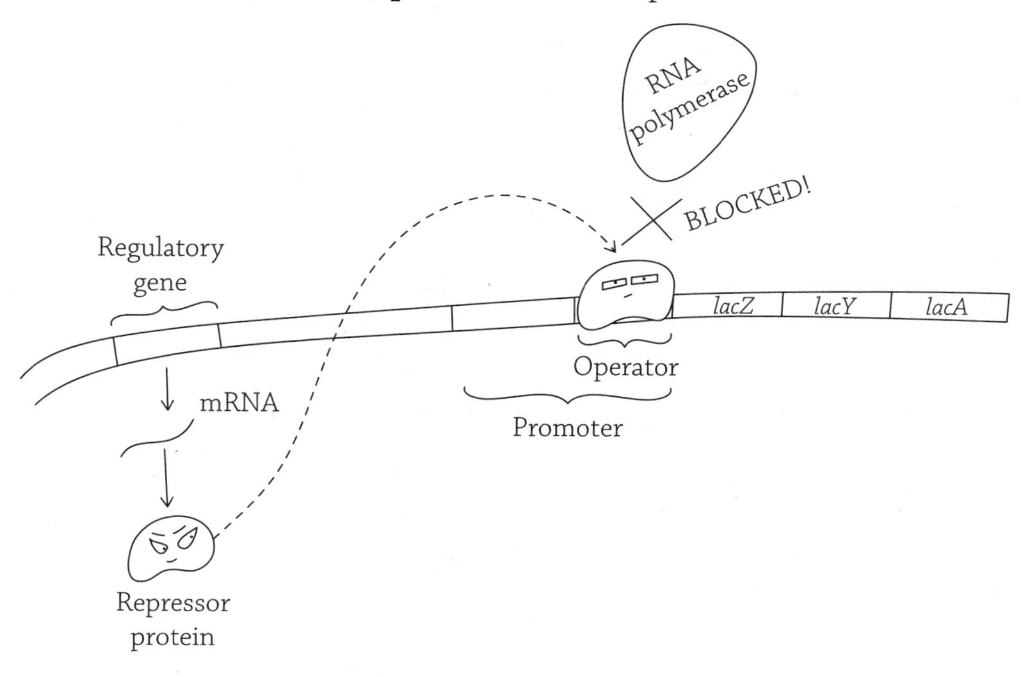

The *lac* operon (in the absence of lactose)

This figure depicts the "default" setting of the *lac* operon: the regulatory gene produces a repressor protein that immediately binds to the operator, effectively blocking RNA polymerase from transcribing (expressing) the structural genes.

The structural proteins of the *lac* operon are normally not being transcribed. Because their expression will only occur if the operon is somehow activated, the *lac* operon is said to be *inducible*. What would be a logical signal from the environment that *E. coli* should allow the operon to begin creating the enzymes to digest lactose? . . . Lactose! If lactose is present, it will bind to the repressor and inactivate it! The inactive repressor is unable to bind to the operator, and therefore RNA polymerase is able to bind the promoter and transcribe the structural genes.

BTW

Okay, based on the lac operon's default setting, you can make a logical guess about whether or not the sugar lactose is normally available in E. coli's environment. You should deduce that lactose is normally NOT a food source for E. coli, because that regulatory gene produces an active repressor that immediately blocks RNA polymerase from expressing the genes. And what do those genes code for? Enzymes that DIGEST LACTOSE. So, it makes sense that if there is no lactose, don't bother creating the genes to break it down.

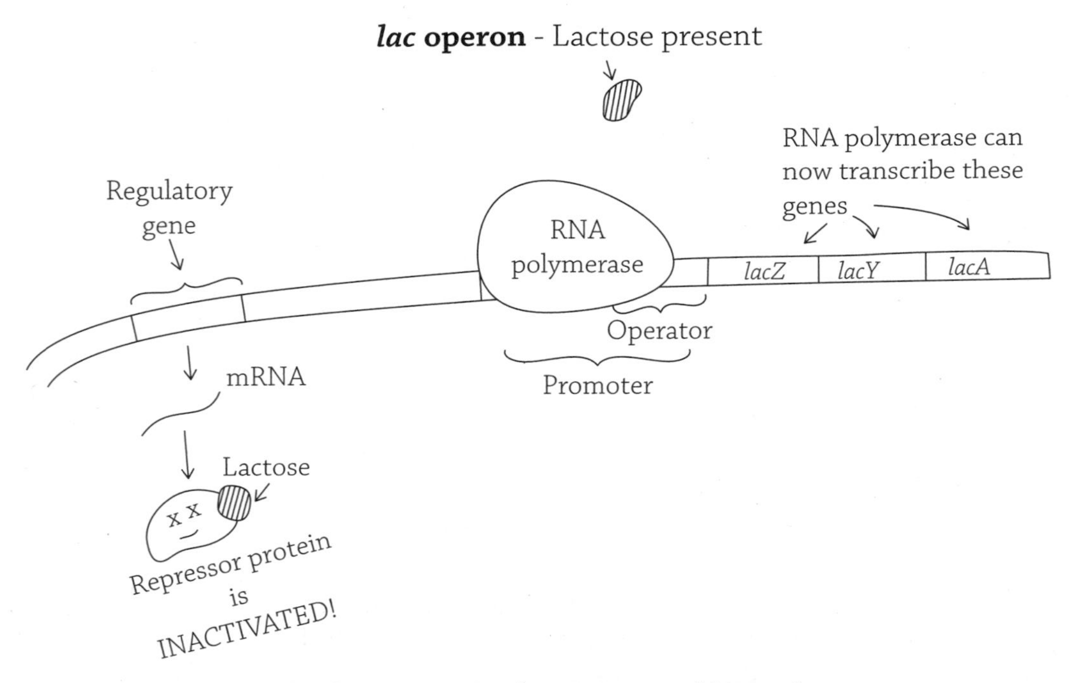

The *lac* operon (in the presence of lactose)

What do you think would happen if all the lactose is then digested? The repressor would no longer be inactivated, so it would once again stop the transcription of the enzymes needed to digest lactose (which is no longer available!).

trp Operon

Here is our second operon example, the *trp* operon. "Trp" is the abbreviation for tryptophan, an essential amino acid that *E. coli* needs to live. The structural genes in the *trp* operon code for five different enzymes. These enzymes work together in the multi-step synthesis of the amino acid, tryptophan. As we said before, it's an all-or-nothing situation: in order to create tryptophan, the cell must go through five steps, and each step requires a specific enzyme; those enzymes are created by the genes *trpA* through *trpE*.

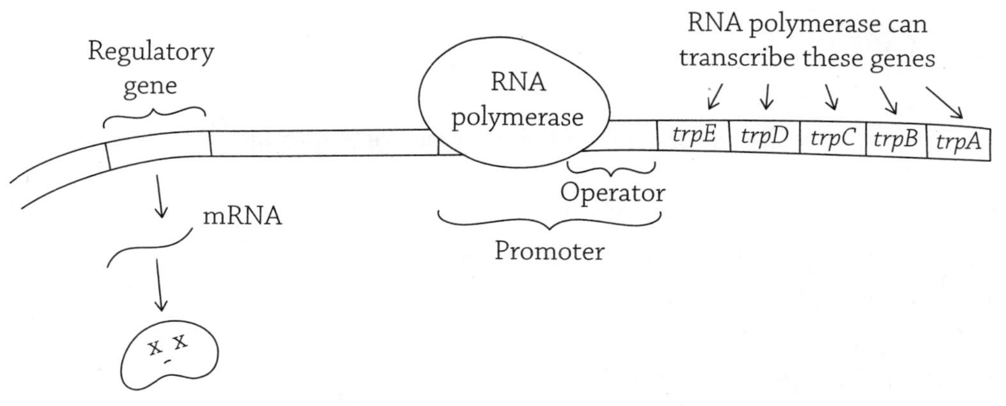

The *trp* operon (in the absence of tryptophan)

Unlike the *lac* operon, the *trp* operon's regulatory gene produces a repressor that is *inactive*. The inactive repressor is not able to bind to the operator and block RNA polymerase. Therefore, the cell is usually producing the genes needed to synthesize tryptophan. This means the *trp* operon is repressible (because it's normally expressing the structural genes and must be told to stop doing so). Now, when would the cell NOT want to create these enzymes? Yes, exactly! When tryptophan is freely present in the environment! Why could *E. coli* waste the time and energy to make the amino acid if it's already there?

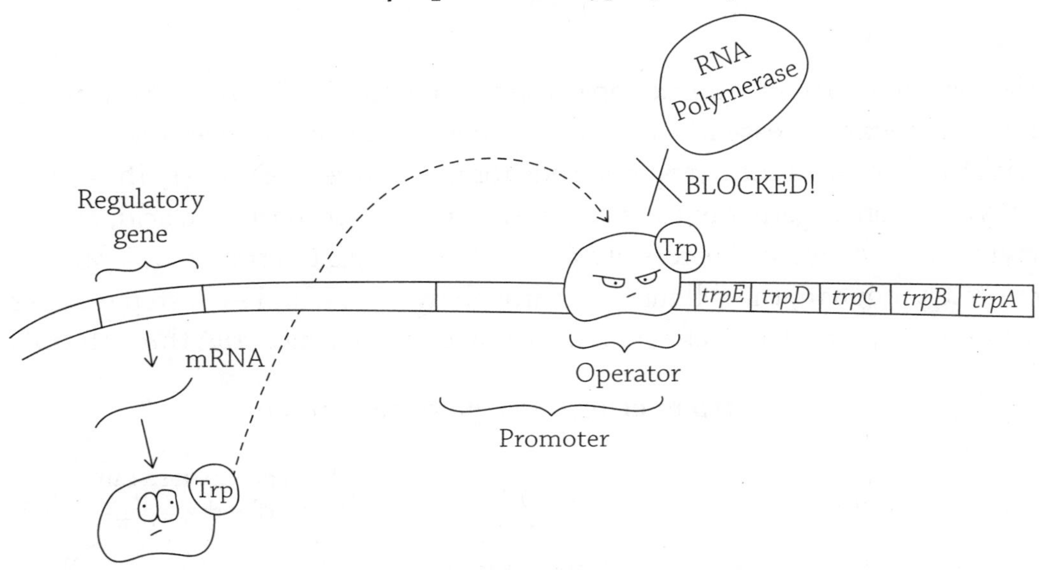

The *trp* operon (in the presence of tryptophan)

Cleverly, tryptophan acts as a **corepressor** for the operon. If tryptophan is available, it will bind to the inactive repressor and activate it! The repressor then binds to the operator, blocking RNA polymerase and the transcription of the tryptophan-synthesizing genes. Brilliant!

As I tell my students, understanding the concept of how an operon works can be tricky. To make it even more challenging, there are these two specific examples that are both

BTW

A thought experiment: Is the amino acid tryptophan usually present in E. coli's environment? Without knowing anything about E. coli's living conditions, you can answer this question by thinking about the trp operon. The default setting is to allow the transcription of the genes needed to create tryptophan; therefore, the amino acid tryptophan is normally NOT available since it needs to constantly make it itself!

exactly the same (the components of an operon and how it functions), but also totally different:

Comparing *lac* and *trp* Operons

Operon	Purpose of structural genes	Default activity of repressor	Default expression of mRNA	Inducible or repressible?
lac	Create enzymes needed to **digest** lactose	Active (able to bind operator)	Transcription not occurring (mRNA not created)	Inducible
trp	Create enzymes needed to **synthesize** tryptophan	Inactive (unable to bind operator)	Transcription occurring (mRNA created)	Repressible

I love the logic behind how these regulatory sequences work. And once you get it, it really clicks in your brain. The **must know** concept stated that operons provide a way to express a collection of genes only if they are needed. The *lac* operon only creates the enzymes to digest lactose if lactose is present; the *trp* operon only creates the enzymes to synthesize tryptophan if there is no tryptophan available. An efficient and clever means to control gene expression.

REVIEW QUESTIONS

1. Fill in the blanks: The process of ________________ is when an undifferentiated ________________ cell begins to selectively express certain genes and turns into a specific tissue type.
2. Why is a promoter important to transcription?
3. General transcription factors are small proteins that bind to the ________________ and help ________________ to land properly. If the cell needs to create a lot of mRNA, however, it needs the help of specific transcription factors called ________________. Unlike the general transcription factors, these specific transcription factors bind upstream of the promoter at sequences in the DNA called ________________ regions.
4. Choose the correct term from each pair of words: By adding methyl groups to **DNA/histones** gene transcription will be **increased/ decreased**. If, instead, acetyl groups are added to **DNA/histones** gene transcription will be **increased/decreased**.
5. General transcription factors bind at the ________________ within the promoter, whereas specific transcription factors (activators) bind at a specific combination of ________________ in the ________________ region.
6. Is the *trp* operon an inducible or repressible operon? What does this mean regarding the availability of tryptophan in *E. coli*'s environment?

7. For each of the following, indicate whether the statement is true or false:
    a. During transcription, the entire chromosome is unwound and unzipped.
    b. The process of transcription involves the enzyme RNA polymerase moving down the entire chromosome and creating a piece of mRNA.
    c. For a given gene, only one side of the DNA double helix contains the correct code for protein synthesis.

8. What is the role of general transcription factors in gene expression?

9. Label the following picture:

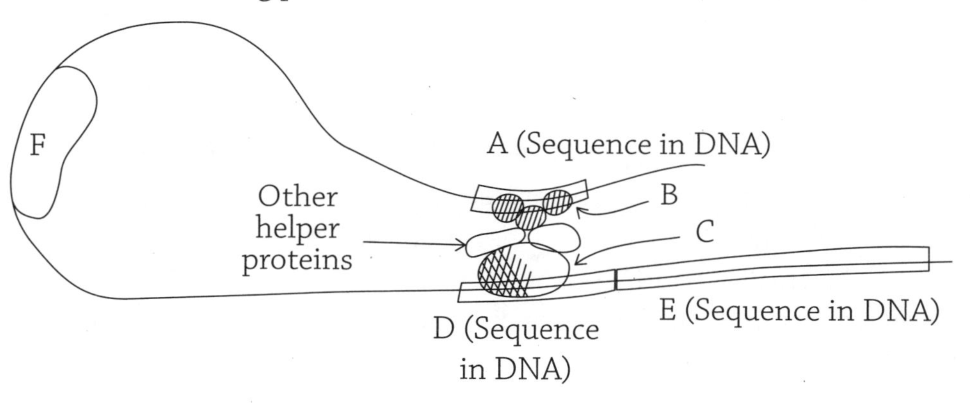

10. Transcribe and translate the following gene (top strand is the template strand):

    GCTACTGATCGACCCCCATAATGAAAATCTTTT

    CGATGACTAGCTGGGGGTATTACTTTTAGAAAA

11. Select the right term from the pair: The *lac* operon is a(n) **inducible/ repressible** operon because it is normally not transcribing the genes for lactose digestion.

12. Fill in the following table about the *lac* and *trp* operons:

Operon	Purpose of structural genes	Default activity of repressor	Transcription occurring or not occurring?	Inducible or repressible?
lac	Create enzymes needed to ______ lactose	______ (able to bind operator)		
trp	Create enzymes needed to ______ tryptophan	______ (unable to bind operator)		

13. Which of the following would lead to decreased levels of gene expression?
    a. Acetylation of histone proteins
    b. Binding of activators to the enhancer region
    c. Methylation of DNA
    d. When lactose is present for a bacterial cell with the Lac operon

The Cell Cycle and Mitosis

MUST KNOW

After the chromosomes are duplicated, the process of mitosis pulls the genetically identical sister chromatids into two separate cells.

The goal of mitosis is to produce cells that are just like the starting cell.

e just learned about DNA and how it's copied. But when, exactly, does that occur in a cell's "lifetime"? Once a cell replicates its DNA, then what? Once the genome is copied, is there more than one way to divvy it up among daughter cells? Those are all important questions that are answered when you take a step back and look at the big picture of a cell's life cycle, summarized with this simple figure:

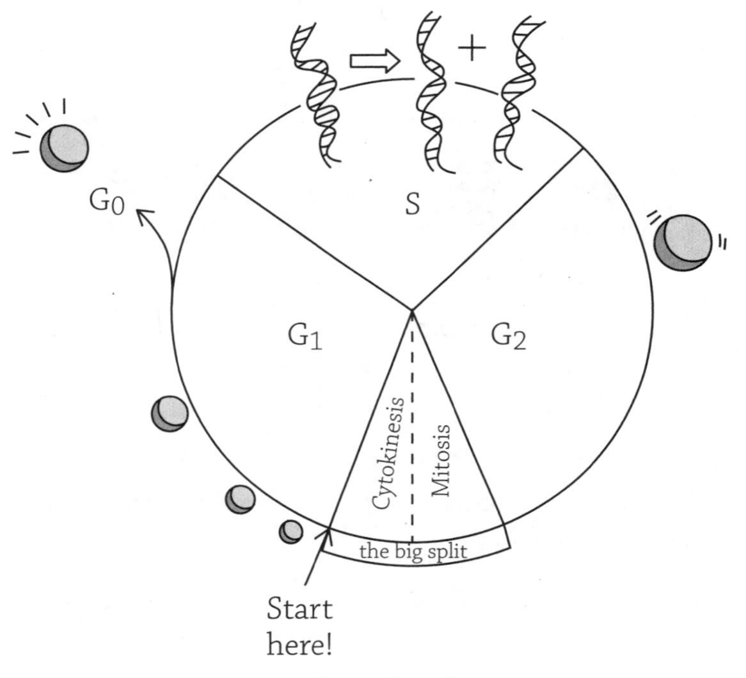

The cell cycle

The Cell Cycle

In the figure above, we can see the stages in a cell's life, starting from when it was "born," meaning it was the product of a mother cell splitting into two daughter cells (at the "start here!" in the figure). This type of cell division produces exactly the same type of cell that you started with (most of the time, this is what a dividing cell has in mind . . . making more cells just like it). If you take anything and split it in half, the resulting two objects have half the volume of the original, right? Therefore, each of the new daughter cells need to grow a bit before it reaches its intended size. This growth stage occurs during G_1 (it helps if you think that G stands for "growth") and allows the cell to increase its cytoplasmic volume. Cells spend most of their lives in the G_1 phase, because this is also when a cell does its "thing," whether it's making insulin (a pancreatic beta cell), helping digestion (a cell lining the small intestine), or capturing images (a photoreceptor cell of your eye). In fact, if a cell is more than happy just to do its job—and it has no intentions of dividing—then it can "exit" the cell cycle and chill out in G_0 phase ("gee naught"). If it gets the signal that it needs to replicate, it will dive back into the cell cycle.

After G_1, the cell progresses into the S phase. Here, the "S" stands for synthesis, because the cell is synthesizing (creating) more DNA. Yes, this is when replication of the chromosomes occurs! All of that gory detail we just learned about DNA replication happens here, in the S phase. At this point, the cell is committed to finishing the process of cell division; if it doesn't, it would end up with twice as much DNA as it should contain, and that is not a good thing. Therefore, a cell that is currently in S will definitely move into G_2.

In the G_2 phase, the cell grows some more, including making more organelles in preparation for the big split. And when a cell splits, there are essentially two parts that need to divide: first the nucleus, and then the rest of the cytoplasm. This is where it gets interesting. In order for a cell to split in two and produce two genetically identical daughter cells, it can't just randomly chop itself in half. The process must be methodical and careful, in order to ensure each daughter cell ends up with a complete set of chromosomes, identical to the original mother cell. The process of nuclear division is called **mitosis**. Shortly thereafter, the rest of the cell divides (meaning the cytoplasm and all the stuff in it), and the cell physically splits into two; this part of the cell cycle is called **cytokinesis**.

BTW

Cyto- means "cell," and *kinesis* means "movement" . . . the cell is splitting!

Mitosis

Mitosis is the process of the nucleus dividing in two. This process must be very carefully coordinated to ensure that one of each chromosome copy ends up in one of the resulting daughter cells. As our **must know** states, the goal of mitosis is to produce cells that are *exactly* like the starting cell. Any differences in genetic makeup would be considered a mutation. In order to ensure perfection, mitosis includes four steps: **prophase**, **metaphase**, **anaphase**, and **telophase**. To help set the stage, let me show you what an example cell would look like in G_1, when it's just chilling out doing its cellular thing.

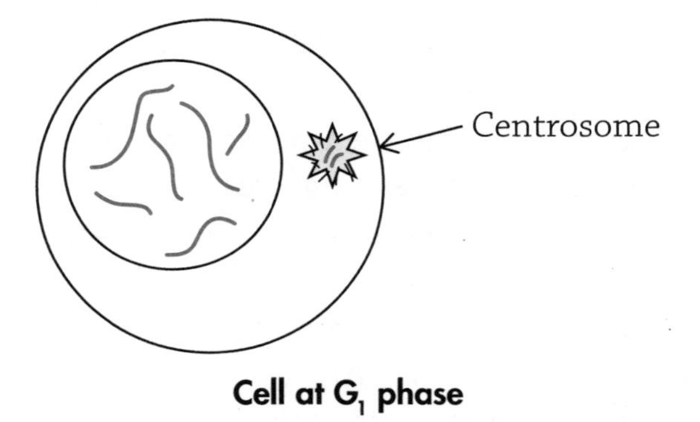

Cell at G_1 phase

This cell has a total of six chromosomes contained within its nucleus. If you look closely, you will notice that they occur in pairs: two large chromosomes, two medium, and two small. There is also a single **centrosome**, a little structure that will soon be very important in divvying up those chromosomes.

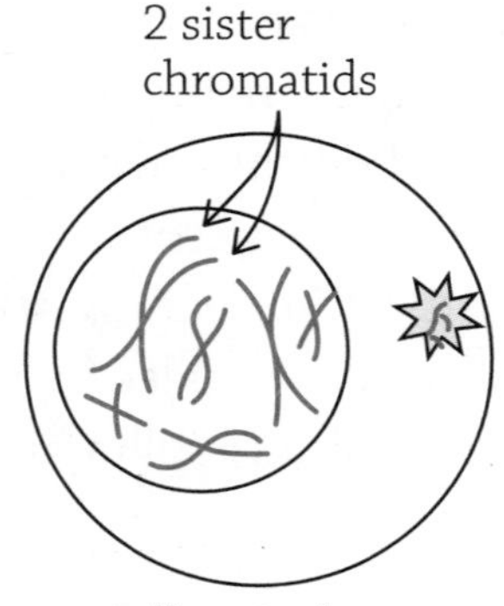

Cell at S phase

Now our cell has completed the S phase, and each chromosome has been replicated (thanks to DNA polymerase). The chromosomes now look like "X" structures because there are two identical copies called "sister chromatids" that are stuck to each other. Sister chromatids are genetically identical copies.

Genetically identical sister chromatids

Region where chromatids are connected (centromere)

Two genetically identical sister chromatids that are attached at the centromere

The next and final stage before the cell divides is G_2. The difference at this point (after completion of G_2) is subtle—the only difference seen in the figure below is there are now two centrosomes instead of one. Remember that in G_2 the organelles are copied. We are focusing on the centrosomes because there needs to be two of them in order for mitosis to occur. Good thing the organelles were copied!

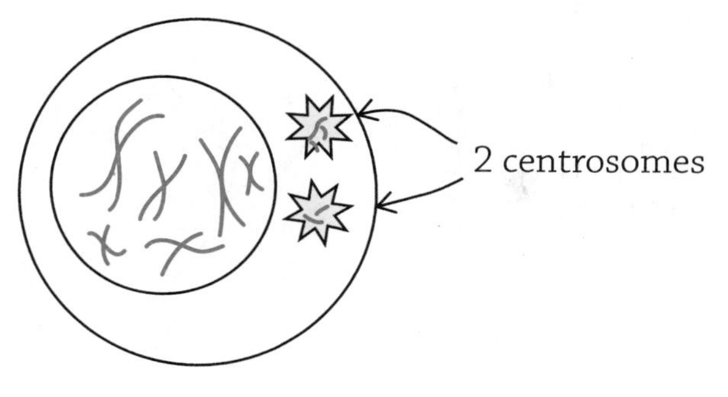

The cell at G_2 phase

After the cell completes G_2, it will begin to split, starting with the nucleus (mitosis). The table below summarizes the major events at each phase of mitosis, plus an example of what the cell would look like.

Stage	Description	Example
Prophase (early)	A. Nuclear membrane disappears B. Chromosomes condense (become tightly wound up) C. The centrosomes begin to move toward opposite ends of the cell and microtubules begin to emerge	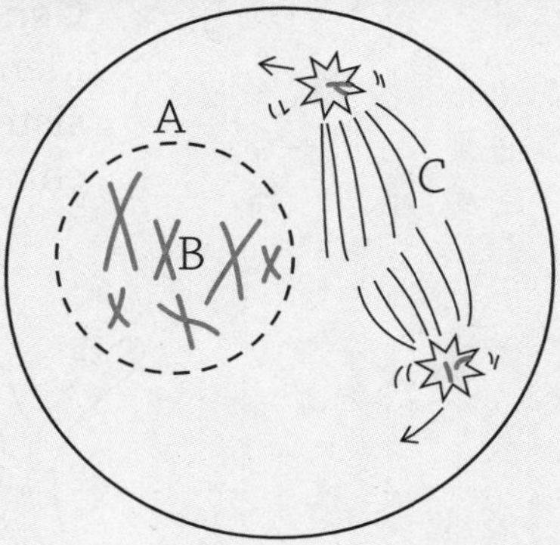 A B C
Prophase (late)	D. Centrosomes reach opposite ends of the cell E. Each sister chromatid becomes attached to a microtubule (from opposite sides of the cell)	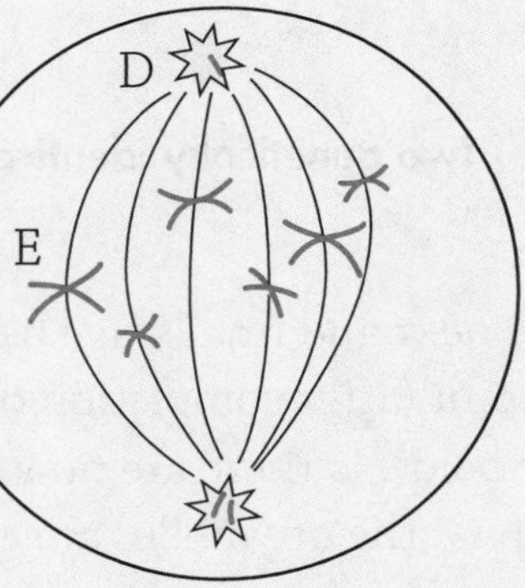 D E
Metaphase	F. The sister chromatids are aligned in the middle of the cell G. Each pair of sister chromatids straddle the "metaphase plate" (imaginary dividing line down the middle of the cell)	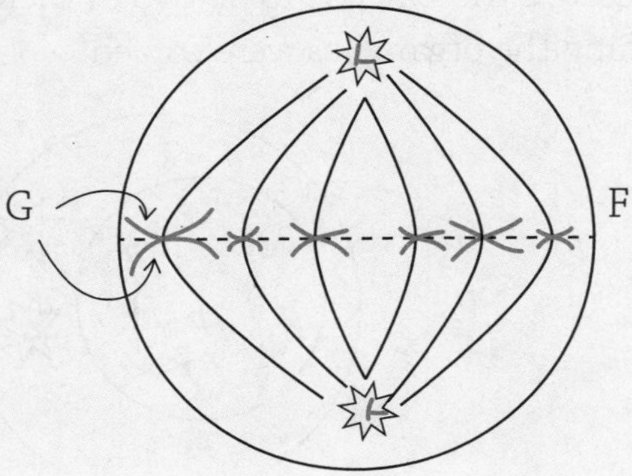 G F

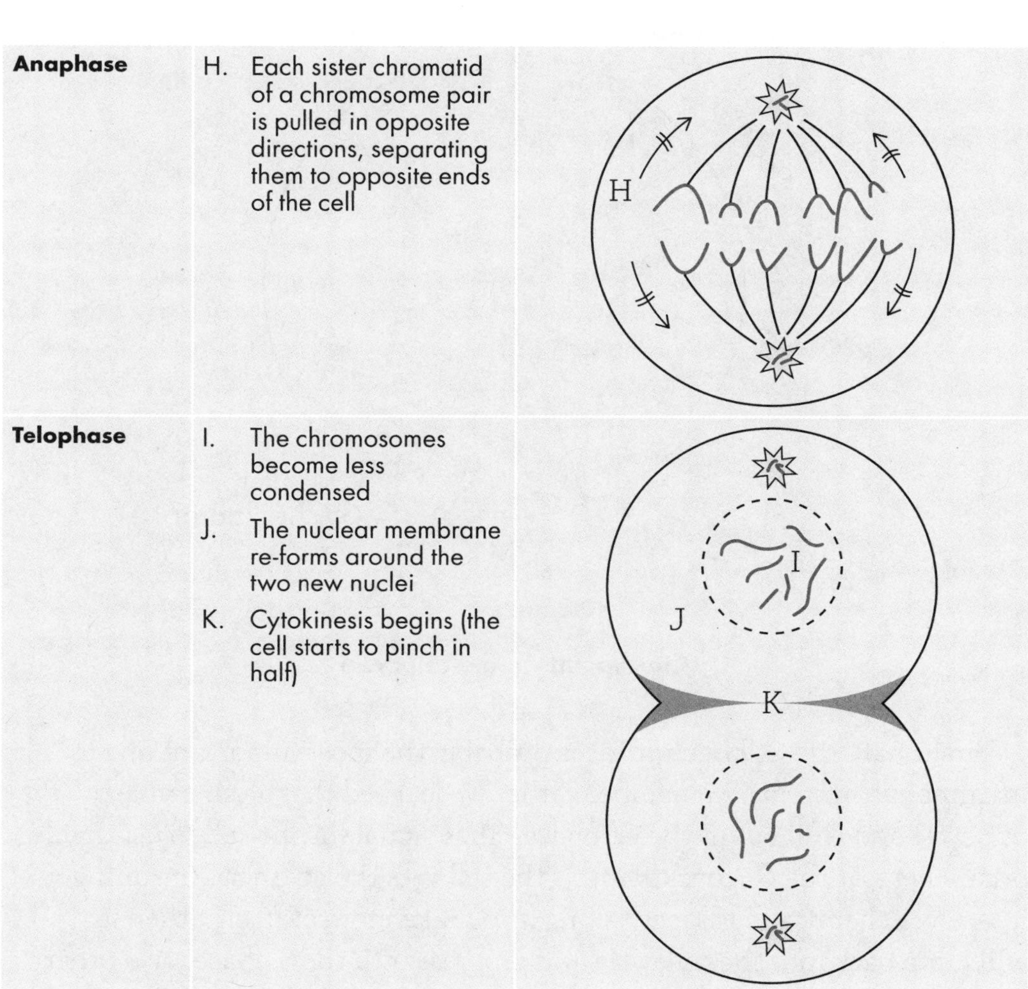

Anaphase	H. Each sister chromatid of a chromosome pair is pulled in opposite directions, separating them to opposite ends of the cell	
Telophase	I. The chromosomes become less condensed J. The nuclear membrane re-forms around the two new nuclei K. Cytokinesis begins (the cell starts to pinch in half)	

When it's all over, you will now have two cells, each with an identical assortment of chromosomes. Any genetic differences are considered to be mutations, which can be dangerous for the cell. Because it is so important to ensure the two resulting daughter cells are perfect genetic copies of one another, there are safety checks embedded in the cell cycle. When a cell progresses through the cell cycle, it will pause at a given checkpoint to make sure conditions are good; if they are, the cell progresses.

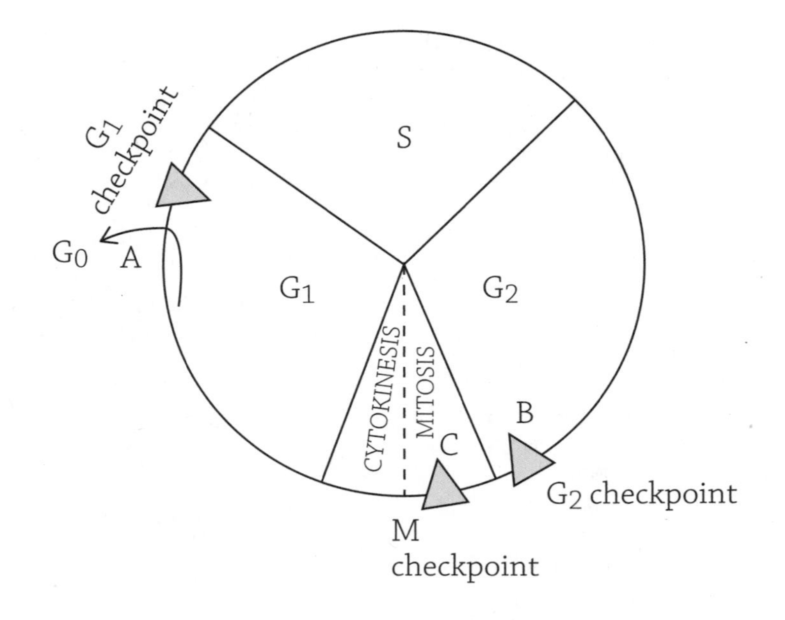

Checkpoints in the cell cycle

First of all, the **G_1 checkpoint** is probably the most important one in mammalian systems. Remember earlier we learned that a cell can "exit" the cell cycle and chill out in the G_0 phase? That signal for the cell to *not* divide occurs here, at the G_1 checkpoint. If the cell gets the go-ahead (meaning it receives chemical messages telling it to make more of itself), then it will jump back into the cell cycle and progress into the S phase. Two other checkpoints occur later, once the cell is in the process of actually dividing. The **G_2** and **M checkpoints** ensure the chromosomes are attached to microtubules, and the chromosomes are properly replicated; if chromosomes are not properly tethered, or if the DNA is damaged, the cell is not allowed to progress until it is fixed—or if the damage is beyond repair—the cell may have to undergo **apoptosis** (programmed cell death). If a damaged cell is allowed to replicate and pass on the damaged DNA to daughter cells, it may result in a cell behaving badly and copying itself unnecessarily.

The cell cycle checkpoints are important, and it's doubly important that cells listen to these checkpoints. Furthermore, a cell needs to hear a certain chemical signal (a **growth factor**) in order to divide. Cancerous cells are normal cells gone rogue; they no longer listen to cell cycle signals that tell them to divide (or not). On top of that, if a cancerous cell does decide to stop dividing, it may stop anywhere in the cell cycle, and not at the proper checkpoints. Cancerous cells grow when it is inappropriate to do so. If the mass of cancerous cells begins to invade neighboring tissue, this can harm the organism.

REVIEW QUESTIONS

1. What is the difference between mitosis and cytokinesis?
2. Fill in the blanks: After the S phase, each chromosome is now composed of two genetically identical ____________ and they are attached at the ____________.
3. How many checkpoints occur in the cell cycle? During what phases do they occur?
4. Which stage of the cell cycle does the following cell depict? How do you know?

5. Is there any time a cell may exit the cell cycle and not progress to the next stage?
6. Briefly describe what occurs in each of the phases of the cell cycle: G_1, S, G_2, mitosis, cytokinesis.
7. True or False: Any genetic differences between daughter cells after mitosis are considered to be mutations.
8. If a cell fails a checkpoint but has already copied its DNA, it may undergo ____________ in order to prevent a defective cell from dividing (and possibly becoming cancerous).

9. What phase of mitosis is the cell below currently in? What structures (labeled G) are straddling the middle dividing line of the cell (labeled F)?

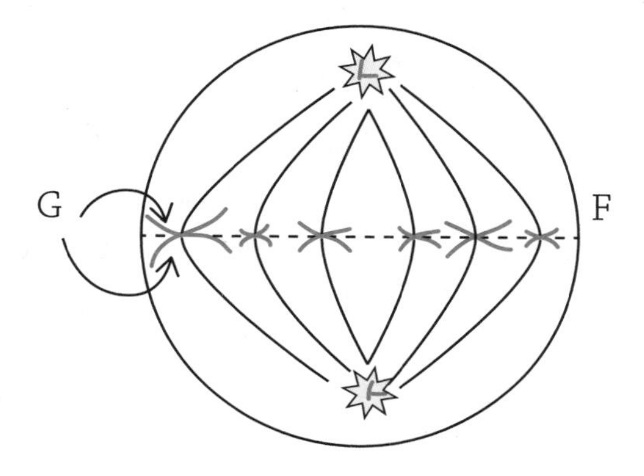

10. What is the name of the chemical signal that tells a cell it needs to divide?

11. Which of the following would not signal a cell to divide?
    a. Successful passage through the G_1 checkpoint
    b. Molecular signals resulting in apoptosis
    c. Presence of growth factors
    d. Passage through S phase of the cell cycle

MUST KNOW

- In meiosis, after the chromosomes are duplicated, the cell will undergo two rounds of cell division.
- The first division (Meiosis I) separates the homologous chromosomes into two separate cells.
- The second division (Meiosis II) separates the genetically idential sister chromatids into a total of four separate cells.
- After Meiosis I (and the homologous chromosomes are separated), the resulting cells are haploid.
- The goal of meiosis is to produce cells that are different from the starting cell.

As we just learned, mitosis is a type of cellular division that creates two genetically identical daughter cells. This is important, because if a single fertilized egg is going to undergo countless rounds of cell division to create the final multicellular life form, it better make sure every cell has the same set of chromosomes!

Now, there is a very important exception to this rule of "if you make cells for a multicellular organism, they better all have the same genes." The **gametes**—egg and sperm—are each unique creations with a different combination of chromosomes. Sure, all the eggs in a female (or sperm in a male) are created using the same source cell, but the process of meiosis ensures that each resulting gamete cell is different in two ways:

1. Meiosis creates cells with half the DNA of the parent cell (specifically, one set of chromosomes instead of the usual two sets).
2. Meiosis shuffles these two sets of chromosomes like a deck of cards and then randomly deals them out into the gametes.

It is a **must know** to understand that meiosis intends to create cells different from one another (unlike mitosis, which aims to create identical cells). Let's talk about these differences in more detail.

EXTRA HELP

This is an important concept that we should pause to reconsider. A multicellular organism is made of trillions of cells, each with the same DNA (thanks to mitosis). But what about the fact that a multicellular organism isn't a huge blob of the same types of cells? We have muscle cells, skin cells, cells that produce hair, and cells that secrete insulin. This wide variety of cells still contain the same DNA, the same chromosomes, the same genes. The difference, however, is what genes they choose to express. A pancreatic cell (alpha cell, specifically) will focus in on that insulin gene and create lots of insulin protein; your skin cell, however, will happily ignore that insulin gene. This is the phenomenon of gene expression (back from Chapter 11).

Meiosis Creates Cells with Half the DNA (only one set of chromosomes)

Let's start with you, shall we? A long time ago, a man and a woman made a baby, and that baby was you. You are a genetic combination of those two people: the man sent over a set of chromosomes in a sperm cell, and the woman sent over another set of chromosomes in an egg cell. This term *set* is significant. Humans have a total of 46 chromosomes, but more specifically, there are *two sets of 23*:

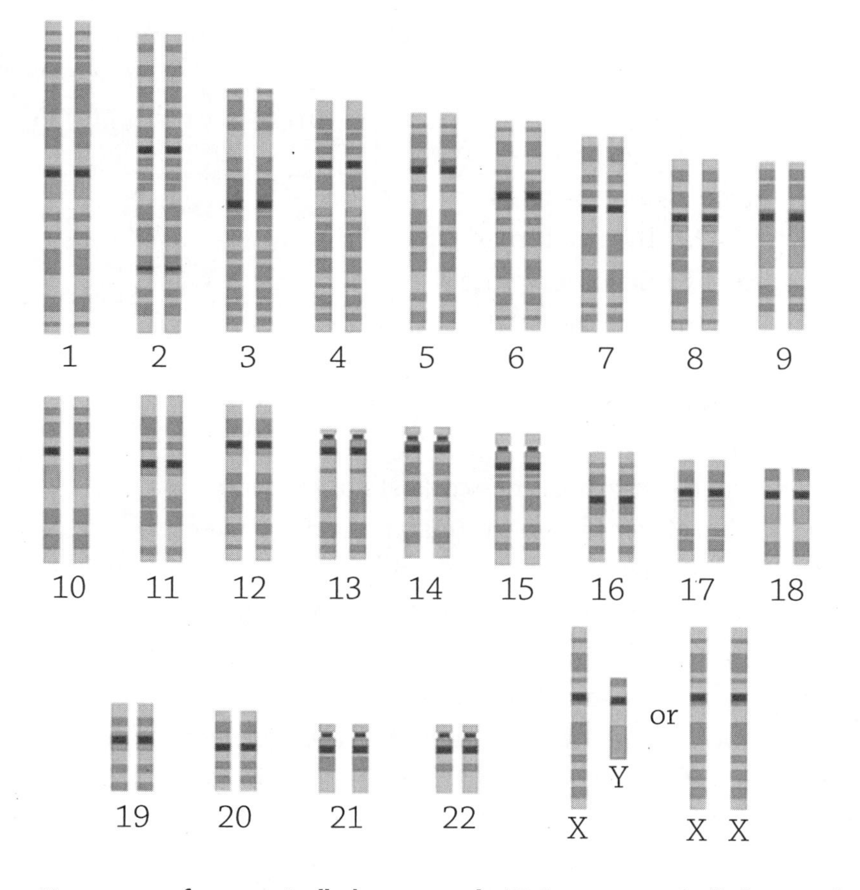

Karyotype of a genetically human male (XY) or a genetically human female (XX).

Author: Mikael Häggström. https://commons.wikimedia.org/wiki/File:Human_karyotype.svg

A **karyotype** is a visual layout of all the pairs of chromosomes in a critter. The previous figure on page 207 is an example of a human karyotype. Look at chromosome number 1 . . . there are, in fact, two of them. Where did you get two number 1 chromosomes? One came from your mom's egg and one came from your dad's sperm! You are literally the genetic composition of your mother and your father. These two chromosomes contain the same genes (about 2,000 of them), and they are the same size and shape. They are not, however, genetically identical (one came from your mother and one came from your father . . . and they are most certainly not genetically identical). These matching chromosomes are called **homologous chromosomes**. If you take one of each homologous pair, you create a single set of chromosomes, numbers 1–22. The final "pair" of chromosomes are the sex chromosomes—the X and the Y chromosome—and they are not considered homologous because they do NOT contain the same genes. If you have one X and one Y chromosome, you are male; if you have two X chromosomes, you are female.

Chromosomes are assigned numbers based on their size. The biggest chromosome is #1, and the smallest chromosome is #22.

EASY MISTAKE

Don't get these two mixed up . . . sister chromatids are genetically identical; homologous chromosomes are NOT genetically identical.

In summary: a single *set* of chromosomes consists of chromosome numbers 1 through 22, plus either an X chromosome or a Y chromosome. The shorthand way to indicate a cell contains a single set is "n," also called **haploid**; the gametes' chromosome count is n = 23. When a sperm fertilizes an egg, the two sets combine into 2n = 46. Another term for having two sets of chromosomes is **diploid**. By far, most of the cells in your body (besides the egg or sperm) have two sets of chromosomes, and are designated 2n. Furthermore, the majority of cell division occurring in your body is through mitosis, by which a diploid cell creates more diploid cells. The ONLY cells in the human body that undergo meiosis are the special cells in the ovaries and testes that are responsible for making eggs and sperm. **Spermatocyte** cells in the testes undergo meiosis to create sperm, and **oocyte** cells in the ovaries undergo meiosis to create eggs.

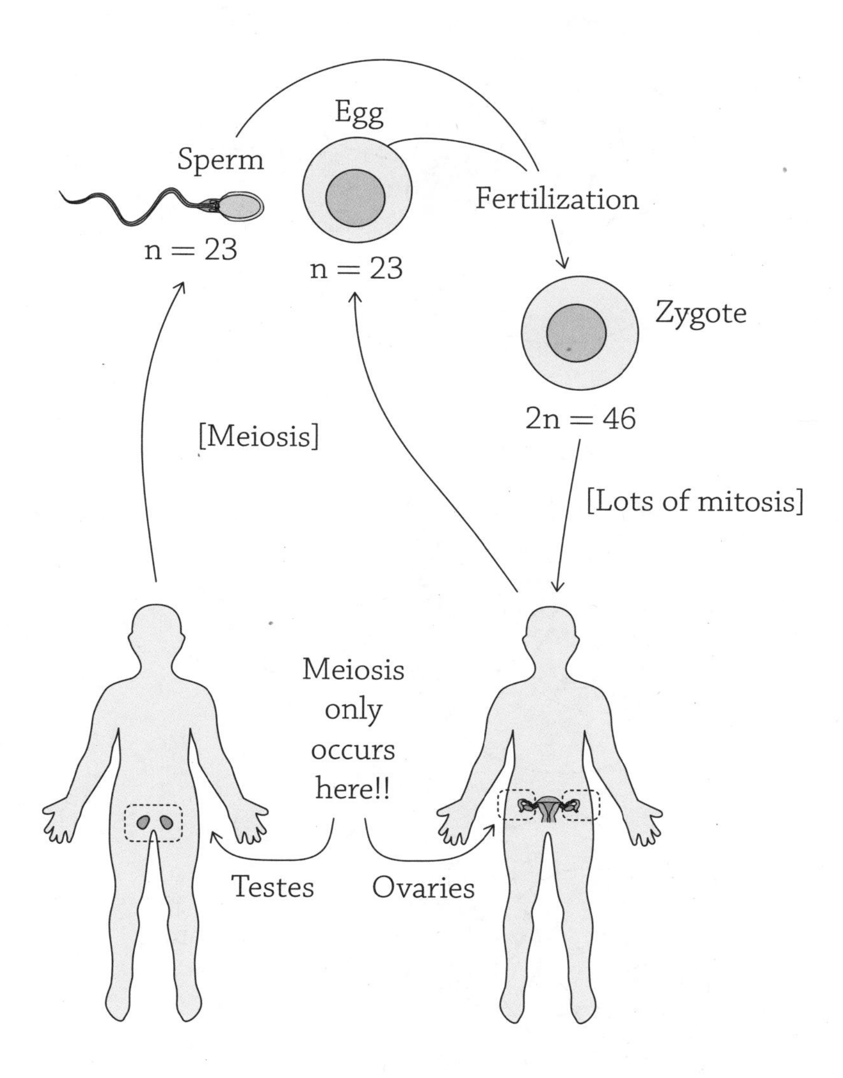

The way meiosis manages to create haploid cells with only one full set of chromosomes is quite clever and involves two division steps instead of one (as in mitosis). I will show you how it's done, starting with the same cell we used in the mitosis example.

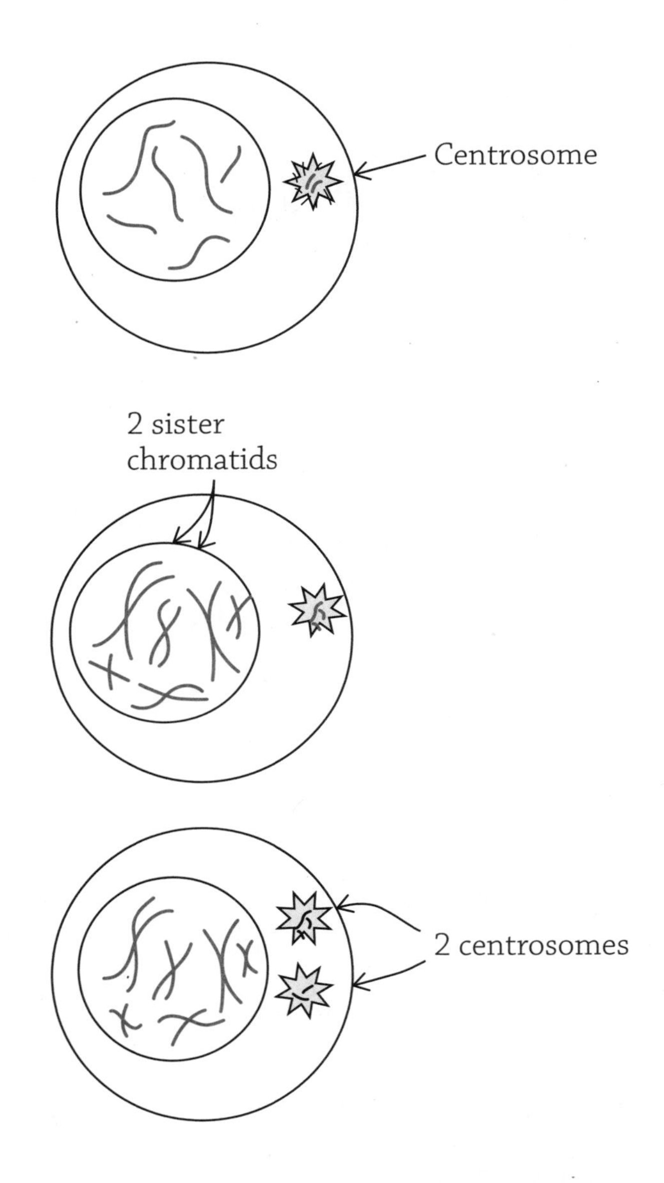

As in mitosis, the cell must first replicate its DNA, creating chromosomes consisting of two genetically identical sister chromatids. If the cell divides by mitosis, it would then split apart the two sister chromatids and be done with it. Meiosis, however, first divides up the homologous chromosome, and THEN it divides up the sister chromatids (just like mitosis).

Meiosis I (dividing up the homologous chromosomes)

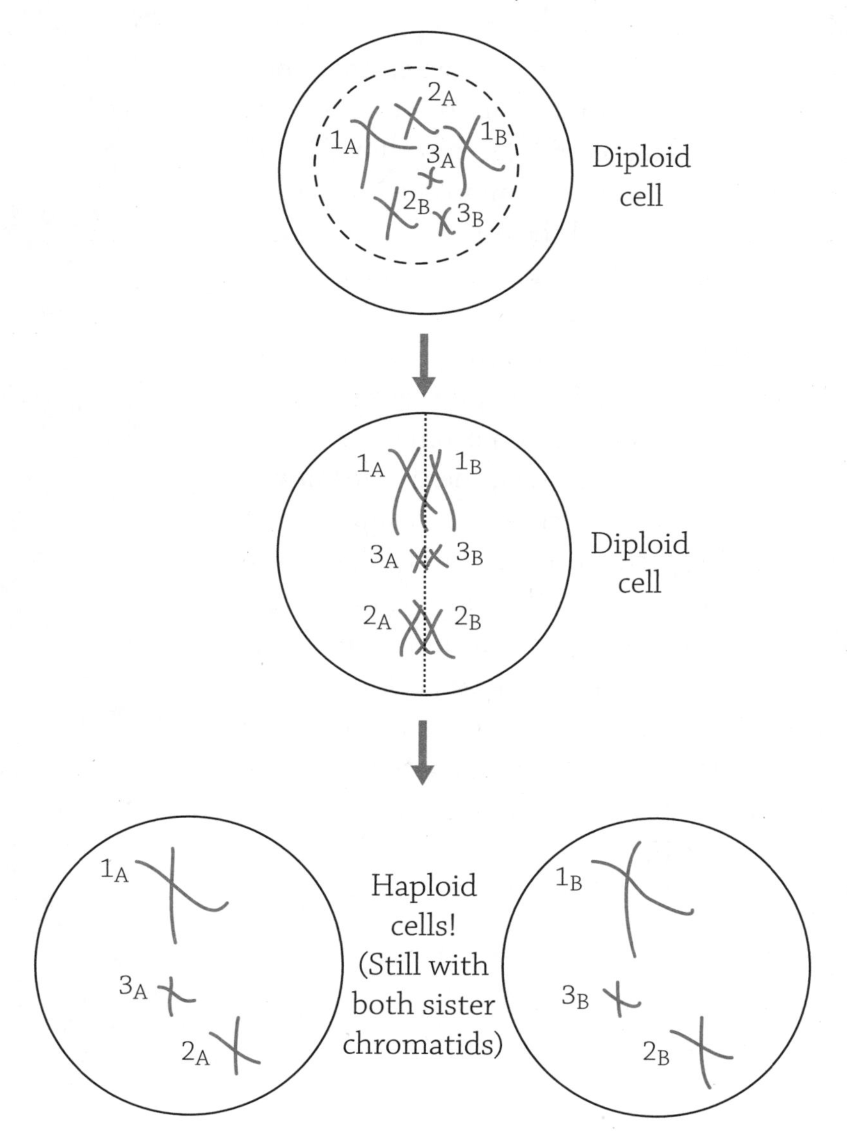

Meiosis I is filled with novel happenings that did not occur in mitosis. For example, the homologous chromosomes seek out their partner and clump together, in a process called **synapsis**. This is necessary, because when these newly formed **tetrads** (tetra- = four, as in four total sister chromatids) line up in the middle, the cell needs to make sure one homologous chromosome is on one side of the metaphase plate, and the other homologous chromosome is on the other side of the metaphase plate.

Once each of the homologous chromosomes go their separate ways (and end up in separate cells), each cell will have only ONE of each chromosome 1, 2, 3 . . . and so on. That means after meiosis I, the two resulting cells are **haploid** cells containing only a single set of chromosomes. Yes, each chromosome still has a sister chromatid attached to it, but that doesn't matter in the definition of haploid (one set of chromosomes) versus diploid (two sets of chromosomes). The cell does, however, need to undergo a second round of cell division in order to separate those two sister chromatids into separate cells. This happens in meiosis II.

To simplify things, we are not going through the detail of prophase → metaphase → anaphase → telophase again. Just understand that each of those steps do occur, in both meiosis I and meiosis II. For example, when the homologous chromosomes line up in the middle during meiosis I, it's called metaphase I; when the sister chromatids line up in the middle during meiosis II, it's called metaphase II.

A common test question tests your understanding of what it means for a cell to be haploid (it must have only one of each homologous pair). A cell is considered haploid at the end of meiosis I, even though the sister chromatids are still attached. A pair of sister chromatids does NOT define the cell as diploid!

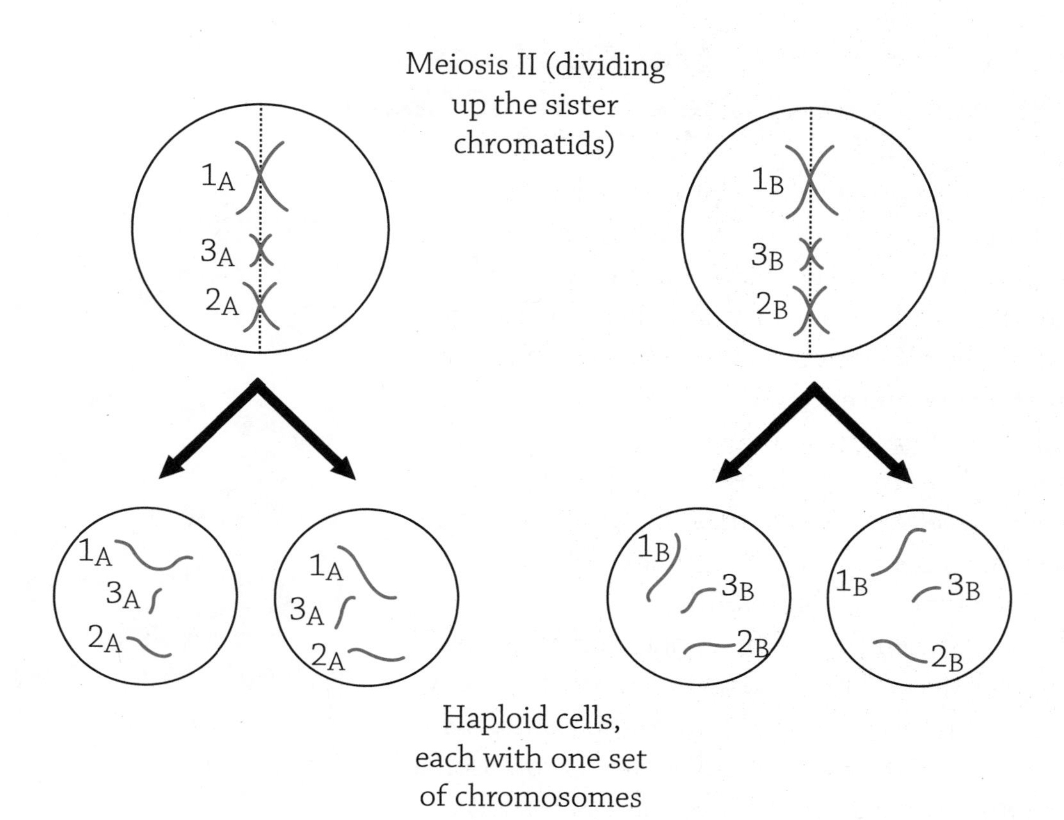

It's important to note that DNA replication did NOT occur again before this second division! If it had, we would end up with too much DNA in the cell. Remember, the point of meiotic division is to cut the chromosome number in half, and undergoing two rounds of cell division (with only one pass through the S phase) accomplishes just that!

Meiosis Shuffles the Two Sets of Chromosomes Like a Deck of Cards

Now we know that mitosis produces diploid cells that are exactly like the parent cell, and meiosis produces haploid cells that are different from the parent cell. This ability of meiosis to produce genetically different cells is very significant, and our **must know** concept. In the process of evolution, it is these differences among individuals of the same species that provide the reservoir by which new traits are selected for as "most fit" and passed on to the next generation.

One of meiosis's big shuffling steps occurs in meiosis I, when the cell first splits and separates the homologous chromosomes. The origin of those two sets of chromosomes are from your parents, right? And recall that a pair of homologous chromosomes contain the same genes but are not genetically identical; it makes a difference, then, which of the homologous pair lines up on which side of the metaphase plate. For example, earlier, when we discussed the steps of meiosis I, you saw this cell:

BTW

Evolution of a species is based on there being variation in a population. In order for evolution to occur, some members of the population are a bit better at survival, simply because they were born with traits different from their not-as-successful brethren. The differences in members of a species allow natural selection to occur, because some traits are "selected" as being more fit in the struggle for survival. These differences in members of a species are in large part due to meiosis's ability to shuffle things up and create a bunch of new genetic combinations in the gametes.

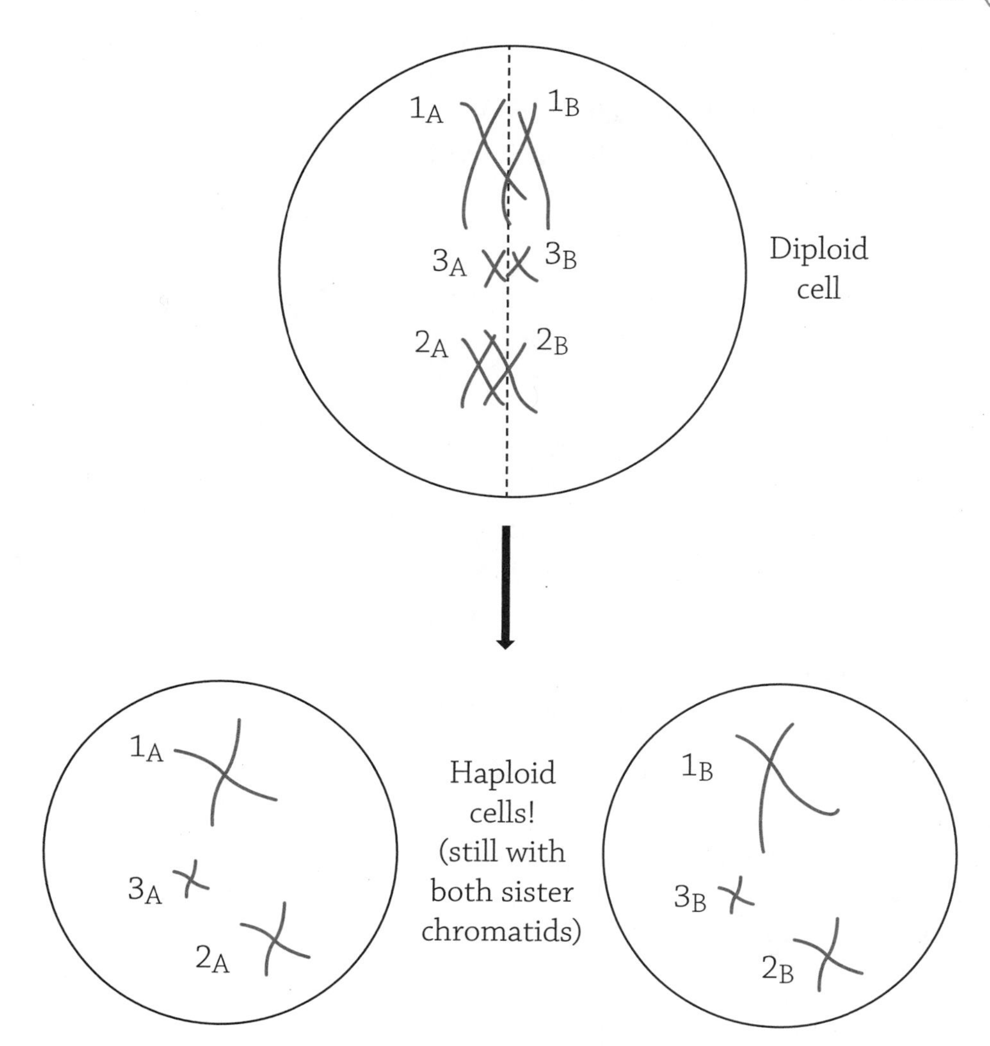

In this scenario, after meiosis I, the resulting daughter cells contain one set of chromosomes (1, 2, and 3) and they each came from the same "source" . . . chromosomes 1_A, 2_A, and 3_A were from your mom, and 1_B, 2_B, and 3_B were from your dad. After meiosis II (when those sister chromatids are separated), the resulting gamete (whether egg or sperm) would also contain chromosomes entirely from maternal or paternal origin. But is there another way this could have played out? Instead of all of your maternal chromosomes lining up on the left side of the metaphase plate, only 1_A and 2_A are on the left; chromosome 3 lined up differently, with the father's copy on the left.

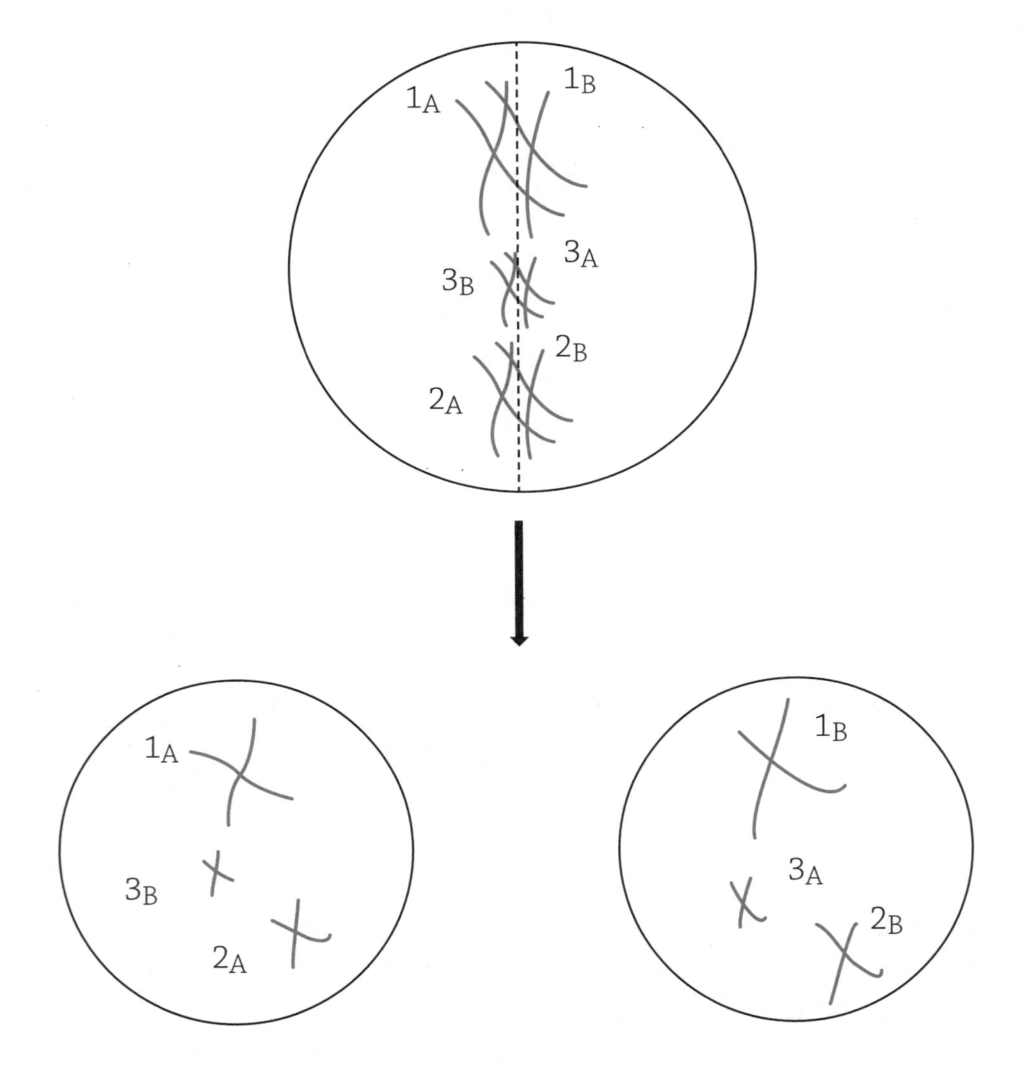

Notice that just because of random alignment of homologous chromosomes, the resulting cells will now have a combination of chromosomes, this time from both sources (your mom's and your dad's DNA).

This random alignment of the homologous chromosomes along the metaphase plate is referred to as **independent assortment**. It's "independent" because chromosome 1_A doesn't really care which side 2_A chooses . . . it's gonna do its own thing.

In this example, we're using a simple cell with 2n = 6, or n = 3. If you want to figure out how many different combinations are possible with this independent assortment concept, you use the equation 2n, where n = haploid chromosome number. Therefore, there are 2^3 = 8 different combinations of gametes possible. Now . . . how many combinations are possible in YOUR BODY? We have a haploid number of n = 23, so the number of different gametes you can make on your own is 2^{23}, or a little over *8 million combinations*! What an impressive selection! But guess what . . . the variations don't stop there. There is a sneaky little bit of shuffling that occurs on a smaller scale, when synapsis occurs and those homologous chromosomes pair up. A phenomenon called **crossing over** occurs, when the homologous chromosomes will swap matching segments with each other (thus swapping the same genes). This increases variation because if you inherit chromosome #3 from your dad, it may actually carry a little bit of your mom's chromosome #3, because crossing over occurred.

EXTRA HELP

This is when my students get really confused. I keep talking about making cells using DNA from either the mother or the father. Take a moment to focus again on what we're actually doing: we are talking about how you, in your body, create your own gametes (whether egg or sperm). This is happening only in the testes and ovaries, and you are creating haploid cells based on your own selection of chromosomes . . . which were provided to you by your mom and dad. So, when we talk about how your own body creates a ton of different combinations of egg or sperm, we have to think back to where you received your own two sets of chromosomes (your parents). Every generation is simply creating the next generation out of the genetic material passed down from our ancestors.

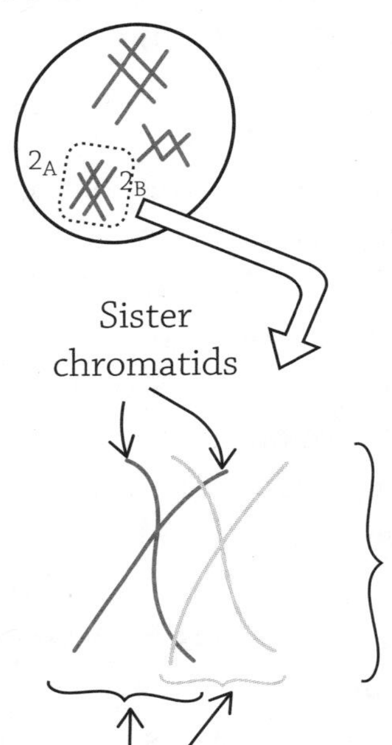

Synapsis between homologous chromosomes 2_A and 2_B

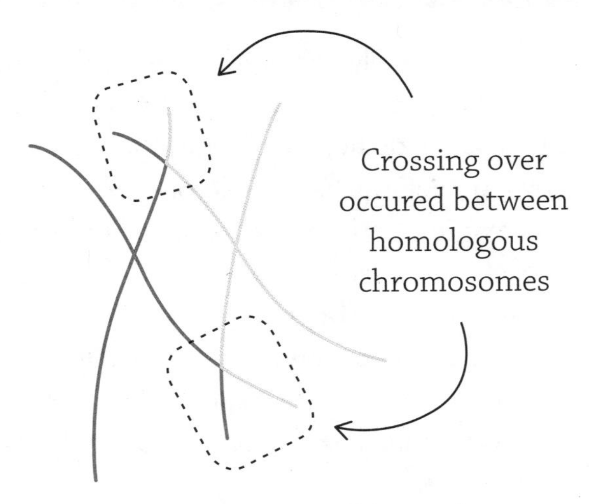

Crossing over occurring between homologous chromosomes 2_A and 2_B

Both of these events—independent assortment of homologous chromosomes during metaphase I and crossing over—are key steps in the **must know** concept of creating a vast array of different genetic combinations in the resulting gametes.

REVIEW QUESTIONS

1. Choose the correct term from each pair: Meiosis creates cells with **half/equal/double** the DNA of the parent cell. Specifically, each gamete contains only **one set/two sets** of chromosomes.
2. True or False: Any cell in your body can undergo meiosis.
3. Choose the correct term from each pair: The specialized cells in the ovaries and testes that produce the gametes start off as **haploid/diploid** cells. Once they divide by **meiosis/mitosis**, they produce the **haploid/diploid** gametes (egg or sperm). When an egg and sperm fuse in sexual reproduction, it creates a **haploid/diploid** zygote cell.
4. What would be the result of crossing over between sister chromatids during meiosis I?
5. Why is it important that meiosis creates gametes that are not genetically identical to one another?
6. Choose the correct term from each pair: The cell that divides by meiosis to create the gametes is **haploid/diploid** because it contains **one set/two sets** of chromosomes (referred to as a **haploid/diploid** cell). The resulting gametes (either egg or sperm) contain **one set/two sets** of chromosomes and are called **haploid/diploid** cells.
7. What three events occurred in meiosis, but did not occur in mitosis?
8. At what stage of meiosis does the diploid cell become a haploid cell?
9. Meiosis creates gametes with different genetic combinations through the process of ______________ (when segments of two homologous chromosomes swap with one another) and ______________ (the homologous chromosomes line up in a different pattern during metaphase I).
10. Which of the following is a correct statement?
    a. Sister chromatids are separated during meiosis I.
    b. Crossing over occurs during meiosis II.
    c. Cells that have completed meiosis are diploid.
    d. At the end of meiosis I, daughter cells are haploid.

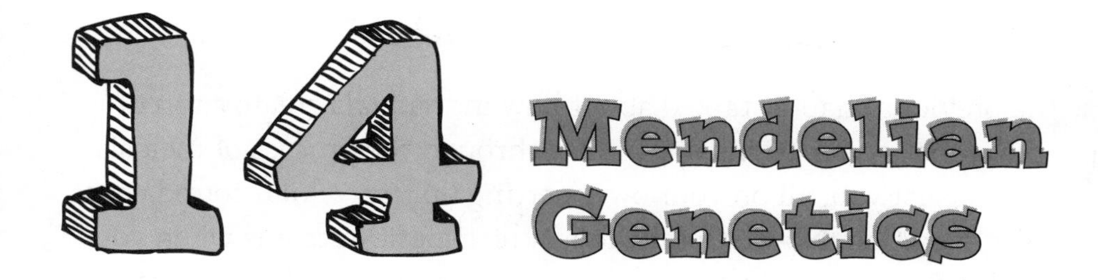

14 Mendelian Genetics

MUST KNOW

- Mendelian genetics enables us to predict what is possible when two organisms create offspring.

- A Punnett square is a model for the different traits possible from a given set of parents.

- In a dihybrid cross that takes two different genes into account, use the rule of multiplication as a quick way to calculate the odds of a particular cross.

- When completing a sex-linked cross, always use XX and XY in your Punnett square.

The previous chapter talked about how individuals create a ton of different combinations of gametes through the process of meiosis. This is just you, all on your own, setting up the cellular foundation for the next generation. Now let's talk about the genetic outcome when you create offspring (with the help of another person's gametes). Welcome to Mendelian genetics and Punnett squares!

The topic of Mendelian genetics is named after Gregor Mendel, a nineteenth-century Austrian monk who studied inheritance of pea plants. Mendel noticed that certain crosses would produce consistent ratios of traits in the offspring, and it appeared that these traits were controlled by some sort of distinct unit. It is impressive that he deduced the basic principles of inheritance and patterns of inheritance well before anyone knew genes (the "distinct units") even existed!

In Mendelian genetics, you use what is called a **Punnett square** to model the different possible genetic outcomes (offspring) when two people have a baby. Punnett squares are very closely related to meiosis, because you decide what goes along one side of the square based on the possible sperm produced from the father, and along the other side, the possible eggs produced by the mother.

Before we jump into some Mendelian genetics, it would help if we look at some helpful vocabulary.

BTW

Keep in mind that each gamete (egg or sperm) will have only ONE of each chromosome from the parent, due to meiosis dividing up the two sets of chromosomes (going from a diploid cell to a haploid gamete). But when using a Punnett square, you're not considering the entirety of a chromosome; instead, you are focusing on one (at the most two) particular gene(s). The same rules apply, though. When forming gametes, only one of each given gene ends up in a sperm or egg.

Term	Definition	Example
Genotype	An organism's actual genetic makeup	When naming a gene, you usually choose a single letter to represent the gene. Let's say there's a particular leafhopper, and its coloration is controlled by a single gene, represented by the letter "G."
Phenotype	What an organism looks like (based on its genetic makeup)	Our example leafhopper can be either green or brown.
Allele	Genes come in different forms; these forms are called alleles	The gene for leafhopper color can come in two forms: the brown allele or the green allele.
Dominant	An allele for a given gene is considered dominant if you see that particular allele's phenotype, even if the diploid organism has only one copy of the allele. The dominant allele's phenotype determines the letter you use to represent the gene, and the dominant allele is a capital letter.	For our leafhopper, the green allele is dominant. That's why the letter "g" (**g**reen) is used, and a capital G stands for the green allele. If the leafhopper's genotype is either GG or Gg, it will be green.
Recessive	A recessive allele's phenotype is "hidden" if there's also a dominant allele. The only way a recessive allele's phenotype is expressed is if the organism only has the recessive allele (gg).	If G (green) is dominant, then g (brown) is recessive. The only way to get a brown leafhopper its genotype is "gg."
Homozygous	The term is used to describe the genotype of an organism that has two of the same allele.	Both GG and gg are "homozygous" genotypes. Specifically, GG is homozygous dominant and gg is homozygous recessive.
Heterozygous	The term is used to describe the genotype of an organism that has two different alleles.	Gg (one of each allele)

Let's dive right into an example.

EXAMPLE

▶ A male brown leafhopper is crossed with a green heterozygous female leafhopper. What is the expected genotypic and phenotypic ratios of their offspring?

▶ It's a good idea to start with writing the genotypes of the parents. A brown leafhopper has the recessive phenotype, so its genotype must be gg (homozygous recessive). The other parent is the green phenotype (which could be either the GG or Gg genotype), but it states that it's heterozygous, so it must be Gg. Now we're ready to create a Punnett square to see all the different combinations of leafhopper babies they can make!

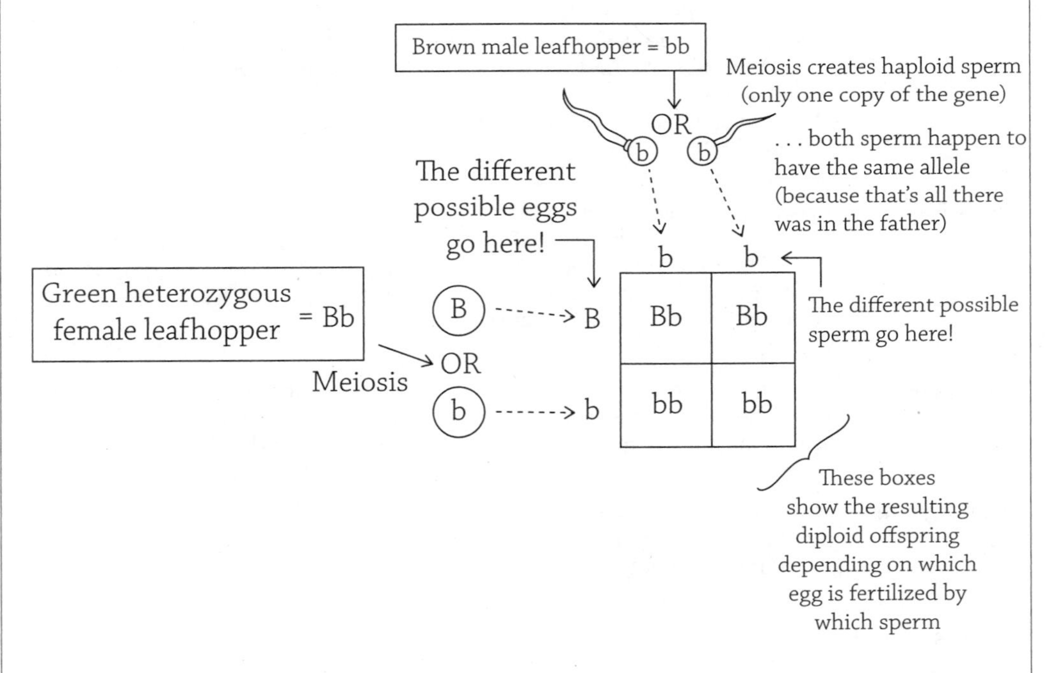

▶ Now we can answer the question! It asked both the phenotypic and genotypic ratios of the offspring, and the answer lies in your Punnett square. Half (or 50%, or 2/4) of the offspring are green, and the other half are brown. That is the phenotypic ratio, because you are describing what the offspring *look* like. The genotypic ratio happens to be the same, but you need to state the genotypes: half are Bb (heterozygous) and the other half are bb (homozygous recessive).

▶ Excellent work! Ready to up the challenge a bit? What if we were tracking *two different genes* in this leafhopper, instead of just one? This is called a **dihybrid cross** (as opposed to a single-gene Punnett square problem, called a **monohybrid cross**). For example, there's a different gene that determines wing shape: rounded wings (R) are dominant to sharp wings (r).

Let's try another example:

EXAMPLE

▶ A female green leafhopper with rounded wings (heterozygous for both traits) crosses with a male green (heterozygous) leafhopper with sharp wings. What are the odds that they will produce a green offspring with sharp wings?

▶ First, do not panic. There's a lot of information thrown at you right there, but I promise, it is no more difficult than our previous example. You simply need to clearly write out the parents' genotypes and then create two separate Punnett squares, one for each gene (either body color or wing shape).

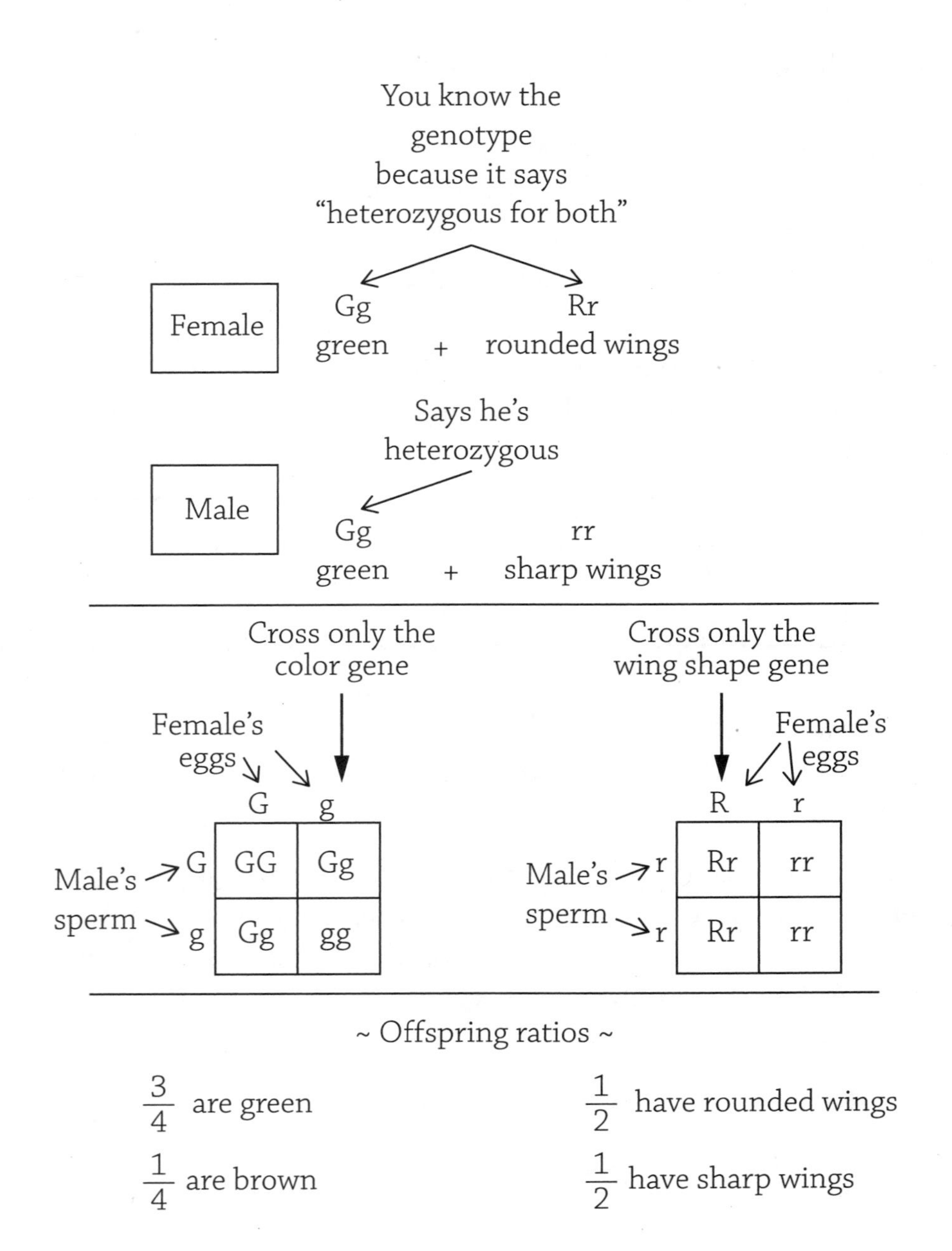
You know the genotype because it says "heterozygous for both"
Female
Gg
green
+
Rr
rounded wings
Says he's heterozygous
Male
Gg
green
+
rr
sharp wings
Cross only the color gene
Female's eggs
G
g
Male's sperm
G
g
GG
Gg
Gg
gg
Cross only the wing shape gene
Female's eggs
R
r
Male's sperm
r
r
Rr
rr
Rr
rr
~ Offspring ratios ~
3/4 are green
1/4 are brown
1/2 have rounded wings
1/2 have sharp wings

- Once you have all the offspring possibilities for both traits, it's a simple matter of determining the odds that a green leafhopper will also have sharp wings. This is solved using the rule of multiplication that states the possibility of two independent events occurring together is determined by multiplying the odds of each event occurring alone. So, for this problem, you multiply the odds of a green offspring (3/4) by the odds of a sharp-winged offspring (1/2) to get 3/8! You have a 3/8 chance of producing a sharp-winged, green baby leafhopper!

BTW

When multiplying fractions, all you have to do is multiply the numerators together to get the numerator of the answer, and then multiply the denominators together to get the denominator of the answer:

$$\frac{3}{4} \times \frac{1}{2} = \frac{3}{8}$$

I don't mean to stress something that may seem obvious to you, but I can't tell you how often my students make a silly mistake at this step, multiply their fractions incorrectly, and get the wrong final ratio!

Sex Linkage

There's another tricky scenario you may run into, and it's a sex-linked gene. Recall that there is a non-homologous pair of chromosomes called the sex chromosomes: X and Y. These two chromosomes have totally different genes from one another. A female has two X chromosomes and a male has one X and one Y chromosome. A sex-linked trait is a gene that is found only on the X chromosome (and is thus "linked" to the X chromosomes). This makes a huge difference, because although females have the "normal" two copies of a sex-linked gene, males only have one copy. This changes things up quite a bit. For example, color blindness is a sex-linked trait, designated by a small superscript "c" on the X chromosomes (X^c). It is super important that when you consider a sex-linked trait, you ALWAYS track its movement on the X chromosome. The small letter c correctly suggests that it is a recessive trait; the normal color vision allele is a capital C (X^C). It is a **must know** that when figuring out a sex-linked trait, *always* take XX and XY into consideration.

If it's a sex-linked trait and you forget to take the X and Y chromosomes into consideration, you will have a ton of trouble trying to solve the cross. If you just can't get a simple cross to work, remember . . . it may be sex-linked!

The odds of finding a color-blind male are much greater than a color-blind female, simply because it is a sex-linked trait. The color-blind allele is recessive, and females have two X chromosomes. Therefore, a color-blind female has two recessive color-blind alleles, one on each X chromosome:

$$X^c X^c$$

In order for a man to be color-blind, however, he needs only one recessive color-blind allele:

$$X^cY$$

Without a second X chromosome, whatever genes are on a man's X chromosomes are the ones to be expressed. A woman, however, could have a single color-blind allele and have normal vision:

$$X^C X^c$$

In this case, she is referred to as a "carrier," because she carries within her the recessive color-blind allele but has normal vision. So, let's try a straightforward sex-linked question.

EXAMPLE

▶ What are the odds that a color-blind man and a carrier woman will have a color-blind child?

▶ First, write the genotypes of each parent:

Man = X^cY

Woman = $X^C X^c$ ← Remember, a carrier is a woman with one recessive sex-linked allele

Then, create a Punnett square:

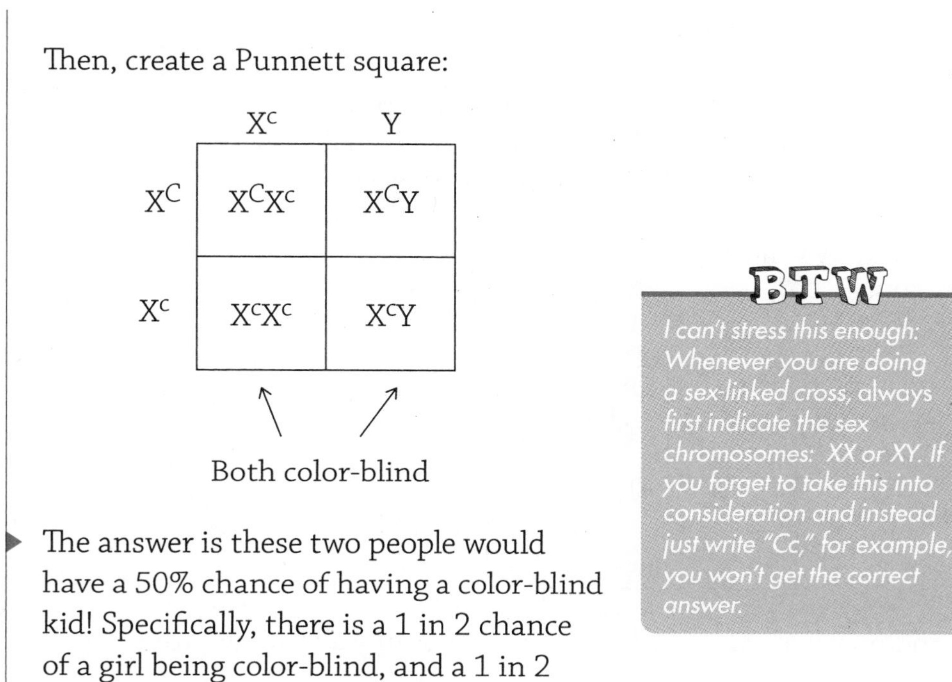

▶ The answer is these two people would have a 50% chance of having a color-blind kid! Specifically, there is a 1 in 2 chance of a girl being color-blind, and a 1 in 2 chance of a boy being color-blind.

BTW

I can't stress this enough: Whenever you are doing a sex-linked cross, always first indicate the sex chromosomes: XX or XY. If you forget to take this into consideration and instead just write "Cc," for example, you won't get the correct answer.

Make sense? Okay, let's try a more difficult example.

EXAMPLE

▶ Hemophilia is a sex-linked trait where an afflicted person's blood does not clot properly. Sunny has normal blood clotting (although her own mother was a hemophiliac) and has a baby with Maynard, also with normal blood clotting. What are the odds that their child will have hemophilia?

▶ In this scenario, both Sunny and Maynard have normal clotting. You know, therefore, that Maynard must be X^BY, where B indicates a normal blood clotting gene. At first, you don't know about Sunny, who is described as also being normal . . . but her own mom has hemophilia! So first you can determine Sunny's genotype . . .

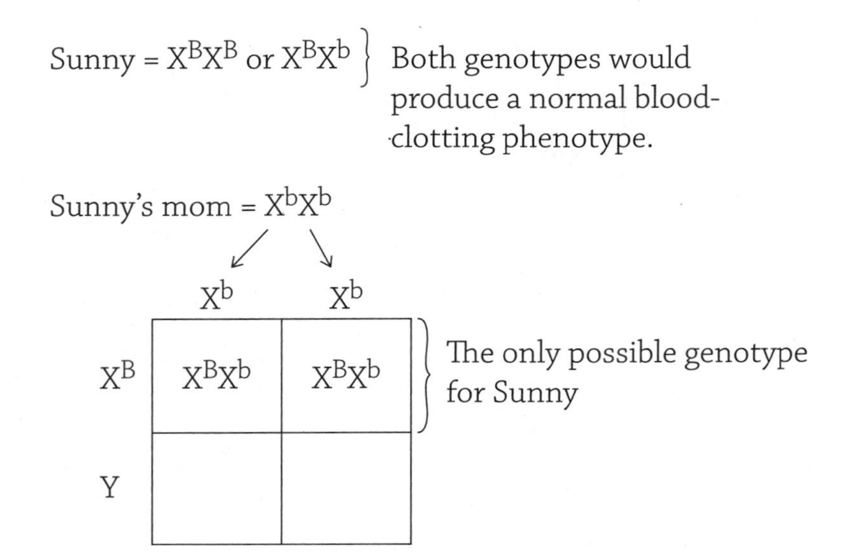

. . . and now you can see what the kids of Sunny and Maynard would be:

Maynard \ **Sunny**	X^B	X^b
X^B	X^BX^B	X^BX^b
Y	X^BY	X^bY

▶ The answer to the question "what are the odds that their child will have hemophilia?" is 25%. One out of the four possibilities (the X^bY boy) would have hemophilia.

REVIEW QUESTIONS

1. Match the following term with the correct definition:

a. Allele	1. Two matching alleles
b. Dominant	2. This allele is seen in a heterozygous combination
c. Genotype	3. What an organism looks like
d. Heterozygous	4. Different forms of the same gene
e. Homozygous	5. In order to see this type of trait, it must be homozygous
f. Phenotype	6. An organism's genetic makeup
g. Recessive	7. One of each allele

2. Fruit fly (*Drosophila melanogaster*) eyes are normally red, but there is a recessive allele that makes them appear dark (ebony). If a heterozygous red-eyed fly is crossed with a fly with ebony eyes, what percentage of offspring will have the recessive phenotype?

3. In cats, brown fur (B) is dominant to white fur (b), and short fur (S) is dominant to long fur (s). What are the odds that a cross between a male cat heterozygous for both traits and a white, short-haired female (who is heterozygous for fur length) will produce a kitten with white, long fur?

4. A man with normal color vision marries a woman who also has normal color vision. Her father was color-blind. What are the chances that their child will be color-blind?

5. A plant with purple flowers crosses with a plant with white flowers. All the offspring produce purple flowers. Based on this description, which flower color allele (white or purple) is dominant and which is recessive?

6. True or False: When a homozygous dominant trait is crossed with a homozygous recessive trait, the phenotypic ratio in the offspring will always be 100% recessive phenotype.

7. In chickens, brown feathers (B) are dominant to white feathers (b) and large combs (L) are dominant to small combs (l). If a white rooster that is heterozygous for large combs mates with a hen that is heterozygous for both traits, what are the odds of producing a white chicken with a large comb?

8. If a color-blind man has a baby with a woman who is homozygous for normal vision, what are their chances of having a daughter who is a carrier for color blindness?

9. If a homozygous dominant trait is crossed with a heterozygous trait, what are the chances of producing an offspring with the recessive trait?
   a. 75%
   b. 50%
   c. 25%
   d. 0%

PART FOUR

Evolution

One of the most important topics in biology is the concept of evolution. This elegant process explains how we ended up with such a staggering variety of life-forms on Earth. It's important to first understand the difference between **evolution** and **natural selection**. The term *evolution* means "change over time"; *natural selection* is *how* this slow change occurs.

15 The Theory of Natural Selection

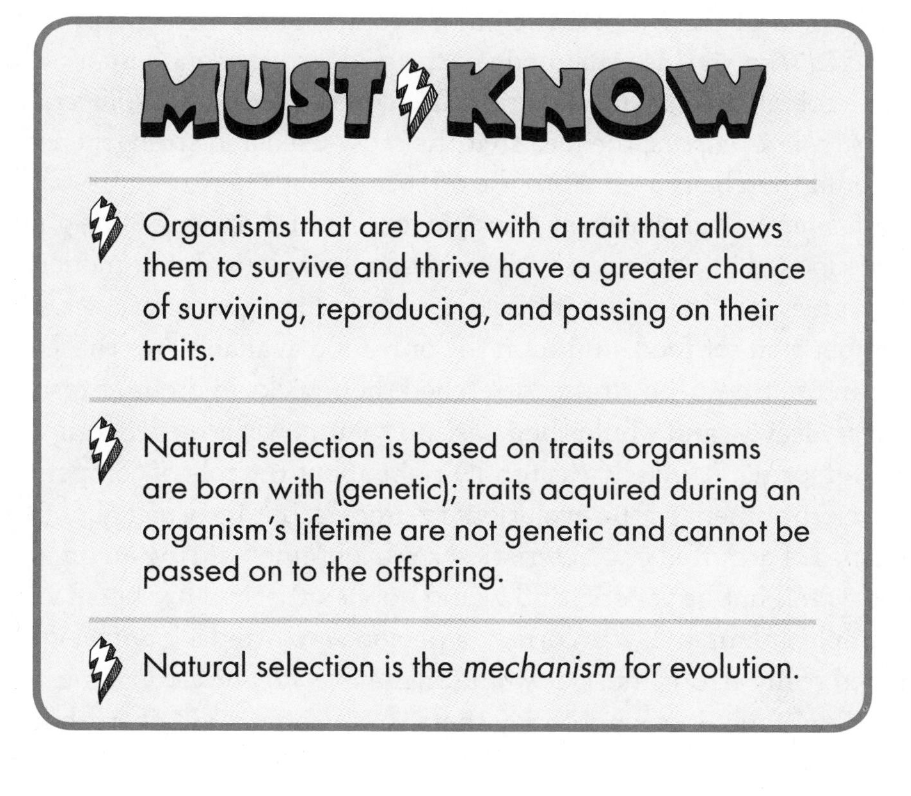

The Development of the Theory of Natural Selection

We give Charles Darwin credit for being the "father of evolution," for explaining how the process of evolution occurs in nature. Any good idea, however, grows from a foundation of previously learned knowledge; Darwin had the help of many other excellent minds who also studied evolution. In 1809 (the year Darwin was born!), a French biologist named Jean-Baptiste Lamarck realized that evolution was a slow process that occurred because of environmental changes. He explained evolution based on two principles: (1) *use and disuse*, the idea that body parts used frequently are made stronger and more robust, and parts that aren't used deteriorate; and (2) *inheritance of acquired characteristics*, the idea that the most-used and strengthened body parts would then be passed on to the offspring.

Giraffes provide an excellent example for Lamarck's theory. The giraffes' ancestors looked like antelope or deer. Lamarck explained that modern giraffes came about based on inheritance of acquired characteristics. There was competition for food, and soon the only food available was the leaves high up in the trees. The giraffes stretched their necks in order to reach those tasty leaves, and would then pass on their newly acquired long necks to their offspring. Lamarck was totally right about the role of competition and the environment in the evolutionary process, but incorrect about the inheritance of acquired characteristics; traits obtained during an organism's lifetime would not be genetic, and would not be inherited by the offspring. For example, if Lamarck was correct, a person who lifted a lot of weights and developed really strong muscles would then have a really strong, muscle-bound baby. Nope. That's not to say there isn't a component of truth in that scenario. An adult may indeed have children who are also strong, but it's because they are genetically predisposed for increased muscle mass. Even with the missteps, Lamarck did an amazing job proposing a logical explanation for evolution that included the importance of the environment and competition.

Two geologists also played important roles in Darwin's formation of a theory of evolution. Understand that the prevailing belief during Darwin's day was that Earth was very young (only a few thousand years old) and populated by unchanging species. In the 1700s, a geologist named James Hutton suggested that Earth was, in fact, very old. Furthermore, geologic features were formed by very gradual mechanisms that could occur because the planet had been around a long, long time. During Darwin's time, another geologist—Charles Lyell—suggested that those slow geological processes described by Hutton were still happening today, as they had in the past. Darwin had the realization that Earth must be significantly older than a few thousand years—otherwise, how could valleys be carved out by rivers, or mountains stretch up into the sky? The timescale of Earth was long enough (billions of years!) to allow gradual changes to happen in geology . . . why not biology?

Another significant insight came from an unlikely source: an economist named Thomas Malthus. Darwin read a 1798 essay by Malthus describing how an increase in the human population would lead to a strain on resources such as food and space. This would lead to disease, famine, and death. Darwin realized that this applies to *all* life, not just humans. All species have the capacity to overpopulate and run out of resources, leading to a struggle to survive.

Artificial selection is the process of changing a species by selectively breeding individuals that have certain traits a person wants to foster. Humans have been benefitting from artificial selection for thousands of years. For example, the different breeds of dogs we have today arose from artificial selection, starting with a gray wolf. Plants such as broccoli, kale, and cauliflower were all created by artificial selection, starting with a wild mustard plant. The big difference between artificial selection and natural selection? In natural selection, the environment applies the selective pressure; in artificial selection, *we* determine what is "most desirable" and select who gets to survive and reproduce.

In 1831, Darwin was invited to sail on the ship HMS *Beagle* as it took a 5-year journey to map the coast of South America and the Pacific Islands. This presented an amazing opportunity for Darwin to collect all sorts of samples of plants, animals, and rocks. He kept an extensive diary of all of his observations, such as insightful comparisons between fossils found in different locations. The best stuff came at the end, when he arrived at the Galapagos Islands. Here, Darwin was blown away by the similarities (and differences!) between the species found on one island compared to another (or even the mainland). The life-forms were so perfectly suited to their environments, yet maintained similarities to comparable species on a neighboring landmass. For example, the finches on each island had beaks perfectly shaped for the primary source of food in their environment: narrow, sharp beaks for catching insects; thick, strong beaks that can crush hard seeds; or beaks able to tear at vegetation. Although all of these finches had remarkable similarities to one another, they each also had adaptations that allowed them to survive and reproduce in their particular environment.

During his voyage, Darwin also found fossils of sea life buried high up in the mountains. He recalled what he learned from Hutton and Lyell, that Earth was very old and underwent slow geological processes that could result in huge changes (such as an ocean disappearing and being replaced by a mountain range!).

One of the coolest things happened earlier, when they visited Argentina. Some of his crewmates brought back an armadillo, which Darwin had never seen before. He had, however, found a fossil of a *giant* armadillo (now named *Glyptodon*) while on a fossil-hunting trip in the area. So, here's this mind-blowingly huge version of an extinct critter that looks eerily similar to a version walking around today. The similarity just screamed "evolutionary ancestor."

Fossil remains of *Glyptodon*

Author: Dellex. https://commons.wikimedia.org/wiki/File:Glyptodon_Skelett.JPG

Once Darwin returned home, it took him more than 20 years to study the significance of his notes and comb through the specimens gathered on his voyage. He came up with a theory explaining how evolution could give rise to such a rich and varied selection of life, and why a species ended up being so perfectly fit for its environment. Darwin published *The Origin of Species* in 1859, explaining how evolution occurs through a process called natural selection. This is our **must know** concept. Even though he saw the results of evolution (different—but related—species), he didn't know how it had happened. The *how* of evolution is the process of natural selection.

BTW

It's a common misconception to think that by using the phrase theory of evolution, *it suggests that it's nothing better than a guess. Please understand that in science, the term* theory *carries a lot of weight. A theory is supported by a large body of evidence and has been repeatedly confirmed through experimentation. I mean, gravity is just a theory . . . and we feel pretty confident about that one.*

The Mechanism of Natural Selection

Although natural selection is responsible for some pretty remarkable things, the process isn't overly complex. Our **must know** concepts can be simplified to these four steps:

1. **Individuals in a population are born with genetic differences (variation).**

Alleles refer to different variations of the same gene (as we learned back in Chapter 14). These variations are the basis for natural selection, because some alleles provide an advantage over other forms. Please understand that the different forms of a gene are not *made* through natural selection, they are *chosen* by natural selection. New alleles are made by random mutations! Another thing to keep in mind is that natural selection acts on *phenotypes*, not *genotypes*. The genes are only specific sequences of nucleotides kept safely in the cell's nucleus. The environment doesn't interact directly with the genes (the genotype); the environment instead interacts with the physical trait the DNA codes for (the phenotype).

Genetic variation comes about by many different means (which is great, because variation in a population means a healthy population). When gametes are formed during meiosis, an organism's genes are lined up and shuffled in different ways. This—along with crossing over—yields a huge number of different eggs or sperm with different combinations of alleles. To mix it up even further, the process of sexual reproduction takes one of those genetically unique eggs and fertilizes it with a genetically unique sperm. The offspring receives two copies of each gene, one from each parent. So many combinations. Finally, what if there happens to be a mutation in one of the sex cells? That would create a brand-new genotype! And if that heritable mutation happened to have a beneficial phenotype, it could go on to create a new adaptation.

IRL Genetic variation is naturally found in populations, unless the population is undergoing *asexual reproduction*. This type of reproduction relies on the "parent" producing genetically identical copies of themselves, without the use of gametes (egg and sperm). Bacteria undergo asexual reproduction by first copying their chromosome and then splitting themselves in half (binary fission), and many plants can clone themselves. Humans have, in fact, cloned plants for thousands of years by cutting off a part of a desirable plant and coaxing it to grow as a new plant. A cloned crop plant can increase food yield and save millions of lives, but a cloned crop can also be more susceptible to disease (the lack of genetic variation limits the population's ability to adapt). Cloning is much more difficult in vertebrate animals, but it is becoming more successful as advances in genetic engineering continue.

2. **There are usually too many offspring for the environment to support, because resources are limited.**

The idea of overproduction rolled around in Darwin's mind after he read Malthus' essay on human population growth. This is a really important thing to consider for the process of natural selection, because if the environment would support unlimited numbers of offspring, natural selection wouldn't occur. As it is, however, exponential growth is only possible with unlimited resources, no predators, and no disease . . . and that's not likely.

3. **There's a lot of competition, and any individual who happened to have been born with a helpful trait is able to outcompete others in the population.**

Environments place limits on population growth because, as we mentioned above, an environment just can't provide unlimited space and resources. So, this leads to an epic struggle for food, water, and shelter . . . and some organisms are more successful than others.

4. **This most-fit individual gets to survive and reproduce. By doing so, the survivor's "most-fit" genes are now passed on to the next generation, resulting in an adaptation.**

   Certain variations allow an individual to outcompete others and win that struggle for resources. An **adaptation** is a helpful trait found in a population that is the product of natural selection. Natural selection accumulates these adaptations because they help an individual survive and reproduce.

As so often happens in science, this simple concept has hidden pitfalls in the form of misconceptions. I will cover three important points that are easy to misunderstand.

- **You gotta be born with it.**

It's super, super important to understand that any trait that helps an individual survive must be genetic and heritable, otherwise it doesn't count (sorry, Lamarck). Only those qualities that can be passed on to the next generation (in their genes) have an impact on the next generation's genetic makeup (gene pool). If, for example, an octopus developed super-stretchy legs during its lifetime and allowed it to better reach a crab hiding in the rocks, that doesn't result in evolution. Sure, that one octopus will be better able to find food and survive, but it cannot pass on its improved stretchiness to its offspring unless it was a genetic trait it was born with.

The reason I keep stressing that "you gotta be born with it" is because even though we accumulate genetic mutations during our lifetime, that doesn't mean your kids will inherit any of those mutations. What if, for example, you got radioactive waste on your hands and it caused genetic mutations in your fingertips so they grew suckers that allowed you to climb walls. Super cool, right? But would this mutation be *passed on to the next generation*? To answer that question, think about this, where are the genes to make the next generation coming from? Yup, the egg and sperm.

In order for you to pass on some helpful mutation, you need to be born with it because it needs to be in your gametes (egg or sperm). And in order for a certain gene to be found in the cells that create egg and sperm, it needs to have been there when you were nothing but a single-celled zygote. Not some random mutation you got because you handled glowing, bubbling radioactive waste (please don't ever handle glowing, bubbling radioactive waste).

- **Populations evolve, not individuals.**

Another important thing to keep in mind is the fact that populations evolve, not individuals. Adaptations occur in populations as most-fit genes (as selected for by natural selection) start to accumulate. An individual can't evolve because you are born with the genes you have and they don't change, beyond a random mutation. The best an organism can do is survive, have lots of offspring, and contribute its excellent genes to the next generation.

- **The story isn't over once the population adapts.**

A population's adaptations that are considered "most fit" can change as the environment changes. The selective pressures of evolution are the conditions in which an organism lives, and if those conditions change, so do the qualities that best help a critter survive. For example, European tawny owls come in two colors: brown and gray. The gray color helps them blend in better when winter brings the snow, providing an advantage when hiding from predators. Because of climate change, however, the winters are becoming milder and there is less snow. Natural selection has shifted to favor the brown color, and there has been an increase in brown owls.

REVIEW QUESTIONS

1. ____________ is change over time. ____________________________ is the mechanism that explains how it happens.

2. Why were the geologists' (Hutton and Lyell) insights into geological processes so helpful to Darwin when forming his theory of evolution?

3. Lamarck had his own theory of evolution, but it was flawed due to the principle of inheritance of acquired characteristics. Why was it wrong?

4. A(n) ____________ is a helpful trait found in a population that is the product of natural selection.

5. A species of woodpecker that evolved to possess brightly colored head feathers is an example of [choose one: **macroevolution/ microevolution**]. The ancestral woodpecker that gave rise to many different species better adapted for their particular environment is an example of [choose one: **macroevolution/microevolution**].

6. Match the person with the correct influential insight:

1. Geologist who suggested Earth was extremely old.	a. Lyell
2. Naturalist who believed species change over time.	b. Malthus
3. Economist who linked population growth to declining resources.	c. Lamarck
4. Geologist who believed slow geological process was still occurring.	d. Hutton

7. How does artificial selection differ from natural selection?

8. Choose True or False for the following statements:
    a. Any beneficial genetic mutation will be passed on to the next generations.
    b. Populations, but not individuals, are able to evolve.
    c. Once a most-fit phenotype is selected through evolution, the population's adaptations are "fixed" and will no longer change.

9. Briefly describe the four steps of natural selection.

16 The Evidence for Evolution

MUST KNOW

- Molecular similarities, anatomical evidence, and the fossil record all provide evidence for evolution.

he argument that evolution is only a "theory" doesn't take into account the overwhelming sum of scientific findings that again and again bolster the concept of species changing over time.

Molecular Similarities

It's safe to say that the more closely related two organisms are, the more similar their DNA and protein sequences. In fact, try ranking the following organisms from having the highest similarity with your own DNA to having the lowest number of nucleotide bases in common: chicken, chimp, fruit fly, another human, and yeast.

If you think about it logically, you can group these organisms based on similarities to yourself. Yeah, good chance that you and another person have a lot in common (genetically speaking). A chimp is a primate (as are you), so that's another good bet. A chicken is a bird, not a mammal like us, but it is a vertebrate, unlike that fruit fly. And a yeast cell? Hm. Not a close relative. So, if you check out the table below, there shouldn't be any huge surprises.

You compared with . . .	DNA sequence % similarity
another person	99.4
a chimp	98
a chicken	65
a fruit fly	60
a glorious and majestic yeast cell	26

A comparison of the sequence of amino acids in a particular protein is another way to establish relatedness. When choosing a protein, it's best to look at one that is found in all forms of life. If a protein is found in many different types of organism, it suggests it evolved early in Earth's history and was so important to cell function that throughout evolutionary time, it was "conserved" for its original use. Its importance means the protein-encoding gene remains intact and relatively free of mutations. That isn't to say there aren't some changes in the sequence of nucleotides (and thus a change in the amino acids of the protein), but the changes are minimal. The protein cytochrome C is an excellent choice because it is found in all eukaryotic cells and plays a very important role in the electron transport chain. It is a chain of about 100 amino acids, and homologous similarities of cytochrome C are suggestive of common ancestry.

Your cytochrome C compared with the same protein from a . . .	Number of amino acids that are different from the human cytochrome C
chimp	0
chicken	13
fruit fly	25
yeast (in all its single-celled glory)	44

Notice that when comparing cytochrome C, the lower the number means a more similar protein because we're counting the number of *different* amino acids.

Anatomical Evidence

If you look at two different species of snake, you can easily tell they're related because they are both snake-shaped; their anatomies are similar. Yet anatomic evidence of common ancestry can be more subtle than comparing the overall body shapes of two different critters. This includes studying **developmental similarities, homologous structures,** and **analogous structures.**

Developmental Similarities

Usually, if critters resemble each other as embryos, they are closely related and have a common ancestor. Surprisingly, a human embryo looks a lot like a fish embryo, and sure enough, we share a vertebrate common ancestor.

As the embryo develops, the initial similar body plan will begin to change and look more like the adult form. This happens because certain genes switch on and direct changes that lead to the formation of specific bits like tails, legs, wings, or gills. Regardless of the critter, all vertebrate embryos have three special structures: a tail, tiny limb buds, and pharyngeal arches. As I said earlier, as development continues, species-specific genes switch on and start to mold the embryo into its final body plan. That tail in a human embryo goes away as development continues, and the pharyngeal arches end up as part of our inner ear (as opposed to gills).

Yes, some people are born with "tails." Please don't make the mistake of thinking it's this wonderful, wagging appendage that could act as a prehensile fifth hand. No, it's simply a vestigial structure due to the embryonic tail vertebrae not being reabsorbed back into the embryo during normal embryonic development. Sorry to disappoint.

Homologous Structures

Homologous structures are body parts that are structurally similar in different species (because they share a common ancestor) but can have very different functions. For example, bat wings, human arms, dog legs, and dolphin flippers are all constructed of the same bones of the arm (humerus, radius, ulna) and hands (carpals, metacarpals, phalanges). This shouldn't be a huge surprise considering we're all mammals and are very closely related. The cool thing about homologous structures is how different they look, thanks to natural selection's ability to carve a structure into a form most fit for the environment. A bat's wing is obviously for flying, and a dolphin's flipper is for swimming. Same bones, different functions! Now, there's a chance that a homologous structure may not be used much in a certain species, and evolution will eventually cause it to wither into a small and useless form.

It's common for structures not actively used in a species to become minimized and "go away." Natural selection chooses adaptations that are useful; non-useful structures can be too much energy to keep and may eventually fade away.

Vestigial Structures

Vestigial structures are homologous structures that no longer perform a function. For example, mammals that eat a lot of plant material have a structure at the beginning of the large intestine called a **cecum**. It's a large pouch that provides housing for mutualistic bacteria that help break down the difficult-to-digest cellulose found in plant cell walls. Our cecum has shrunk down to a small pouch, including the **appendix**, a little finger-like extension that doesn't do much besides cause some excitement when it becomes inflamed in some unlucky individuals and needs to be immediately removed (an appendectomy).

New research suggests that the appendix may still serve an important function as a home for beneficial gut bacteria!

Analogous Structures

Analogous structures are interesting because they're like the opposite of homologous structures: they *look* similar, but *don't* have a common origin. Bird wings and insect wings look similar (they both look like . . . uh . . . wings) but have developed from totally different tissues: bones and feathers versus membranes. These two animals—birds and insects—are not close animal relatives, because one is a vertebrate and one is an invertebrate. But as a testament to the power of evolution, these two distant relatives both independently evolved to have wings because of the similar environment in which they live.

Geological and Fossil Evidence

Geological evidence of Earth's age was a key component of Darwin's understanding of evolution. During the early 19th century, the common belief was that our planet was formed only a few thousand years earlier, and since that time, neither Earth nor the life-forms on it have changed. The theory of evolution, however, not only states that life-forms have indeed changed, but the timescale necessary for this slow process to occur is possible because of the true age of Earth.

This crucial geological evidence was provided by the geologists Hutton and Lyell, both mentioned earlier in this unit. Hutton's theory of **gradualism** stated that geological structures (layers of sediment, or great canyons) were formed through slow changes over long periods of time. Later, Lyell added to the theory by also discovering that those slow changes occur at a constant rate and are still occurring today. His theory was called **uniformitarianism**. This changed the understanding of Earth's age (which is 4.6 billion years old!) and provided a timeline in which the slow process of natural selection could occur.

Fossils provide a picturesque physical record of evolutionary history. It documents the presence of organisms that are no longer on Earth and proves that life is, indeed, changing. Otherwise you'd be stopping your car to let the occasional giant sloth cross the street.

The giant ground sloth of over 2 million years ago grew 17 feet tall. That. Is. Awesome.

Even though we uncover the skeletons of critters that are no longer here, they often bare an uncanny resemblance to current life-forms. A *Mesohippus* fossil looks very much like modern-day horses and, sure enough, *Mesohippus* is the ancestor of your friendly neighborhood equine! It's also cool to see the extinct species change as their fossils are uncovered in shallower and shallower sediment layers. Because of Hutton's and Lyell's theories, we understand that the slow geological processes formed the layers of Earth by depositing sediment. Newer sediment layers are higher up; the lower you dig, the older the timescale.

If you take a cross section of the fossils found in all these layers, it provides a really cool snapshot of how species change over time. Horses, in particular, provide an intriguing example. Their early ancestor *Hyracotherium* (50 million years ago) looked more like an antelope than a horse, and it had toes instead of hooves! Because of natural selection, many toes became vestigial structures and the middle toe became bigger, eventually forming the modern-day horse hoof. A horse is actually walking on only its middle toes and the hoof itself is a huge, tough fingernail!

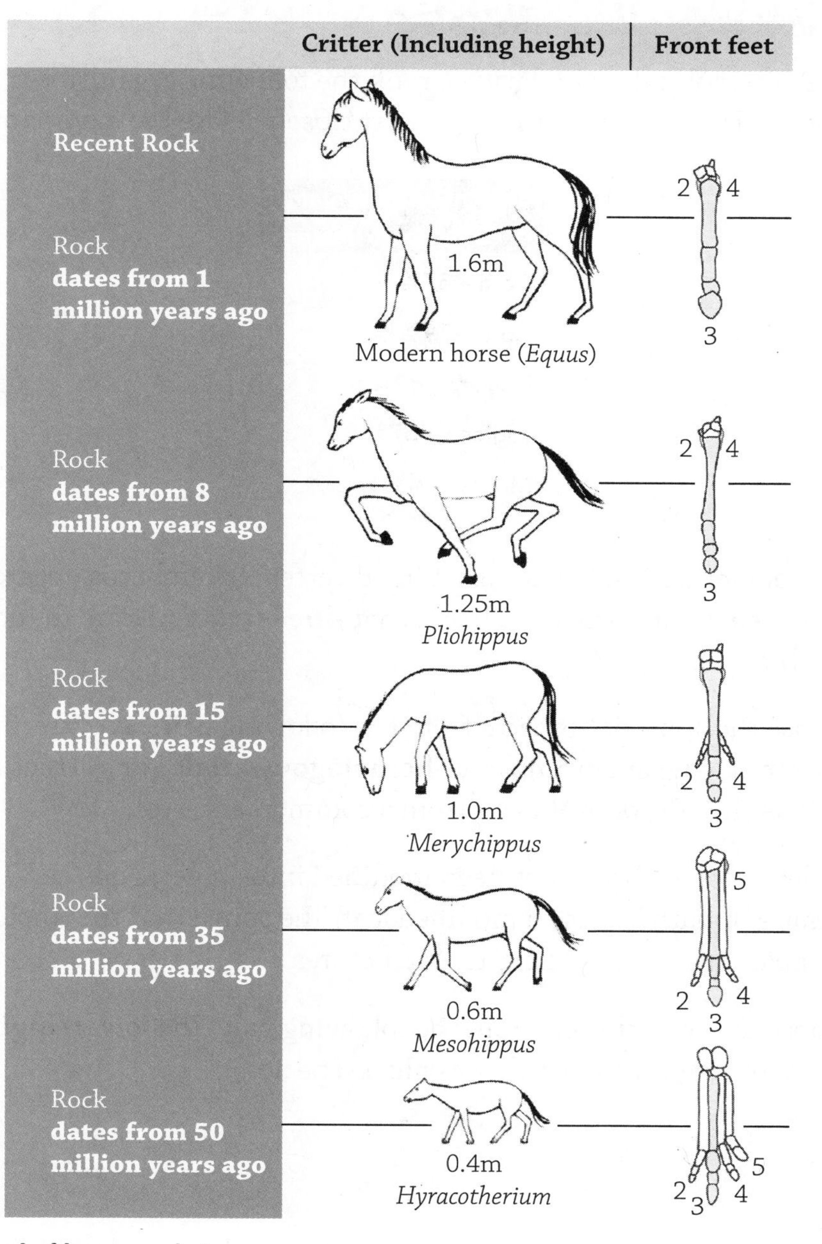

Fossil record of horse evolution

Author: Horseevolution. https://commons.wikimedia.org/wiki/File:Horseevolution-es.png

REVIEW QUESTIONS

1. Based on molecular similarities, rank the following organisms from most closely related to a human to the least closely related to a human:

Cytochrome C protein – % similarity to humans
Corn – 67%
Carp – 79%
Donkey – 89%
Euglena – 57%
Mouse – 91%

2. Developmental similarities exist in all vertebrate embryos, regardless of what their adult form looks like. What three structures are found in all vertebrate embryos?

3. Choose the appropriate term from the following pairs: A butterfly wing and a bird wing are examples of **homologous/analogous** structures because they **do/do not** arise from a common ancestor.

4. The leaves of a cactus have been modified into spines, and the Venus flytrap's "mouth" is also a modified leaf. The spines and the "mouth" are examples of ________________ structures.

5. Choose the correct term from the following pair: The **lower/higher** the sediment layer, the older the fossilized species.

6. A previous question listed the percent similarities between different organisms and a human:

Cytochrome C protein – % similarity to humans
Mouse – 91%
Donkey – 89%
Carp – 79%
Corn – 67%
Euglena – 57%

Explain why these results make sense. Don't use quantities (e.g., percent similarity) in your answer. Instead, use logic to explain why the ranking seems "right."

7. What is the relationship between homologous and vestigial structures?

8. What is the driving force behind the development of analogous body structures?

9. Fill in the blank and choose the correct term from the pair in bold: Hutton's theory of ________________ stated that geological structures were formed through **fast/slow** changes over long periods of time.

10. When comparing two different species, which of the following types of evidence would provide the most reliable evidence of common ancestry?
    a. Fossil evidence
    b. Molecular similarities (DNA)
    c. Anatomical comparisons
    d. Analogous structures

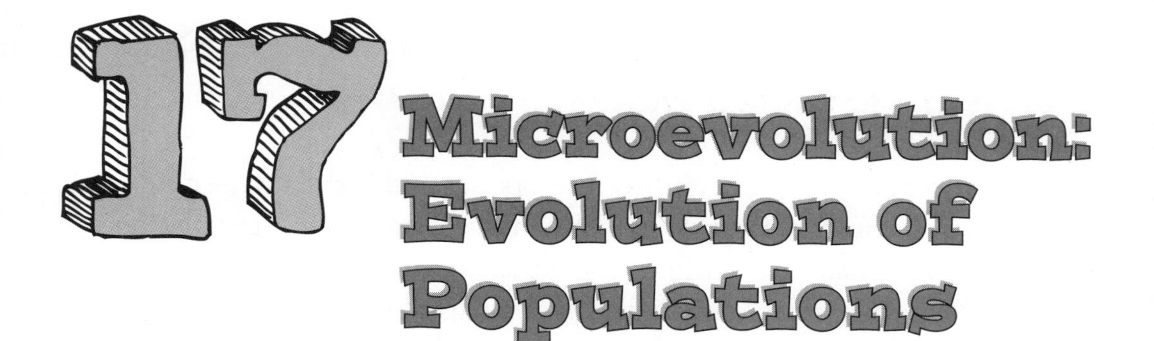

- Evolution can be measured as a change in allele frequencies.

- Hardy-Weinberg provides a "snapshot" in time of allele frequencies, so you can see if allele frequencies are changing over time.

Genetic Variation in a Gene Pool

A **population** is a group of interbreeding species and is the smallest biological unit that is able to evolve. The genetic variation in a population means everyone has slight differences in their phenotype due to slight differences in their genes (genotypes). If one particular phenotype allows that lucky individual to better adapt to their environment, their chances of surviving and reproducing are increased. More reproduction means more babies, which means more chances to pass on their more-fit genes to their offspring. This leads to an increase of that particular allele in the next generation . . . and that means the population is EVOLVING! Our **must know** concept is that evolution can be measured as a change in a population's allele frequencies. An **allele frequency** means the proportion of one allele compared to all alleles for that trait.

Microevolution refers to the evolution of a single population, as seen by a change in the allele frequencies in that population. The term **gene pool** refers to all the alleles held by all the members of a population. When studying a gene pool, it helps to focus on a particular **locus** (meaning one specific gene). Each member of a population has two copies of each gene, right? So, if there were 50 critters in a population, and each has two genes for feather color, then there would be a total of 100 alleles for the feather-color trait in this population. That is the population's gene pool for the feather-color gene.

Now, let's take it one step further. In simple genetic examples, an allele has two options: the dominant allele (which we'll call A) and a recessive allele (a). When considering the possible *genotypes* for each individual, there are three options:

AA homozygous dominant

Aa heterozygous

aa homozygous recessive

Let's jump into a specific gene pool example. Consider a leafhopper.

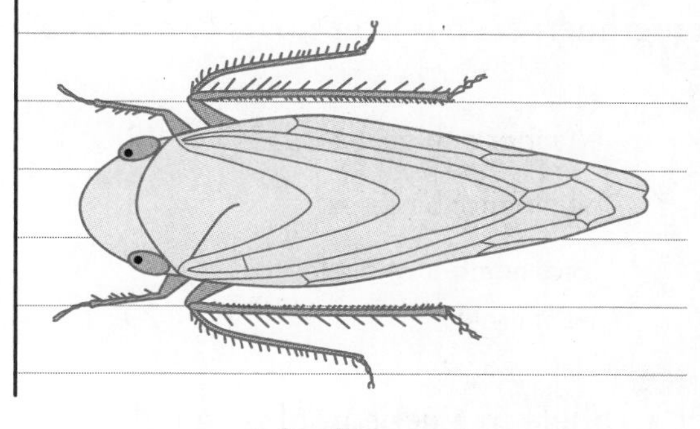

A leafhopper

This little arthropod specializes in eating plants and avoiding predators, so it has adapted to jump far distances with its tiny bug legs. We will say that the leafhopper has two alleles for leg muscles (yes, leafhoppers have muscles. By definition, all animals have muscles. Well, except for sea jellies . . . but I digress). The dominant allele is for big muscles (B) and the recessive is for small muscles (b). The population we're studying consists of ten leafhoppers, and each little arthropod has two genes for leg muscles.

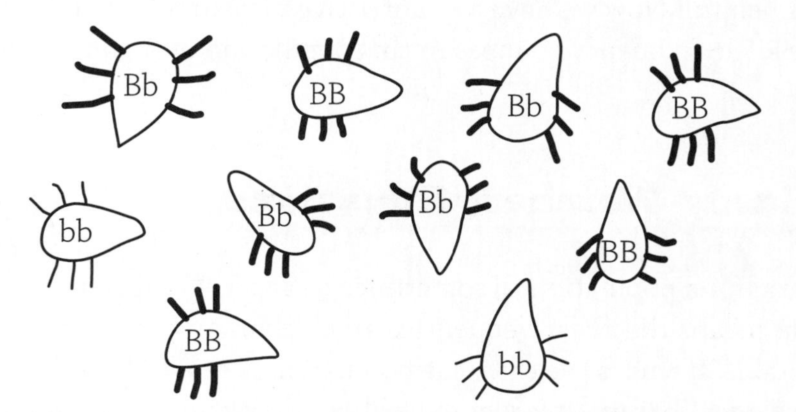

Leafhopper gene pool. B = allele for big leg muscles; b = allele for small leg muscles. Notice the only leafhoppers with small legs are homozygous recessive (bb).

When studying the evolution of a population, it's helpful to calculate the allele frequencies. To do so, count up all the individual dominant and recessive alleles (remember: each critter holds *two* alleles):

Number of B alleles	12
Number of b alleles	8
Total number of alleles for leg muscles	20

The frequency of an allele in a gene pool is calculated by:

$$\frac{\text{Number of one allele (B or b)}}{\text{Total number of alleles (B + b)}}$$

That enables us to determine the frequency of B:

$$B = 12/20 = 0.60 = 60\%$$

As well as the frequency of b:

$$b = 8/20 = 0.40 = 40\%$$

This is helpful! Now we have a quantitative measurement and a way to see if the allele frequency changes in this population over time . . . *which is evolution!*

The Hardy-Weinberg Equations

It is natural for a population to sometimes be at an equilibrium for a given gene. That means the allele frequencies aren't changing and everything is fairly stable. If this is the case, the population is said to be at **Hardy-Weinberg equilibrium**. In order to be at Hardy-Weinberg equilibrium (and keep those allele frequencies stable), five conditions must be in effect:

1. **No mutations!** If there were mutations, you would make a new allele and everything would get thrown outta whack.

2. **Only random mating allowed.** You want all the alleles in the gene pool to be randomly shuffled. No picking and choosing allowed.
3. **Large population.** When you're doing any sort of statistical analysis, the bigger the sample size, the better!
4. **No gene flow.** You can't have new individuals waltzing their way into (or out of) your population, bringing their fancy new alleles with them.
5. **No natural selection (aka evolution).** Because, well, that would certainly disrupt equilibrium.

This all seems unlikely. How could a population *not* be evolving? If you only consider a single gene in that population, however, it makes it easier to imagine. The Hardy-Weinberg equations are a super helpful tool in determining if a population is evolving. You can look at one single gene when the population is at equilibrium, and figure out the fraction of alleles that are dominant and the fraction that are recessive. Then, in a couple years, do the same thing again! Any change in the gene pool's allele frequencies means your population is evolving. It is a **must know** to realize the Hardy-Weinberg equilibrium provides a "snapshot" of allele frequencies, so you can see if those allele frequencies are changing over time.

Two equations are used when calculating the Hardy-Weinberg equilibrium:

$$p + q = 1$$

and

$$p^2 + 2pq + q^2 = 1$$

Not bad, right? The "p" stands for the frequency of the dominant allele and the "q" stands for the frequency of the recessive allele. There are only two options available—dominant and recessive—so when you add them up it has to equal 100%:

$$p + q = 1 \text{ (meaning 100\%)}$$

The previous equation focuses entirely on the alleles in the population. The second equation focuses on the genotypes of the members of the population. Keep in mind that each individual has two genes for each allele. To understand the equation, consider a cross of two big-muscled leafhoppers, both of which are heterozygous (Bb). Instead of using our letter designations specific to this one gene, use Hardy-Weinberg's "p" and "q":

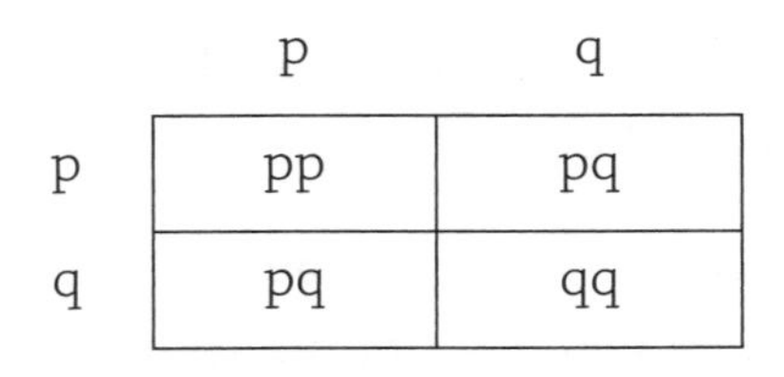

	p	q
p	pp	pq
q	pq	qq

Now the second equation makes sense, because it's showing you the different kinds of genotypes you can get: homozygous dominant (pp), homozygous recessive (qq), and heterozygous (pq):

$$p^2 + 2pq + q^2 = 1 \text{ (meaning, as before, 100\%)}$$

The 2 before the pq is because if you look at our Punnett square, we generated two different heterozygous genotypes (pq).

▶ You are studying fur color in a population of proud and majestic wild hamsters. This population is currently in the Hardy-Weinberg equilibrium. The fur allele comes in two forms: brown (dominant) and gray (recessive). In this population of 250 hamsters, 175 are brown and 75 are gray. What are the frequencies of the brown and gray alleles? How many hamsters are heterozygous for fur color?

▶ When given a Hardy-Weinberg problem, start with the phenotype for which you *know* the genotype: gray. It's recessive, so it has to be q^2. Always start by finding the frequency of the recessive allele "q." Once you get that, everything else is cake! The dominant phenotype, on the other hand, could be p^2 or pq . . . and you're not sure how many individuals are homozygous dominant or how many are heterozygous. Just leave it alone and focus on your recessive numbers:

q^2 = (75 gray) / (250 total) = 0.3

$q = \sqrt{0.3} = 0.55$ ← the allele frequency of the recessive (gray) allele

And since $p + q = 1$:

$p = 1 - q$

$p = 1 - 0.55 = 0.45$ ← the allele frequency of the dominant (brown) allele

BTW

Just because a gene is recessive doesn't mean it's rare. Most (55%) of the alleles in this fictitious hamster population are for the recessive allele, and this often occurs in nature, too.

- The question also asked us how many hamsters are heterozygous for brown fur. This is tricky because it's easy to forget to actually *finish the problem:*

  Heterozygous genotype: $2pq = (2)(0.45)(0.55) = 0.50$

- As many of my students have found out, it's tempting to stop at this point (but you would be wrong!). Read the question carefully. It asks how many hamsters are heterozygous. You just calculated the frequency! So, to finish the problem:

Remember: if the question asks for the total number of individuals of a certain phenotype, do not stop after you calculate the frequency! You must then multiple the frequency by the number of individuals in the population to find the answer to the question.

Number of heterozygous hamsters: (0.50) (250 total hamsters) = 125 hamsters are heterozygous

How Populations Can Change

As we learned earlier, there is always variation in a population (it would be *so boring* otherwise). A good way to model variation is by relating the different phenotypes to how many individuals have each phenotype. The perfect type of graph to use is a **histogram**, where the x-axis shows the different possible variations and the y-axis shows the number of individuals with a given variation. Because traits are selected for by natural selection, the highest number of individuals possess the most-fit phenotype. Flanked on either side are the rarer "extreme" members of the population. There are not as many of these guys because their phenotypes are not as advantageous as the most-fit phenotype. This creates a bell-shaped curve (also called a "normal distribution") where the majority of individuals have the average measurement:

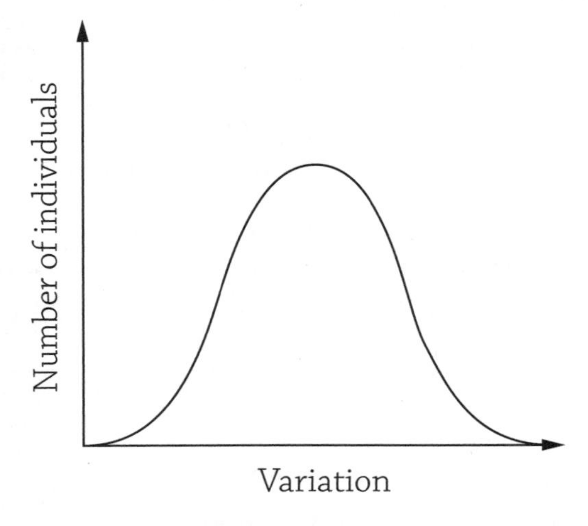

Normal distribution in a population

When conditions change, so does this curve. What if something happened to the environment and one of the extreme phenotypes was now the most-fit phenotype? The individuals with the not-frequently-seen extreme phenotype are now living the happy life of having a selective advantage that results in their increased survival and more successful reproduction. Meanwhile, the majority of the population now are stuck with a phenotype that is no longer so helpful. They don't survive as well anymore, and they don't leave as many offspring. This creates a change in the gene pool of the population, and the graph will now shift.

This small-scale evolution within a single population is referred to as **microevolution**. The differing patterns of microevolution include stabilizing, directional, and disruptive selection. **Stabilizing selection** does what it says: it stabilizes what you already have by reinforcing the current average most-fit phenotype. Extreme phenotypes continue to be selected against and you maintain the status quo! Robin birds lay an average of four eggs (their clutch size). If their clutch is too big with too many eggs, there may not be enough food to go around and all the chicks will be malnourished. Too small a clutch size and there's a possibility of no surviving offspring. Stabilizing selection ensures the extreme phenotypes (too many or too few eggs) are selected against.

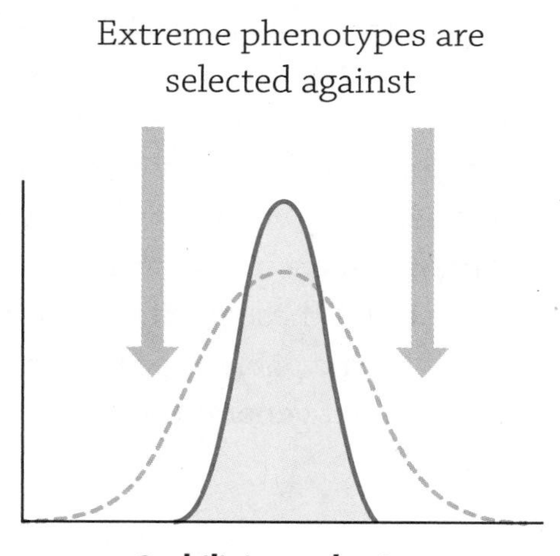

Stabilizing selection

Directional selection occurs when one extreme phenotype is favored over the other, causing a shift toward one end of the spectrum. For example, a new insect-eating predator moves into an ecosystem currently populated with dragonflies with medium-length wings. The new predator easily chases down the small-winged, slower dragonflies, and the faster insects with larger wings are better at evading the new predator. The few insects that were the extreme phenotype are now the most fit and are better able to survive, reproduce, and pass on their long-winged genes. The next generation (the solid line in the graph) now has a new (longer-winged) most-fit average.

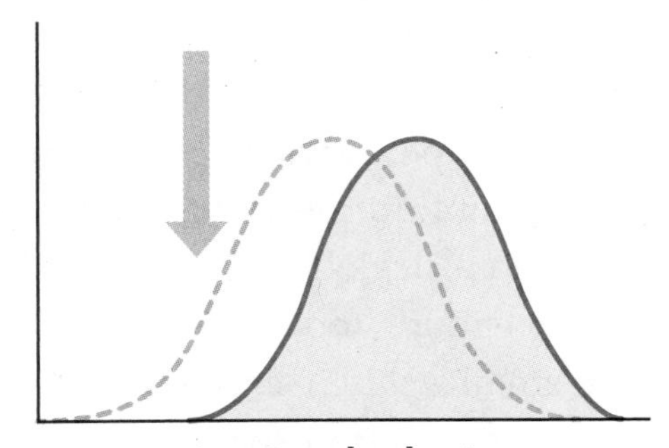

Directional selection

Finally, there's the chance of **disruptive selection** (and it's quite, uh, *disruptive* to the current population). Imagine a population of flowering plants that produced numerous clusters of flowers on each plant. A new species of bee began to pollinate the flowers, but it preferred plants with fewer flowers (easier to position itself correctly to reach the nectar) or plants with tons of flowers (which were very visible to the flying bees). What used to be the most-numerous phenotype with an average number of flowers is now being passed over by the bees! These plants aren't being fertilized, and instead, the plants with the extreme phenotypes are getting pollinated and are producing the seeds for the next generation.

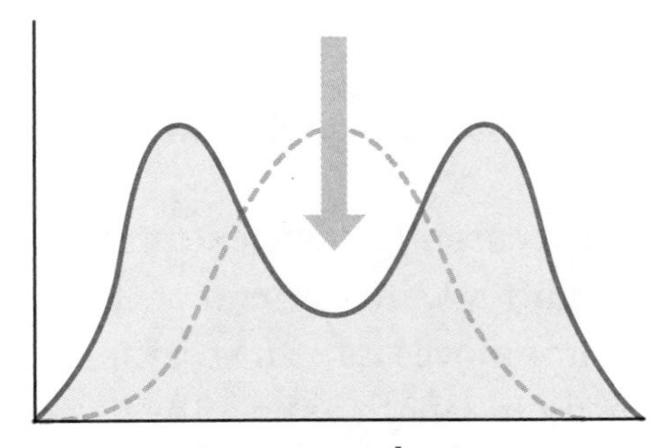

Disruptive selection

Gene flow and genetic drift

It's logical to assume evolution always occurs because some phenotype is better than another, and natural selection is in the biological driver's seat. But it's a bit more interesting than that, because sometimes evolution can occur because of totally random events, and natural selection isn't involved at all! This is called **genetic drift,** and any change in allele frequencies is due to *chance*. In order for genetic drift to occur, the original population size has to be drastically reduced, and the remaining population randomly has a different allele composition compared to the original population. There are two ways in which population size could be decreased enough for genetic drift to happen: bottleneck effect and founder effect.

The **bottleneck effect** happens when a population is drastically cut, and only a few survivors are left. By random chance, some alleles may be completely lost from the surviving population, or some alleles are now overrepresented. When the population increases again, its gene pool is created from the survivors and the new gene pool can look drastically different from the original. Bottlenecks often cause a loss of genetic diversity.

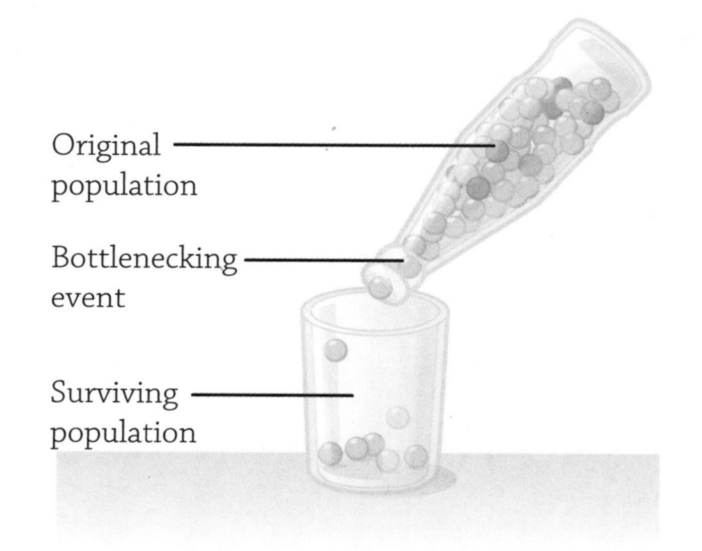

Bottleneck effect

Author: OpenStax, Rice University. https://commons.wikimedia.org/wiki/File:Bottleneck_effect_Figure_19_02_03.jpg

The rare Siberian tiger was hunted nearly to extinction in the 1940s, with their lowest numbers dropping to a devastatingly small 40 tigers. That small handful of tigers contained the only alleles left from their original population. Due to decades of conservation and protection, their numbers have since increased to 900. Unfortunately, this new tiger population has very little genetic diversity because of their having undergone the bottleneck effect.

If a handful of members of an original population move away and begin their own population, this is called the **founder effect.** A loss of genetic variation occurs when any new population is established by a very small number of individuals. If the founding individuals happen to carry rare alleles, then the newly created population will have a higher frequency of those alleles.

If you are asked for an example of a bottleneck effect, don't make the mistake of accidently describing natural selection. In natural selection, a large portion of the population dies but those who are more fit survive. The survivors of a bottleneck event have to just randomly survive (and not because of some advantageous trait).

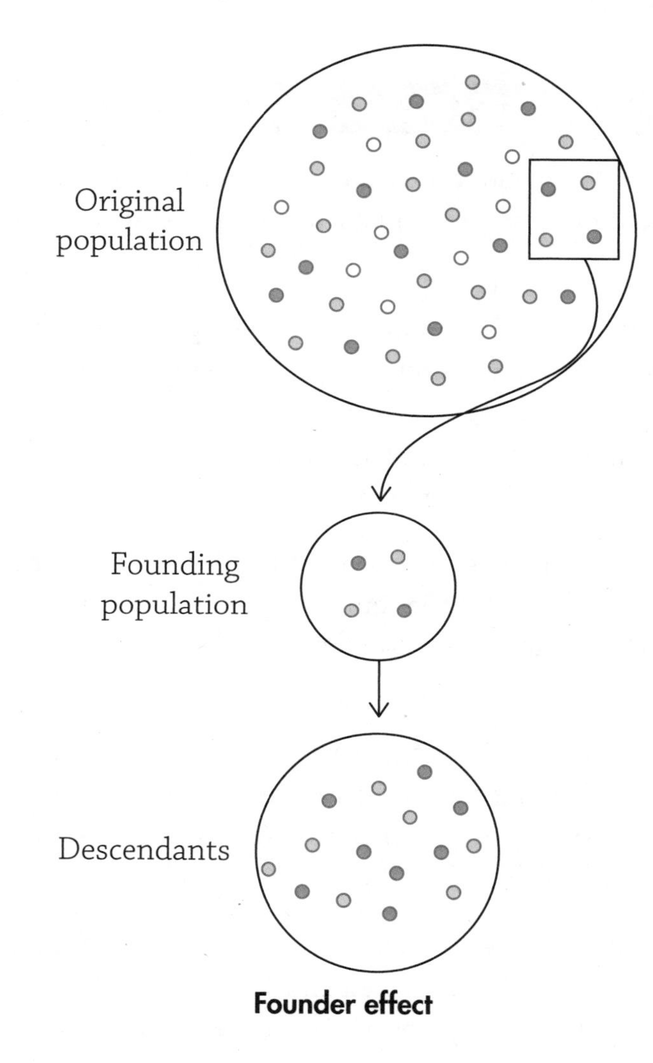

Founder effect

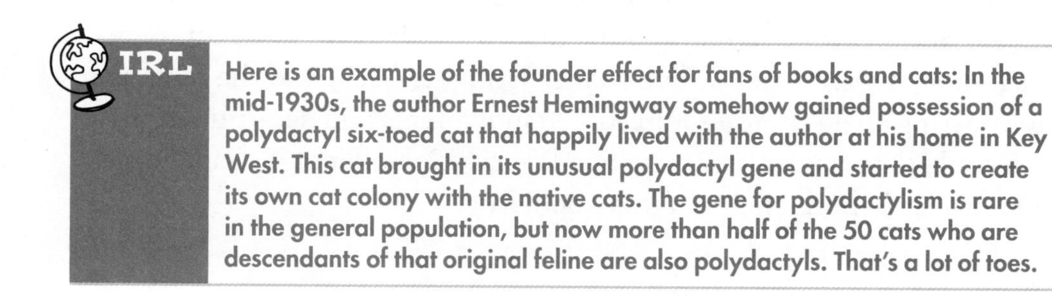

Here is an example of the founder effect for fans of books and cats: In the mid-1930s, the author Ernest Hemingway somehow gained possession of a polydactyl six-toed cat that happily lived with the author at his home in Key West. This cat brought in its unusual polydactyl gene and started to create its own cat colony with the native cats. The gene for polydactylism is rare in the general population, but now more than half of the 50 cats who are descendants of that original feline are also polydactyls. That's a lot of toes.

REVIEW QUESTIONS

1. The Hardy-Weinberg equation is a way to check the allele frequency of a population, but only if the population in question meets five conditions. List them.

2. Within a population of 251 snails, the shell color brown (B) is dominant over the shell color tan (b), and 40% of all snails are tan. Given this information, calculate the following:
    a. The number of snails in the population that are heterozygous.
    b. The number of homozygous dominant individuals.

3. If a population suddenly suffers a drastic reduction in numbers due to a catastrophic lava flow, the remaining population will eventually repopulate. The reestablished population has very little genetic diversity due to the ________________.

4. A population of stinkbugs range from slightly smelly to extremely smelly, with the majority of bugs with the medium-smelly phenotype. If a predator began to only eat the medium-smelly bugs, explain what kind of selection this describes.

5. Write the Hardy-Weinberg equation that tallies the different alleles for a given gene: ______________________. Next, write the equation that summarizes all the given genotypes possible: ______________________. Explain what each component of the equation stands for.

6. The ability to taste the chemical phenylthiocarbamide (PTC) is due to a gene that codes for a receptor in the taste buds. The two most common alleles for this gene are the dominant tasting allele (T) and the recessive non-tasting allele (t). You sampled 220 individuals and determined that 150 could detect the bitter taste of PTC and 70 could not. Calculate the frequencies of both alleles in this population.

7. Evolution in a population that occurs because of a chance change in allele frequency is referred to as ________________. For example, a small number of individuals may leave their original population and start their own. If these colonizing individuals happen to carry an assortment of ________________ that are not similar to their original population, their new population will have a different ________________. This is an example of ________________.

8. What type of selection would strengthen the status quo of what is currently the most-fit phenotype?

9. Some 35% of the critters in a population (in Hardy Weinberg equilibrium) has the recessive phenotype. What is the percentage of heterozygous individuals in this population?
   a. 59%
   b. 41%
   c. 48%
   d. 24%

18 Macroevolution: Evolution of Species

The process of speciation can only occur if there is a separation that divides a gene pool.

If microevolution is the change in the gene pool of a single population, then **macroevolution** is the creation of entirely new species. Macroevolution is a huge process that created all life-forms on Earth today and is defined as the rise of two or more species from an existing species (the process of **speciation**). In order to create a different species from a current population, there needs to be some sort of separation within the population. It is our **must know** to understand that the process of speciation can only occur if there is a barrier that divides the ancestral species' gene pool. The two groups of individuals are no longer able to mate and exchange genes, and the gene flow is interrupted; you now have two separate gene pools.

This, however, isn't enough to say speciation has occurred! Just because the populations are separated doesn't mean they're different species. If you recall, a species is defined as a group of individuals who are able to interbreed and produce viable offspring. The isolated gene pools need to *change* enough so that if brought back together, the members can no longer interbreed. This can happen just by random genetic drift, as we learned about in the previous chapter.

Furthermore, what if the two populations lived in slightly different environments? Natural selection would take over! Random mutations would be selected for or against (depending on the environment) and soon enough, different adaptations would accumulate. Would your two separate groups now be considered different species? Once again, it all has to do with their ability to mate and produce offspring. Eventually, the adaptations prevent mating between individuals of these two separate groups (even if they did come back together). Now speciation has occurred!

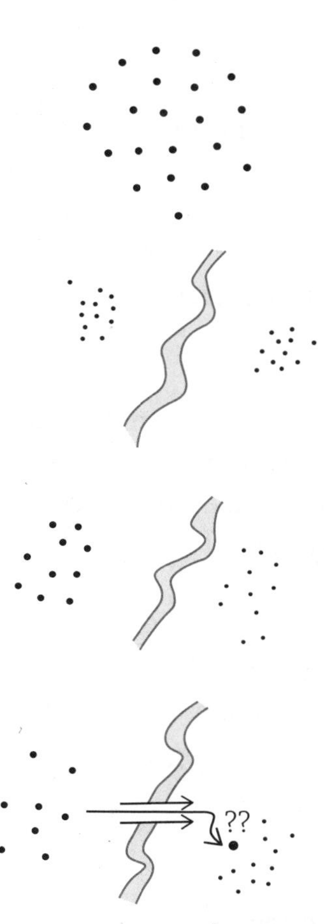

1. Original population
2. Gene pools become isolated
3. The 2 separate gene pools begin to change because of **genetic drift** or **natural selection**
4. Adaptations occur that prevent mating if the opportunity arises

The process of speciation

An adaptation that prevents the production of offspring is referred to as a **pre-zygotic barrier** or a **post-zygotic barrier**. Recall that a **zygote** is the cell that forms immediately after fertilization. A pre-zygotic barrier would, therefore, stop the egg and sperm from fusing and creating a zygote in the first place. If fertilization occurs, the process of creating an offspring can still be stopped by a post-zygotic barrier. In this case, the zygote is created, but it may not develop further, or the offspring may not be able to reproduce (it's sterile). Following are a few examples.

- Temporal isolation (pre-zygotic)

The term *temporal* means "time," so an example of this type of isolation would be a difference in mating seasons. If two species' mating seasons don't align, breeding can't occur. For example, some species of frog may live in the same habitat, but because they reach sexual maturity at different times in the spring, they won't interbreed.

- Behavioral isolation (pre-zygotic)

Two closely related species can have significantly different courting behavior, and thus the other species just would not look very attractive. For example, different species of firefly will use a different flash pattern to attract a mate. Wrong light signals, no interest!

- Mechanical isolation (pre-zygotic)

Mechanical isolation means that the reproductive bits just don't fit together. Many species of dragonfly are mechanically isolated because they have incompatible structures used in mating and things just can't line up properly.

- Offspring viability (post-zygotic)

In this case, mating can occur, but the offspring do not survive or are themselves sterile. A horse and a donkey can mate, they can produce an adorable little offspring (a mule), but that mule is itself sterile.

A **liger** is the offspring of a male lion and a female tiger, and a **tigon** is created by a male tiger and a female lion (yes, they're different critters). These offspring were long thought to be sterile, but there have been documented cases of the offspring producing babies of their own. This shows how the concept of "species" can be a bit muddled and not as clean-cut as what we learn. Other interesting hybrids: wholphin (whale-dolphin), beefalo (bison and cow), and a pizzly (polar bear and grizzly).

REVIEW QUESTIONS

1. Imagine a hypothetical population of beetles that are suddenly isolated from one another (for example, an earthquake creates a fault that the insects cannot traverse). One side of the fault has a lot of dark soil and the other side of the fault is covered mostly by grass. Using natural selection, explain how speciation could occur between these two isolated groups of beetles.

2. One species of flower can be fertilized by another species of flower, but once the egg and sperm fuse, their differing chromosome numbers prevent the formation of a viable zygote. This is an example of ____________________, a type of ____________________ barrier that prevents the production of offspring between two different species.

3. About 200 years ago, an ancestral population of flies laid their eggs on tree fruits called hawthorns. Once domestic apple trees were introduced, some flies in the population chose to lay their eggs on apples, instead. The maggots that developed in the hawthorn fruit grew into adult flies who preferred to mate with other hawthorn-reared flies; maggots from apples developed into flies who preferred to mate with other apple-grown flies. This is an example of ____________________, a type of ____________________ barrier that prevents the production of offspring between two different species.

4. Compare and contrast microevolution with macroevolution.

5. In order for speciation to occur, there needs to be some sort of barrier that divides the current population's ______________, thus interrupting the gene flow between the two resulting groups. Next, the isolated gene pools need to ____________________ in a way that renders them unable to interbreed if brought back together.

6. One species of field cricket mates in the spring, whereas a second species of field cricket mates in the fall. This is an example of ________________ isolation, a type of ________________ barrier that prevents the production of offspring between two different species.

7. Which of the following would not prevent formation of a zygote?
   a. Mechanical isolation
   b. Temporal isolation
   c. Offspring viability
   d. Behavioral isolation

19 Phylogeny and Vertebrate Evolution

MUST KNOW

- Phylogenetic trees provide a visual representation of the relatedness of species.

- A phylogenetic tree grows from a common ancestor, and as branches occur, new species arise.

- Traits lower on the phylogenic tree are more common, and traits higher up are shared by fewer species.

You often hear the statement "humans evolved from chimpanzees," but what does that mean? When you say something like that, you're referring to the fact that a long, long time ago, before there were any humans or chimpanzees, there existed a *common ancestor* to both humans and chimps. There was a population of mammals that most likely looked chimp-like, and this population split into two divergent evolutionary pathways: one led to modern-day chimps, the other led to *Homo sapiens* (humans). Speciation occurred (woohoo! we just learned about this in the previous chapter!), and in brief, it looks like this:

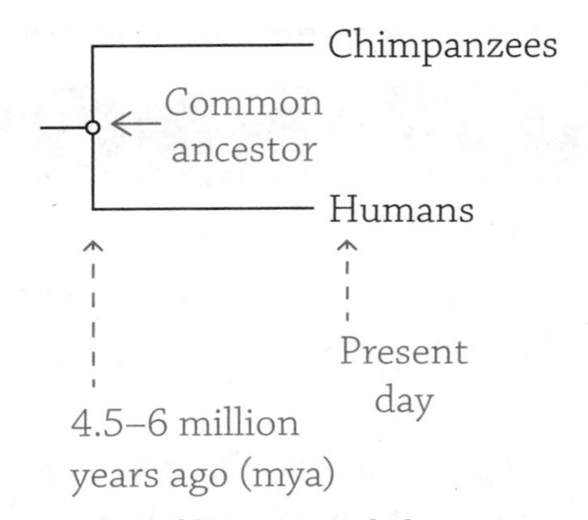

Phylogenetic tree of human and chimpanzee evolution

The common ancestor that existed ~5 million years ago is no longer present because speciation occurred and created the current species (such as humans and chimps). The phrase "we evolved from chimpanzees" is all sorts of wrong because it suggests: chimps → humans . . . and if that happened, there would be no chimps around, *because they all evolved into people!*

The branching diagram below is a **phylogenetic tree** and the basis of our **must know** for this chapter. Phylogenetic trees (and their close cousin, the cladogram) provide a visual representation of the relatedness of species. The point where two branches are connected indicates a common ancestor between the two separate species. When an event caused isolation of gene pools in that ancestral population, it led to speciation (as shown by the branches leading off of the common ancestor).

A phylogenetic tree is a hypothesis about the evolutionary relationships between groups of organisms; it's a hypothesis because humans weren't around to witness the actual evolutionary event occur. These branching diagrams allow us to study evolutionary history by grouping species by shared, heritable traits. These traits, or **characters**, include physical, behavioral, or molecular characteristics.

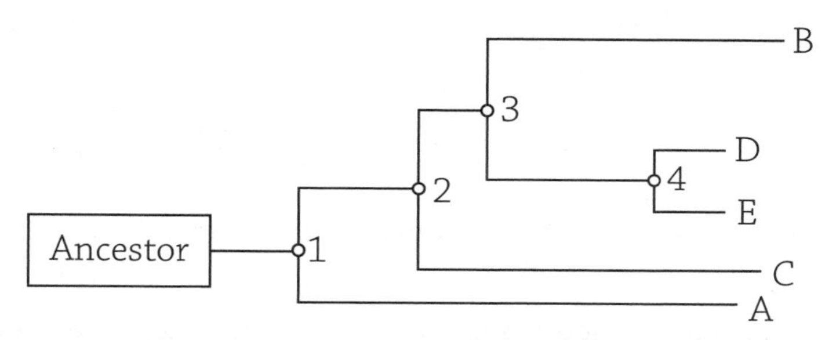

General phylogenetic tree with common ancestors (numbers) and species grouped by common derived characteristics (letters). Each species group can also be referred to as a taxon (or taxa, plural).

A branch point (or "node") is a divergence of two lineages from a common ancestor. It shows where two groups separated due to some mutation and acquired trait. Node 1, for example, represents the common ancestor for all the species shown (A–E). Notice that species A didn't change much from that original common ancestor. Species A is called the **basal taxon**, meaning it was the first to diverge from the ancestral species and has the fewest number of adaptations acquired since diverging. Branch point 2 represents the common ancestor of all the following taxa (B, D, E, and C).

A **shared character** is one found in all taxa under consideration. For example, in the figure below, a backbone is a character shared by fish, amphibians, reptiles, and mammals (but not the outlier, the invertebrate). The character of hair only applies to the mammal taxon.

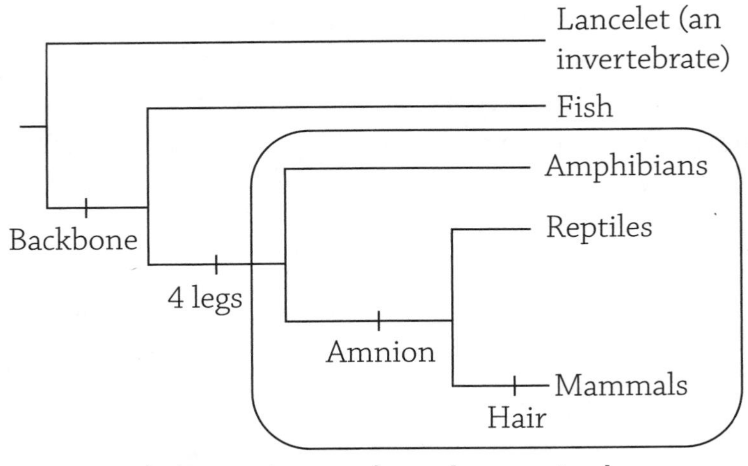

Phylogenetic tree of vertebrate animals

A **derived trait** is a character present in a taxon or taxa, but not in the taxon's common ancestor. A group that shares some derived trait is called a **clade**, and all species after a derived characteristic make up the clade. For example, amphibians, reptiles, and mammals are all of a clade defined by having four legs (the circled portion of the above figure). Again, as our **must know** concept suggested, phylogenetic trees help us visualize how groups of species are related, and how they share similar characteristics. For example, as animals evolved, species acquired traits that helped them better survive and reproduce on land. The large evolutionary events depicted in the previous phylogenic tree outline the momentous process of life moving from the oceans (where it began) onto the land. For example, if you are going to become mobile outside of an aquatic environment, you're going to need legs.

IRL Sometimes, a species' next evolutionary step may involve *losing* characteristics that were acquired earlier in their evolutionary history. Snakes, for example, might have evolved from burrowing lizards. Whales are mammals that decided to head back to water and evolutionarily lost their rear legs (though they maintain a vestigial pelvis that acts as a reminder of their four-legged past).

As species colonized ecological niches that are farther removed from water, they gathered adaptations that helped them fight against constant water loss, either from their own bodies (amphibians lack the water-retaining scale layer that reptiles acquired), or from their eggs (amphibian eggs don't have shells and must remain in water, whereas reptile eggs are shelled and can survive on land). If you look at that phylogenic tree, an "amnion" means a membranous sac that surrounds and protects the embryo. Mammals take it one step further and retain the amnion (and the embryo) inside their bodies during gestation (thanks, Mom).

REVIEW QUESTIONS

1. The saying "humans evolved from chimpanzees" is horribly wrong. Rephrase the statement so it correctly describes the relatedness between humans and chimps.

2. Answer the questions based on the following phylogenetic tree:

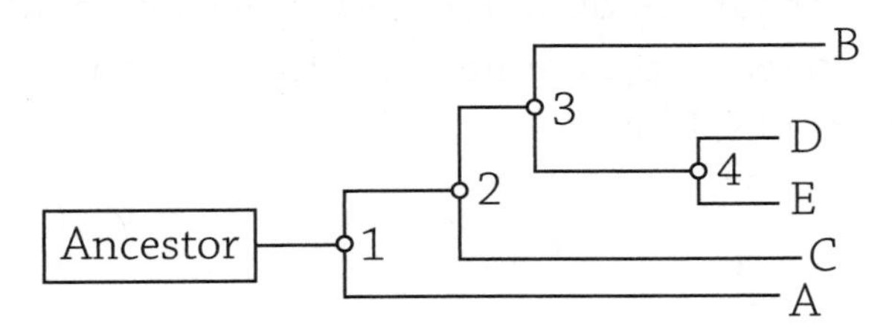

   a. Which node is the common ancestor of species E and B?
   b. Which species is most similar to the original common ancestor for species A–E?

3. Circle the node on the following cladogram that represents the common ancestor for species B and C.

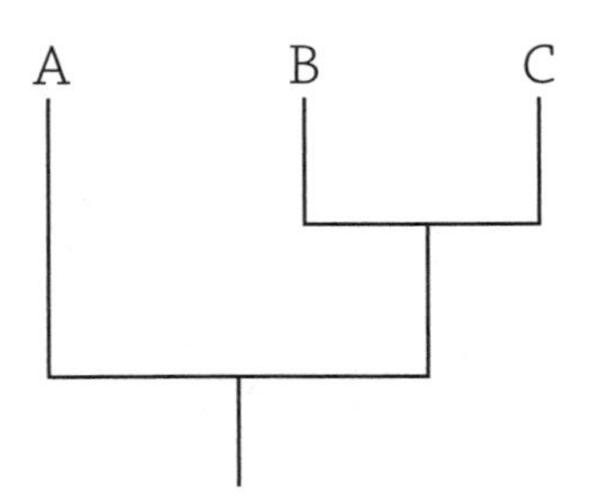

4. Based on the following cladogram, which species is most closely related to species D?

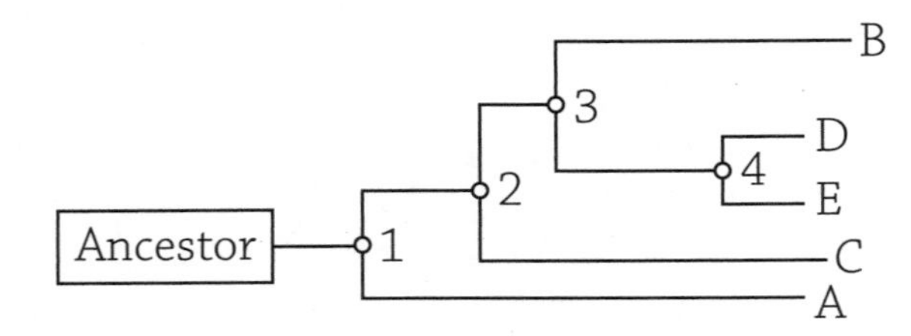

5. Which derived trait(s) in the following phylogenetic tree is(are) only found in the taxa including amphibians, mammals, and reptiles?

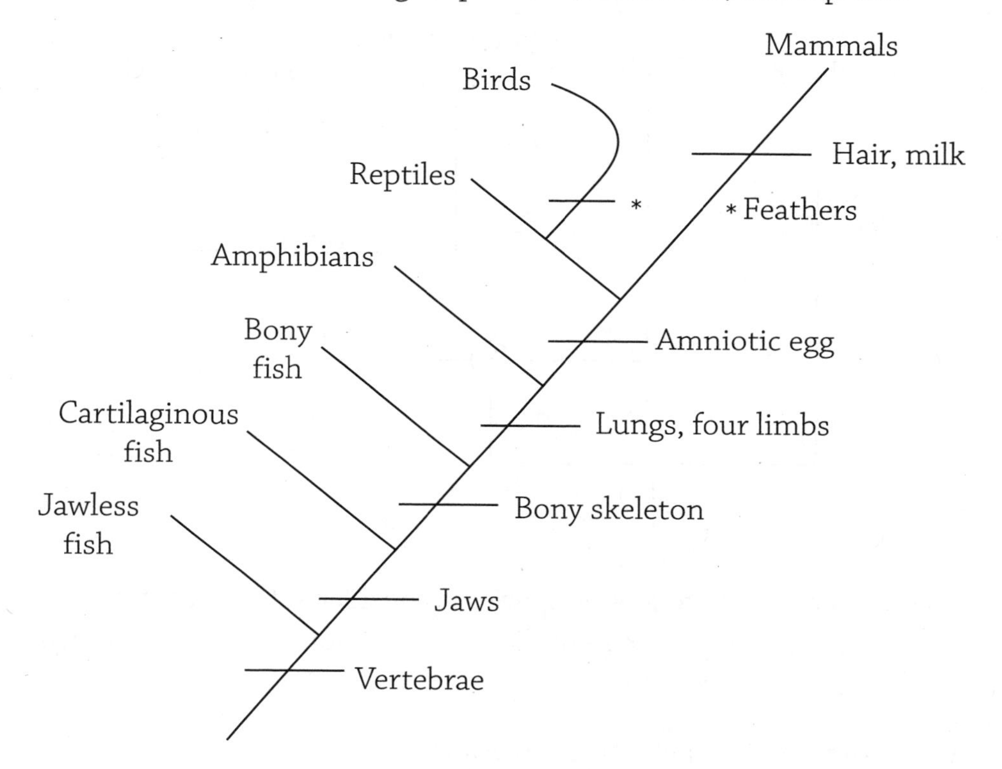

6. Answer the questions based on the following phylogenetic tree:

Salmon
Frog
Bearded dragon lizard
Sloth
Gorilla
Y
X

    a. Which of the following species is considered the basal taxon?
    b. Which node—X or Y—indicates a common ancestor from earlier vertebrate evolution?

7. What can't a phylogenetic tree show you?
    a. Relatedness of different species
    b. Time taken for an evolutionary event
    c. Presence of common ancestors
    d. Species grouped by shared traits

PART FIVE

Forms of Life

If you recall from way back in Part One, one characteristic of all life was the cell as its basic building block. A living thing can be made of a single cell (such as bacteria and many protists), or it could be multicellular. Multicellular organisms include fungi, some forms of protist, plants, and animals. Each of these multicellular life-forms are fascinating and worth studying in great detail. In this book, however, we will focus on two major players: the plants and the animals. I mean no offense to fungi and bacteria—bacteria are actually my favorite biological topic—but plants and animals provide the perfect opportunity to learn how a complex organism relies on smaller systems within it to function properly. As cells divide in the process of growth, they begin to differentiate (Chapter 11), meaning certain genes turn on and the cell adopts a specific shape and function. These specialized cells form tissues, and multiple tissues come together to create the organs that perform specific tasks for the organism. This process of creating small systems in the larger organism is key to the success of both animals and plants.

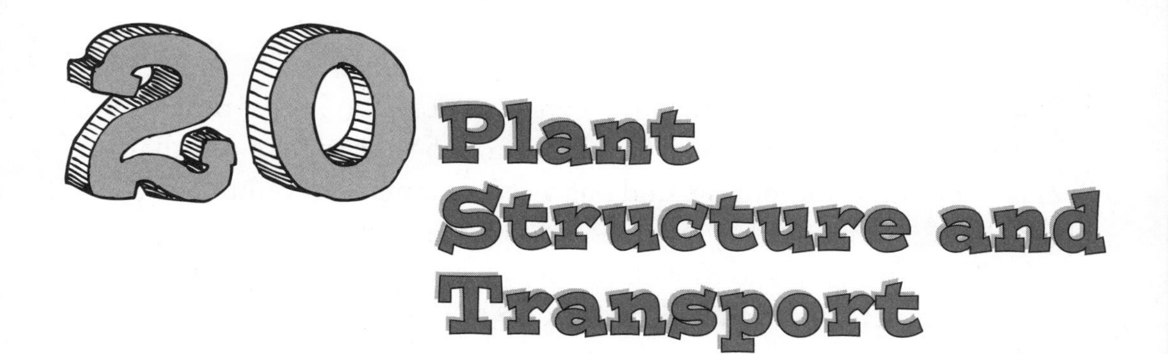

20 Plant Structure and Transport

MUST KNOW

- Plants have specialized tissues capable of performing different functions.
- Transport in plants is driven by water potential.

Plants are easy to define: they are autotrophic (photosynthetic), multicellular organisms made from a specific type of cell that contains chloroplasts and has a cell wall made of cellulose. A plant's body is composed of organs such as leaves (for photosynthesis), a stem (for structure), and roots (for water absorption). It becomes even more fascinating when you look a bit closer at these organs. Like, how exactly is a leaf's form fit for its function of photosynthesis? How does a huge tree move water into its roots and pull it all the way up its body?

Just like in animals, plants' organs are composed of a number of different tissues: dermal, ground, and vascular. This is, in fact, our **must know** when learning about plant structure: plants have different tissues, just like animals do. The **dermal tissue** is what it sounds like: the epidermis (or skin) of the plant. The dermal tissue often secretes a waxy coating to help reduce water loss. The **ground tissue** comprises the major "filling" of the plant and often specialize in photosynthesis and storage. The **vascular tissue** is composed of **phloem** (the transport system for sugar produced through photosynthesis) and **xylem** (the transport system for water). No matter what plant organ you are studying—stems, roots, or leaves—they all contain these three tissue types.

BTW

We will be focusing on flowering plants, the most evolutionarily advanced type of plant. Please understand, however, there are other categories of plants that do not necessarily possess all the traits that we are talking about. Moss, for example, doesn't have vascular tissues and relies on flagellated sperm to swim over to the egg (weird, right?). Ferns have vascular tissues (which enables them to grow much larger than the more primitive moss) but lack seeds to protect their embryos. Conifers (such as pine trees) have vascular tissues and seeds, but the seeds are said to be "naked" because they're not encased and protected in flowers!

Not meaning to be animal-centric, but once again I want to point out that, like animals, plants' organization follows the same hierarchy of organs → tissues → cells. Plants' organs are made of specific types of tissues, and those tissues are made up of specific types of cells! Parenchyma cells are considered the stereotypical plant cells—they are the models for the plant cells you see in your biology book. Parenchyma plant cells have the usual cell wall composed of cellulose, are chock full of chloroplasts, and house a large central vacuole.

If a cell is going to do a lot of photosynthesis, it is most likely a parenchyma cell; this cell type tends to do all the work (storage, metabolism, etc.).

IRL **It's possible to grow an entire plant out of a single parenchyma cell! These cells retain the ability to differentiate into other types of plant cells (with the correct chemical coaxing). If you have ever propagated a plant by cutting off a stem and stimulated root growth—often simply by placing the cutting in water—the parenchyma cells found in the stem are responsible for differentiating and forming the new roots.**

The cool thing about these different cell types, however, is their structure starts to specialize a bit to help in its function, and a major function besides photosynthesis is support. Recall that plants rely on their cell walls to provide structure, so the next two cell types are modified to provide a bit more support. Can you probably guess what part of the plant cell is modified? Yup, the cell wall. For example, **collenchyma** cells are found in young parts of a plant and provide structure without restraining growth. Their cell walls are thicker than in parenchyma cells, yet are still flexible.

IRL **You are familiar with collenchyma cells if you have broken a piece of celery in half and noticed those long strings. Those are cylinders of collenchyma cells. It makes sense, because it's the stalk of a young celery plant that needs to keep elongating and growing. The collenchyma cells allow this growth while providing structure.**

One step along the specialized-for-structure chain involves the **sclerenchyma** cells. These guys are like tiny little bricks: very strong and inflexible. Once sclerenchyma cells have finished growing, they produce a second cell wall that is chock full of a polymer called lignin, which provides even more rigidity. These cells cannot elongate and thus are found in areas of the plant that are no longer growing in length.

A specific type of sclerenchyma cell called a sclereid is responsible for the slightly gritty texture when you eat a pear!

The vascular tissue of the plant is also composed of specific cell types. The water-conducting xylem is made of **tracheids** and **vessel elements**, both of which are tube-shaped hollow cells that are quite dead at functional maturity. The sugar-conducting phloem is composed of **sieve-tube elements** that, unlike the cells of the xylem, are still alive and functioning (though the contents of the cell are reduced to help provide more space for transport). These cells must retain a bit of function because movement of phloem relies on solute gradients and active transport, and a dead cell is unable to actively do *anything*.

Now that we have covered the building blocks for a plant (cells and tissues), we can zoom out a bit and focus on the plant's organs and their functions.

Leaves

A leaf's form is adapted for its primary function: photosynthesis. The driving force behind the creation of glucose is the sun, so leaves tend to have a high surface area to increase the light absorption. There are exceptions to this rule, and they are quite interesting. The following table shows some leaves that have been modified to the point of no longer looking like leaves!

Plant	Modified leaf structure	Purpose
Cactus	Spines	Protection
Poinsettia	Big red "flower petals" (they're actually leaves)	Attract pollinators to the tiny actual flowers in the middle of all those pretty red pseudopetals
Venus flytrap	Death maw	Capture and digest insects
Succulents (like a jade plant)	Big, fat leaves	Water storage

The dermal layer of the leaves has a waxy coating to help reduce water loss by evaporation. The underside of the leaf, however, has special holes called **stomata** (singular, *stoma*) that are very important for photosynthesis. If you think back to the process of photosynthesis, you can probably guess what those holes are for (hint: they provide an entryway for an atmospheric gas to enter the leaf airspaces). Yes, stomata are for carbon dioxide to move into the plant leaves! The carbon dioxide gas enters these pores and then moves throughout the leaf until it gets to the cells that are specialized to make glucose. The holes are flanked by special cells called **guard cells** that are in charge of either opening or closing the stomata. The organization of the leaf tissue is logical, if you think about the function:

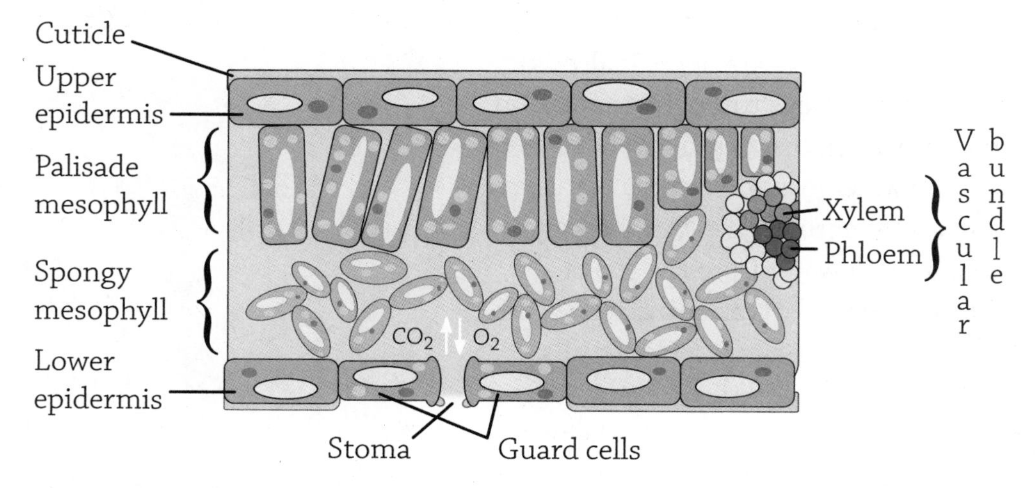

Cross section of a leaf

Source: https://commons.wikimedia.org/wiki/File:Leaf_anatomy.jpg

The top layer of the leaf is filled with column-shaped cells packed as tightly together as possible (the palisade cell layer). This ensures that there are a ton of chloroplasts hanging out on the sunny (upper) side of the leaf, grabbing as many photons of light as possible. These cells use the carbon dioxide that comes in through the stomata, and the water that is brought up from the roots, in order to create glucose. The gaseous CO_2 can get to that tightly packed upper layer because the lower layer is composed of loosely packed cells (the spongy cell layer) that has a bunch of airspaces. Once the glucose is created, it needs to get all throughout the plant's body. The glucose superhighway—the phloem—is conveniently branched and dispersed all throughout the leaf. Alongside the sugar highway is the water conduit (the xylem), and both of these vascular tissues are bundled together to create the veins you see when you hold a leaf up to the light.

The leaf is clearly an important structure for photosynthesis, but it also plays an important role in transport of water up the plant. The movement of a column of water molecules up the xylem is no easy feat. The main driving force is a "pull" that originates from the leaves, specifically, all those open stomata that allow CO_2 to enter. The truth is, when those tiny doorways are open, it also allows water to leave through evaporation. Sure, this isn't necessarily a good thing; too much water loss will cause a plant to wilt and eventually die. But the evaporation of water from the top of a plant (referred to as **transpiration**) is the driving force behind how a plant pulls water up from the soil through the roots. Imagine the column of water in the plant's xylem as a chain of water molecules, each holding on to the other through hydrogen bonds **(cohesion)**. The chain of water molecules is stabilized by the **adhesion** to the inside of the xylem conduit. If something introduces a bubble into the water column, this disrupts the flow of the xylem because the continuous cohesive chain of water molecules is broken. This phenomenon is called **cavitation**.

BTW

Don't forget that cohesion and adhesion are both possible because of water's ability to form hydrogen bonds:

When a water molecule bonds with itself, it is called cohesion. If a water molecule hydrogen bonds with something else, it is called adhesion.

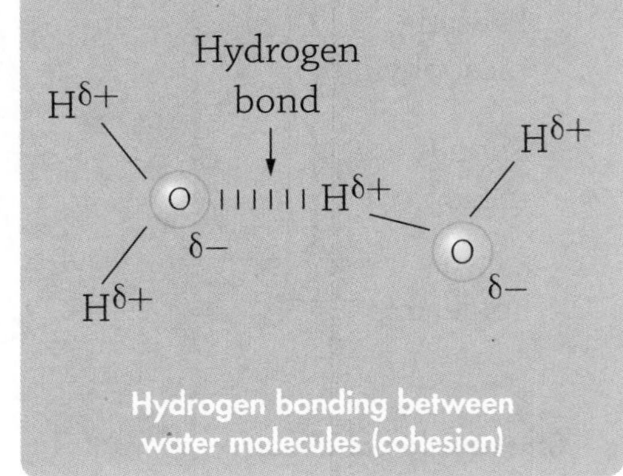

Hydrogen bonding between water molecules (cohesion)

IRL **You are supposed to cut the stems of flowers under water to prevent the introduction of an air bubble into the xylem (cavitation)!**

As transpiration occurs, it draws the column of water from the roots all the way up the plant body. Imagine this happening in a giant tree! There's also a bit of help from the roots (a slight "push" of the water column), but it's not enough to have an effect on larger plants. We'll talk about this a bit more when we get to roots.

Stems

In a typical plant, the leaves are attached to the stem. The stem helps the leaves reach out and grab as much sunlight as possible, and the stem will also grow tall to outcompete neighboring plants for sun. The stem will also push the reproductive bits as high as possible to help disperse pollen and facilitate fertilization. Stems provide the highway through which sugar (in the phloem) and water (in the xylem) will flow. Some plants have modified stems, the most surprising of which is the glorious potato. Even though potatoes grow underground, each tuber is actually a stem modified to store large amounts of starch.

Roots

The roots of a plant extend into the soil and absorb water and nutrients. Some species of plants have one strong major root (the taproot), whereas others have a bunch of smaller, shallower fibrous roots. If you have ever tried to pull up a dandelion from the soil, you have battled the strong grip of a taproot. On a related note, if you have ever ripped up a piece of sod (such as a clump of grass from your lawn), the roots on the underside are perfect examples of fibrous roots. Some plants use their roots to store food and water, such as beets and carrots.

If you have ever pulled a carrot from the ground, the part that gets all the glory is the carroty part (the orange root) that you eat. Sure, it's delicious and good for you, but what was the purpose for the plant? What would a carrot plant do with all that stored starch if we hadn't sauntered along and so rudely ripped it from the soil and ate it? If a carrot plant is left to its own devices, it would eventually use all that stored glucose (starch is the polymer a plant creates to store tons of glucose) and convert it into energy (ATP) to use toward flowering! A carrot flower is actually quite pretty.

The flow of the water in the xylem originates in the roots and moves upward into the rest of the plant's body. As we mentioned, the movement of the water is mostly due to transpiration (evaporative water loss) from the top of the plant, which pulls the cohesive chain of water molecules upward. The root, however, can provide a tiny bit of "push" due to the difference in water potential in the root tissue and surrounding soil. Recall from Chapter 7 that water will diffuse from a region of high water potential to low water potential, and the more solute that is dissolved in water, the lower the water potential. Therefore, the roots of a plant need to have a lower water potential than the surrounding soil in order to facilitate osmosis *from* the soil *into* the roots. The cells of the roots work to make this happen by actively transporting solute (ions and salts) from the soil into the root cells. This lowers that water potential in the plant's tissues, and water will diffuse in after it! This increased volume of water creates a slightly higher pressure in the roots compared to the rest of the plant, but the impact the pressure has on water movement is minimal. Yet if you have ever noticed droplets of water emerging from the edges and tips of the leaves, you have witnessed the effect of root pressure on xylem flow (called **guttation**). The bulk of the xylem movement up a plant is due mostly to the transpiration from the top.

REVIEW QUESTIONS

1. List the three plant tissue types and briefly explain their function.

2. ________________ cells are the "typical" plant cell, whose main roles are photosynthesis and storage. ________________ cells provide structure without restraining growth. ________________ cells are specialized for structure and are no longer able to elongate.

3. Explain why a leaf's structure is perfectly adapted for photosynthesis.

4. How does water potential play a role in movement of water up a plant?

5. Compare and contrast phloem and xylem.

6. For each of the following descriptions, indicate whether it applies to parenchyma (P), collenchyma (C), or sclerenchyma (S) cells:
   a. Has two cell walls
   b. Found in strings of celery because they provide structure without restraining growth
   c. Has lignin in its cell walls
   d. Primary function is photosynthesis
   e. The reason a pear has a gritty feel

7. The process of ________________ is the major force that pulls water from the roots up a plant's body. As water ________________ from the top of the plant, it pulls a chain of water molecules up the xylem. Each water molecule creates a chain through the process of ________________, and the entire chain sticks to the sides of the hollow tracheids and vessel elements through the process of ________________.

8. Which of the following is an incorrect statement regarding plant structure and transport?
   a. Plant organs are composed of dermal, ground, and vascular tissues.
   b. A plant draws water in through its stomata.
   c. Increased evaporation from the leaves would increase movement of water through the xylem.
   d. Plants rely on their cell walls for structure.

MUST KNOW

- Flowering plants' evolutionary success is a result of double fertilization.

- Double fertilization refers to one sperm fertilizing the egg (forming the embryo) and the other sperm fertilizing the two polar nuclei (forming the endosperm).

Plants create a lot of offspring. They have to, considering the low odds of their seeds finding conditions conducive to germination and seedling survival. The simpler plants such as moss rely on water to help their flagellated sperm swim to a neighboring moss plant's egg. Clearly, little moss sperms cannot swim long distances in order to propagate the species. This is why a patch of moss grows slowly by spreading outward; only nearby eggs can be fertilized by neighboring sperm.

As plants became more advanced, they became better adapted for sending their progeny farther away, and this helps them survive the dangerous and dry conditions that may await them. The flowering plants have evolved the most clever means to protect their embryos for the dangerous journey to unclaimed patches of soil. Our **must know** idea for this chapter is that flowering plants' success is rooted (pun intended) in a reproductive mechanism called **double fertilization**.

Although many plants can clone themselves, it behooves any species to mix up their genetics in order to create variation in the population. Plant sex—just like animal sex—involves the fusion of two gametes (egg and sperm) to create a new organism. All plants, from the primitive moss to the fancy flowering plants, rely on eggs and sperm to create an embryo. Types of plants differ, however, in how much protection they provide for this developing embryo. In order for the embryo to survive, it must avoid desiccation (water loss) and live long enough to take root and grow in the soil.

Flowering plants have these awesome seeds that are brilliant adaptations intended to protect the embryo from the dangers of the environments. A seed, at its most basic, is temporary housing for a plant baby. The flowering plants take it one step further and provide a food source (endosperm) inside the house (seed coat) for the baby (plant embryo). But I am getting ahead of myself—that's the end of the double-fertilization story. Let's first consider how it all began in an unfertilized flower.

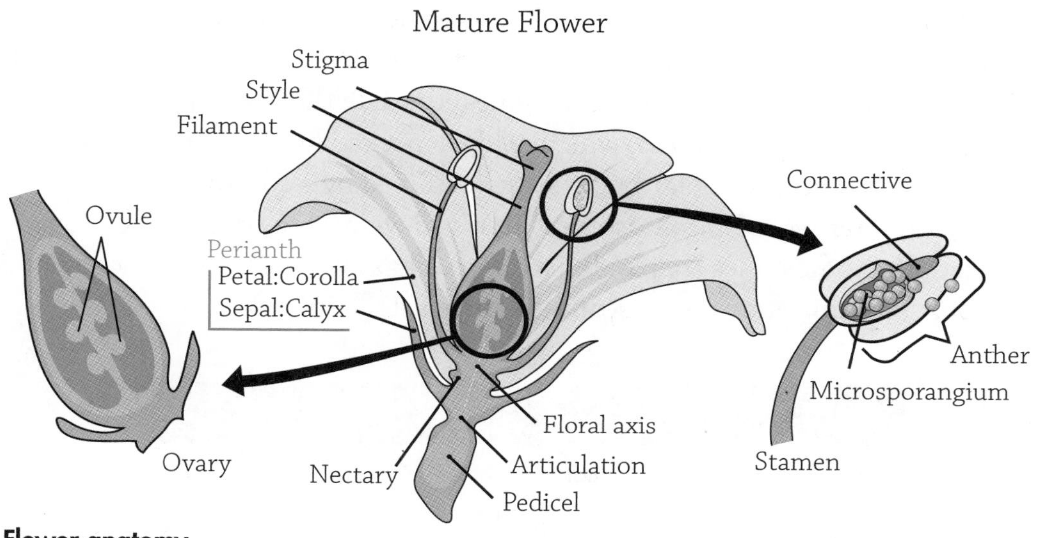

Flower anatomy

Author: Mariana Ruiz LadyofHats. https://commons.wikimedia.org/wiki/File:Mature_flower_diagram.svg

A flower is the reproductive organ of a plant. A flower is categorized as a "perfect" flower if it has both the sperm-producing **anthers** and the egg-containing **ovary**.

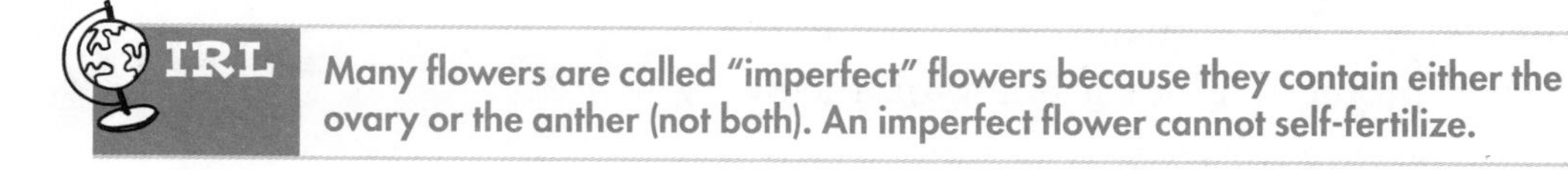

Many flowers are called "imperfect" flowers because they contain either the ovary or the anther (not both). An imperfect flower cannot self-fertilize.

Pollen grains are produced in the anthers, and each mature pollen grain contains two sperm and a third cell called a **tube nucleus.** A pollen grain will catch the wind (or catch a ride on an animal) and find its way over to the sticky **stigma** at the top of the **style**. Once it lands on the stigma, the tube nucleus will grow into a tube that tunnels its way down the style, providing a pathway for the two sperm. Once the sperm reach the ovary, one will fertilize an egg cell that is awaiting inside each ovule. Here is when our **must know** concept of double fertilization comes into play! There were two sperm traveling down the pollen tube, right? One obviously has to fertilize the egg to produce an embryo. The other sperm fertilizes these other two cells called **polar nuclei** that are waiting in the ovule alongside the egg. This forms something called **endosperm**, which is like a packed lunch for the growing embryo. And since two polar nuclei were fertilized by a sperm, that creates a cell with *three sets* of chromosomes! That's so weird! Sperm and eggs have one set of chromosomes (haploid), an embryo has two sets of chromosomes (diploid), and this endosperm stuff has three sets (triploid)!

If a sperm fertilizes the...	It forms the...	Which has a chromosome count of...
Egg	Embryo	2n (diploid)
Two polar nuclei	Endosperm	3n (triploid)

Ta-da! Two sperm, double fertilization, and an embryo (and its food) are formed! Now a cascade of developmental changes are occurring in the flower, starting with the petals falling off (they're no longer needed to attract pollinators). The embryo and the endosperm reside within the **ovule**, which eventually develops into the seed coat (protection for the embryo). The ovary will develop into the fruit (which animals eat and poop out the seeds somewhere else, helping to dispense the embryos).

BTW

The example I am providing is of a perfect flower with the traditional development of seeds and fruits. Please be aware that there are many exceptions to this scenario and the different ways fruits can develop is fascinating and varied. To get the true, full story, you need to immerse yourself in a botany class.

REVIEW QUESTIONS

1. What defines a flower as "perfect"?

2. In a flower, the ________________ produces the ________________, each of which carries ________________.

3. Fill in the following table:

If a sperm fertilizes the...	It forms the...	Which has a chromosome count of...
Egg		
Two polar nuclei		

4. Once fertilization occurs, the ovary develops into ________________, and the ovule develops into the ________________ (protecting the embryo). Since the embryo is stuck inside a protective coating, it needs a packed lunch to keep it alive until the embryo grows into a photosynthesizing plant; this food source is the ________________.

5. What flower type cannot self-fertilize, and why?

6. Pollination is when a pollen grain lands on the ________________. Fertilization is when the egg in the ________________ is fertilized by one of the two sperm that travels down the ________________ through the tube created by the third cell in the pollen grain, called a ________________. The second sperm will fertilize the ________________, turning into food for the embryo, called ________________.

7. For a typical flower, list these structures in order from outermost layer to innermost layer: ovule, egg, ovary, petals.

8. What is the benefit of an ovary developing into fruit?

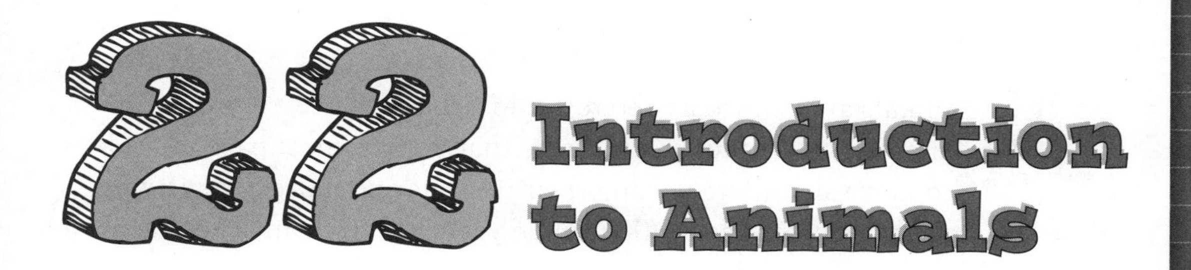

22 Introduction to Animals

MUST KNOW

- Each branch along the animal phylogenic tree introduces new traits.

- Invertebrates (animals without spinal columns) vastly outnumber all vertebrates (animals with spinal cords).

nimals are defined as heterotrophic (they need to eat to obtain energy), multicellular organisms that, for the most part, are able to move about freely using their nervous and muscular systems. The first animals evolved about 700 million years ago from an ancestral protist. Like animals, protists can be heterotrophic multicellular organisms made from non-photosynthetic eukaryotic cells, so it's a logical common ancestor. There are numerous ways to classify animals, but below is one hypothesis of animal phylogeny. One of our **must know** concepts is to notice that each branch in the tree introduces a new trait to that animal group. One thing to marvel about—the only phylum that includes critters with backbones is Phylum Chordata. All other animals are invertebrates, or organisms without a spinal cord. It's pretty amazing how outnumbered we are.

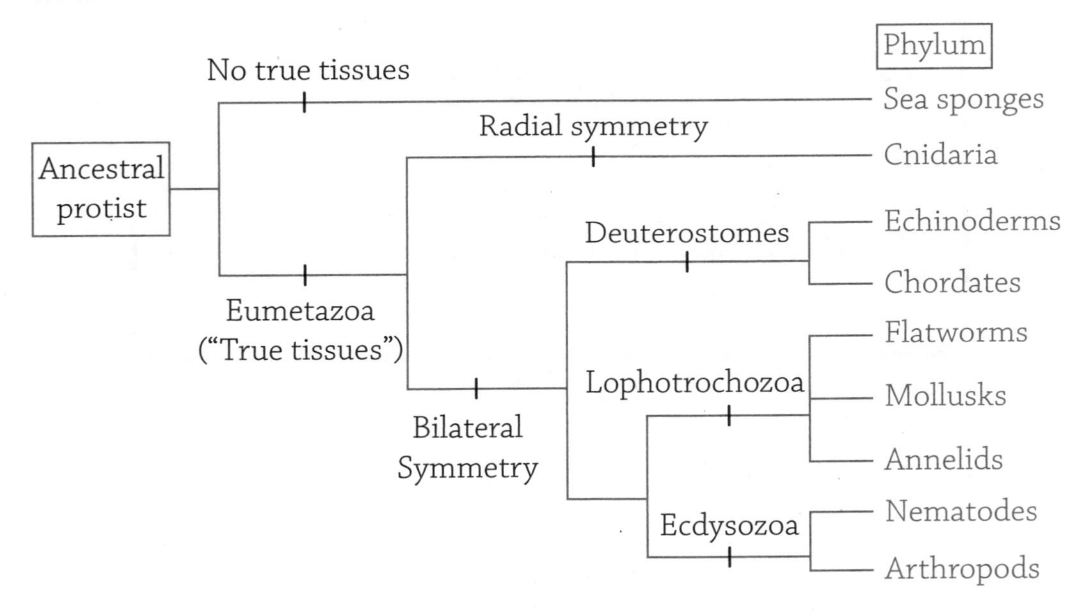

Notice that the outlier is the sponge, a critter that most people wouldn't automatically think of as an animal (but it is!). Sponges are the weirdos in the group because they don't have true tissues, which is more similar to the ancestral protist than to the rest of the animal phyla. The remainder are in the clade Eumetazoa, defined as animals with true tissues (groups of similar cells that work together as a unit).

If you have ever loofahed yourself in the shower with an expensive bath sponge (not the plastic variety) . . . it may literally be the skeletal remains of a sponge animal. Just FYI.

The next division is based on an animal's overall body plan. If it's radially symmetrical, that means it's shaped like a donut or a cup—there's a difference between the top and the bottom, but all the way around it's the same. Bilateral symmetry, on the other hand, means that there's only one way you could cut the shape in half for it to produce mirror images (for us, it would be a line drawn between the eyes, right down the body). The vast majority of critters have bilateral symmetry, and for good reason. This type of symmetry correlates with sensory input being focused in one area: eyes, ears, nose, taste, and the brain to make sense of it all.

The Phylum Cnidaria includes jellies (can't call them jellyfish because they aren't fish) and are from an earlier lineage that had not adopted this more "advanced" bilateral body style. Cnidarians do not have a brain but instead possess a "nerve net," which can detect and respond to stimuli from all directions at once (the benefit of being radially symmetrical!). Everyone else is of the bilateral clade, which can be divided up further based on differences in development. The lophotrochozoans have a specific type of larval stage called a trochophore, but don't stress about what, exactly, a trochophore looks like. It's more important to simply understand that this branch of the phylogenetic tree is based on what the animals look like when they're developing.

BTW

This is a good review of our previous evolution unit, and the importance of common ancestry. We learned that species are often defined as having a common ancestor if their embryos look similar. The different species categorized as lophotrochozoans are being grouped in a similar way, based on their developmental structures (trochophore larva), suggesting a common ancestor and a close relationship!

The lophotrochozoa clade includes flatworms, mollusks, and annelids. Flatworms are . . . well . . . *flat.* This is actually significant because their bodies have a very high surface-area-to-volume ratio and can rely on diffusion to move oxygen and nutrients throughout their tissues (no need for a circulatory system!). Flatworms you may be familiar with are the cute planaria and the arguably not-so-cute parasitic tapeworms. Another phylum is the mollusks, an amazing variety of critters, including clams, snails, slugs, cuttlefish, and the smartest of the invertebrates—the octopi. The final crew in the lophotrochozoa clade is the segmented worms (annelids), including our beloved and underappreciated earthworm.

IRL **There is an animal called a sea slug, an ocean-dwelling mollusk that feeds on algae and comes in an impressive variety of colors, shapes, and sizes. One species in particular—*Elysia chlorotica*—is particularly fascinating because it breaks our very definition of an animal . . . this special slug can photosynthesize! *E. chlorotica* feeds on algae, steals their chloroplasts, and embeds them in their skin, turning the slug a beautiful green. The slug is even leaf-shaped! Animals are defined as heterotrophs, eating to obtain nutrients, but this guy can go for months without food and photosynthesizes instead. Sadly, these amazing little invertebrates are becoming harder and harder to find.**

The next group of bilateral critters reside within the ecdysozoa clade, meaning their life cycles all include a step of ecdysis (molting) of their external covering. This includes nematode worms, tiny guys who have adapted for life in almost every ecosystem (including living a parasitic existence in vertebrate bodies). There is only one other phylum in this clade, but it is the most numerous phylum of animals by far: the arthropods. Arthropods all have exoskeletons, a segmented body, and jointed legs—this includes crustaceans (crabs and lobster) and all the insects. There are approximately a *billion billion* arthropods living on Earth!

The deuterostome clade includes organisms that have a common pattern of development during early embryo formation. There are only two phyla included here, but one of them is the only vertebrate group in the entire animal phylogenic tree: chordates! It is also interesting to note that our closest invertebrate relatives are the residents of Phylum Echinodermata, including the sea stars, sea urchins, sea slugs, and sea cucumbers. I would like to think that we are more like the intelligent and clever octopus, but nope, our close cousin is the glorious sea cucumber. I guess you don't get to pick your relatives.

As we move forward in our investigation of animals, we will focus on the most well studied of the vertebrates: humans.

REVIEW QUESTIONS

1. Choose the correct term from the following pair and then fill in the blank: Animals are defined as **autotrophic/heterotrophic** organisms, meaning that they can only obtain their nutrients by ____________________.

2. Based on the phylogenic tree from the beginning of the chapter, which animals are members of the clade that undergoes ecdysis (molting) during their life cycles?

3. Why don't flatworms need a circulatory system?

4. Based on the phylogenetic tree on page 306, the phylum that includes ____________________ is the closest relative of our phylum, the ______________.

5. True or False: Animals are classified as ONLY being multicellular (and NOT unicellular).

6. Which two special tissue types are unique to the animal phylum?

7. List two characteristics that define members of Phylum Cnidaria (jellies).

8. What is the most numerous phylum of animals, and why?

Animals Need Homeostasis

MUST KNOW

Negative feedback maintains homeostatic conditions.

Positive feedback amplifies the stimulus/response until a goal is achieved.

A hypotonic environment means too much water is absorbed, whereas a hypertonic environment leaches water from tissues.

Aquaporin proteins regulate water reabsorption in the nephrons.

The Importance of Maintaining Homeostasis

One of the qualities of all life is the need to maintain a stable internal environment, a process called **homeostasis**. Homeostasis regulates all sorts of things in the body, such as temperature, blood glucose levels, and salt concentration. In order to maintain ideal levels, the body needs to be able to counteract conditions that stray beyond a set point. The stimulus (whether it's low blood glucose, the body overheating, or becoming dehydrated) is detected by a sensor, which triggers a response to offset the stimulus. This is the process of **negative feedback** and it is so important, it is the **must know** for the chapter.

If you are out in freezing cold weather without enough warm clothes, a sensor in your hypothalamus will notice that your internal body temperature has dropped below the ideal set point (which is approximately 98.6°F or 37°C). The brain-thermostat sends a signal to the blood vessels in your skin, and they respond by restricting and diverting the blood into deeper tissues to reduce radiant heat loss. Your brain also sends signals to your muscles, and they respond by shivering and generating heat (thank you, Second Law of Thermodynamics). These responses help increase internal temperatures, and the brain-thermostat happily goes back to stand-by once homeostatic temperatures are achieved.

If, by chance, you step from those cold surroundings into an overheated room, the thermostat will swing too far in the other direction. This time, signals tell your capillaries to dilate and radiate heat from the skin, and sweat glands start to ooze liquid, providing evaporative cooling. This cools off your core temperature, and the thermostat is once again happy (homeostasis).

There is such a thing as **positive feedback** in the human body, but it's not nearly as common as negative feedback. In a positive feedback loop, once the sensor detects a stimulus, it triggers a response to increase that stimulus even more, until some final goal is achieved. A perfect example is childbirth. The release of the hormone oxytocin causes the uterus to contract and push the baby toward the cervix. The pressure of the baby's head on the cervix causes the pituitary gland to release more oxytocin, which in turn triggers stronger contractions. Stronger contractions mean more forceful pressure on the cervix . . . and more oxytocin . . . and more contractions, until . . . baby!

Hormones

The maintenance of homeostasis oftentimes involves hormones. **Hormones** are chemical messengers produced by the endocrine glands in your body. They travel throughout the bloodstream to cause an effect at a different location. There are many hormones needed to ensure proper functioning of the human body. The table below gives examples of some of the major players:

Endocrine gland	Hormone example	Function
Thyroid	Calcitonin	Lowers blood calcium levels
Parathyroid	Parathyroid hormone (PTH)	Raises blood calcium levels
Ovaries	Estrogens and progestins	Uterine lining growth and development of female secondary sex characteristics
Testes	Androgens	Sperm formations and development of male secondary sex characteristics
Adrenal glands	Epinephrine and norepinephrine	The fight-or-flight hormones
Adrenal cortex	Glucocorticoids	Reduces inflammation
Pancreas	Insulin	Lowers blood glucose levels
Pancreas	Glucagon	Raises blood glucose levels
Posterior pituitary	Oxytocin	Uterine contractions and milk secretion
Hypothalamus	Antidiuretic hormone (ADH)	Water retention by kidneys
Hypothalamus and anterior pituitary (they work together)	Thyroid-stimulating hormone	Tells the thyroid gland to release its own hormones

In order for a hormone to elicit a response, the target must have the correct receptors. The binding of the hormone will cause a cell response through that clever process of signal transduction we learned about in Chapter 4. This is how a chemical can travel throughout the entire body (in the circulatory system), but only have an effect at a specific location. Only cells with a complementary receptor will "hear" the message.

Here is a perfect example of hormonal regulation of homeostasis (and a reminder of our **must know** concept). The endocrine system uses the hormones insulin and glucagon to maintain even levels of blood sugar (glucose). Levels that are too high (hyperglycemia) or too low (hypoglycemia) can be dangerous, as are diseases that wipe out the body's ability to regulate properly (diabetes).

The levels of glucose dissolved in the bloodstream are regulated by the pancreas. If you were so busy you forgot to eat lunch, your blood glucose levels would begin to dip below the homeostatic levels (which is about 90 mg/100 mL).

IRL **If you've gone to your doctor for a checkup and they ask you for a urine sample, they are doing a quick check of your blood glucose levels. They dip a specially treated test strip in the sample for a quick quantitative analysis.**

The pancreas receives the signal that your body cells do not have enough glucose for cellular metabolism and your body needs to release some glucose stored in the liver. Specific cells in your pancreas (the alpha cells) release the hormone glucagon. Glucagon travels to the liver where it frees glucose monomers from the storage polymer glycogen. The freed glucose jumps into the bloodstream, restoring level conditions.

EXTRA HELP

Keeping these terms straight in your head is half the battle!

glucose: the sugar monomer that is being regulated in the bloodstream

glycogen: how glucose is stored in the liver (a polymer of glucose molecules)

glucagon: the hormone released when blood glucose levels are low

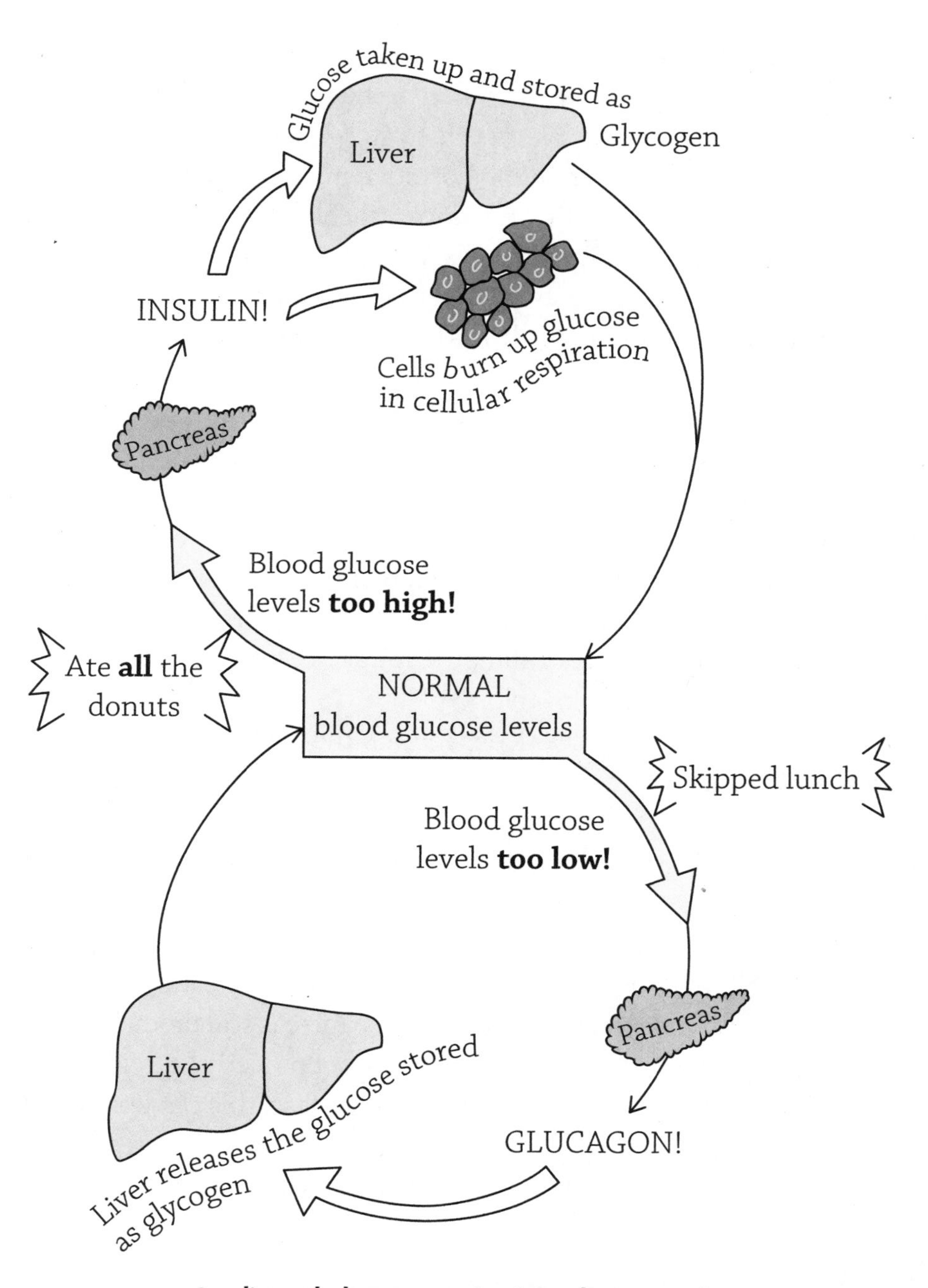

Insulin and glucagon maintaining homeostasis

If, instead, you decided to eat an entire chocolate cake, your blood glucose levels would be too high; cells need to take in the free-floating glucose and use it in cellular respiration, or the liver needs to store it as glycogen. Different cells of the pancreas (beta cells) secrete the hormone insulin. The cells of your body have specific receptors for insulin and, once attached, they take up glucose and use it in cellular respiration. The liver also hears the call to arms, and it brings in glucose and creates the polymer glycogen (for long-term storage of glucose). This drops the blood glucose levels back down to where they should be.

Disruption of Homeostasis

Diabetes mellitus is a disease that renders the insulin signal ineffective. If your body cells cannot hear the signal to take in glucose for cellular respiration, blood glucose levels remain high and a lot of glucose is excreted in the urine (which is why urine is tested for glucose). The body cells resort to using fat as the substrate for cellular respiration, leading to acidic metabolites and dangerously low blood pH.

Why, exactly, insulin fails to work depends on the type of diabetes (I or II). Type I diabetes is called "early-onset" because people are born with this condition. The immune system attacked the beta cells and the pancreas can no longer produce insulin. A diabetic person must rely on properly timed insulin shots in order to control their glucose levels. Type II diabetes (or, **insulin resistance**) is referred to as "late-onset" because the likelihood of developing type II diabetes increases drastically after age 45, after decades of unhealthy eating habits. Type II diabetes is "insulin-independent" because the pancreas still produces insulin, but the cells' receptors for the hormone are no longer properly functioning. If the body cells cannot "hear" the signal to take in glucose, this can lead to health problems. Insulin shots would not help a person with type II diabetes (making the insulin isn't the problem); the only treatment is a carefully controlled diet.

Type II diabetes used to be extremely rare in youth, but it has become much more common in recent years. There has also been a disturbing increase in type I diabetes in people 20 years old and younger.

Osmoregulation When You Live in the Water

The process of **osmoregulation** maintains healthy levels of water and solutes (salts, ions) in tissues. This is important for body cells to function, including neurons (that rely on proper levels of Na^+) and muscle cells (that rely on proper levels of Ca^{2+}). Aquatic critters live in the water and must work to maintain homeostatic levels of both salt and water. A fish living in freshwater (such as a lake or a river) is surrounded by a hypotonic solution; their tissues have a lower water potential. Therefore, osmosis will occur *into* the fish's body (recall that water moves from an area of HIGH water potential to LOW water potential).

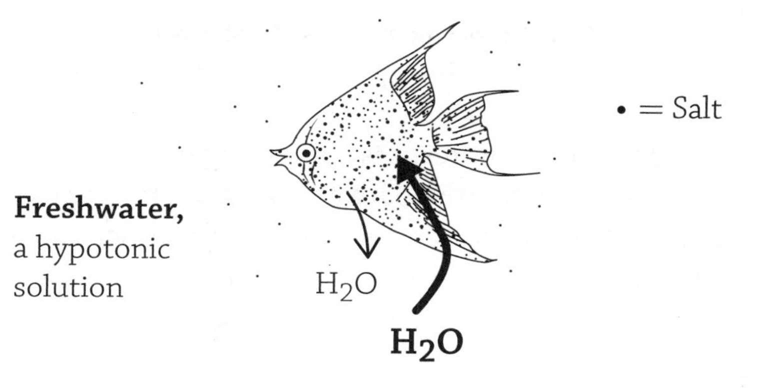

Water movement in a freshwater fish

More water will move into the fish's body than leave it. To maintain healthy internal levels of salt and water, the fish needs to excrete large amounts of diluted (watery) urine. It must also retain as much salt as possible in its body to maintain a healthy osmotic balance; their kidneys reabsorb salt and bring it back into the circulatory system before it can be lost in the urine.

In contrast, a fish living in saltwater (the ocean) has to constantly battle water loss because the surrounding environment is hypertonic and has a lower water potential compared to the tissues.

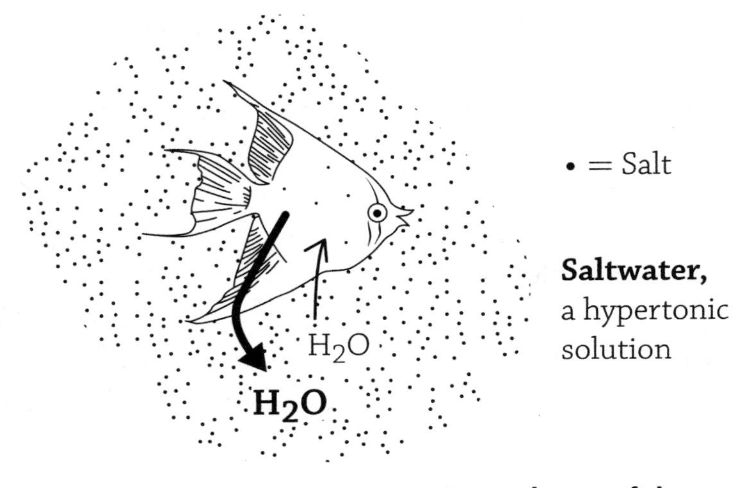

Water movement in a saltwater fish

Not only is the fish losing water, it is also taking in a ton of salt in its diet. The kidneys are going to work hard at saving water by reabsorbing it back into the circulatory system, and getting rid of as much salt as possible. Marine fish, therefore, produce a concentrated, very salty urine.

Excretion

The liquidy insides of animals need to remain free of the dangerous chemicals produced when proteins and nucleic acids are broken down. Refer back to what we learned about the four major macromolecules and consider what element is found exclusively in nucleic acids and proteins . . . nitrogen! Unfortunately, when nitrogenous molecules are dismantled, toxic things such as ammonia (NH_3) are released. The excretory system is specialized to help remove (or, *excrete*) these nasty metabolic waste products (as urine). The excretory system also plays a role in osmoregulation and maintaining healthy levels of salt in the body tissues, as we just learned about.

The excretory system is not about excrement (poop). The digestive system deals with poop, which is nothing more than indigestible food and beneficial gut bacteria. The excretory system is in charge of the momentous task of cleaning your blood and removing toxic chemicals that could kill you. Our kidneys are the key organs of this process, though a tiny bit of nitrogenous waste is also excreted in our sweat. Since this nitrogen-containing waste is so toxic, different critters have evolved to produce a form of this waste that is perfect for the environment in which they live:

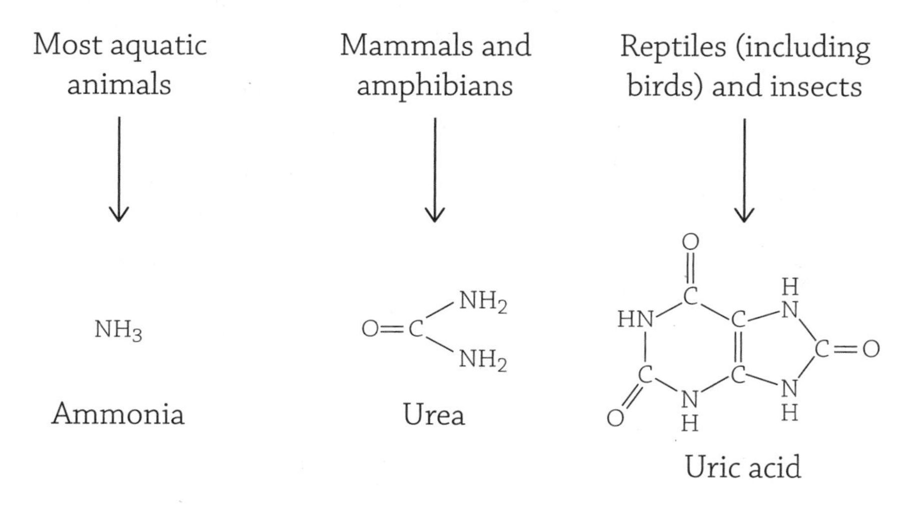

Fish have the luxury of excreting the most toxic and most concentrated form of waste: ammonia. If animals formed and collected ammonia in our bodies, we would toxify ourselves! Ammonia is, however, extremely water-soluble and is quickly whisked away from the water-dwelling animal's body. Land animals can't handle ammonia as our main form of nitrogenous waste, so we have evolved the less-dangerous form of urea: not as concentrated, and not as dangerous.

The most interesting adaptation, however, is excretion of uric acid. This waste product is the only form of nitrogenous waste that is a solid! If you consider the critters that evolved to use this form of nitrogenous waste, you may see something they have in common . . . all their young gestate in eggs. If you're a little baby bird, bearded dragon, or bug, you are stuck in the egg's watery environment while developing. Any water-soluble waste product would get stuck in the egg with the baby, creating a dangerously toxic solution. Therefore, the best waste product is a solid that settles on the bottom of the egg, out of harm's way.

Ever see a bird pee a liquid? That would be *no*, because as we just learned, it's a solid. If you have had a run-in with bird poop, it's actually only half poop; the other half is a white, solid blob of uric acid.

The human kidney is an organ the size of your fist, and we have two of them nestled against the back of our body cavity. The kidneys adjust levels of water, salts, and ions in our blood, along with removing the toxic urea created by the liver. Anything to be excreted is moved from the blood into the **nephrons** of the kidneys. There are millions of these little structures embedded all throughout the cortex (outer layer) of the kidneys. The nephrons collect excess water, ions, and waste and send it all down the ureters, into the bladder for storage, and eventually expelled through the urethra (as urine).

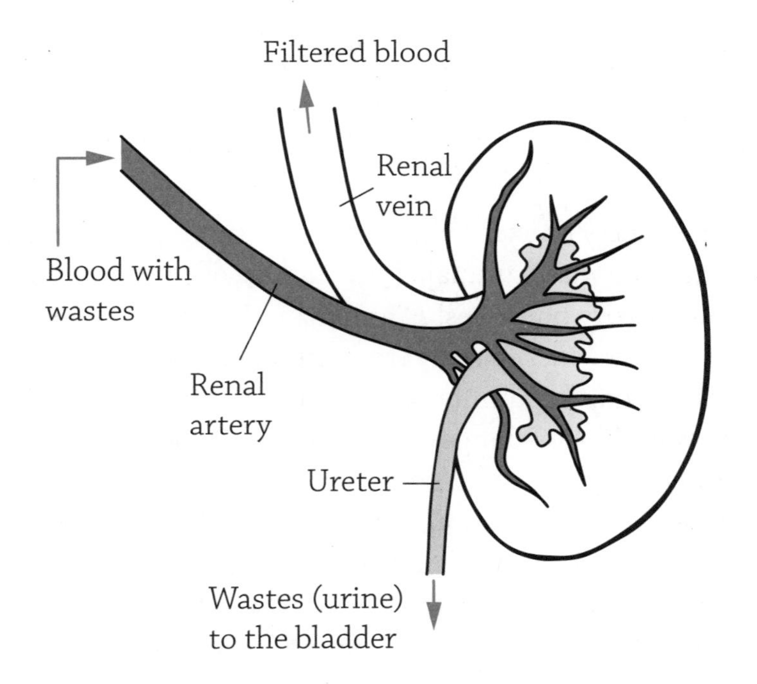

Blood flow through a kidney

Author: The National Institute of Diabetes and Digestive and Kidney Diseases. https://www.niddk.nih.gov/health-information/kidney-disease/kidneys-how-they-work

While you sit there casually reading, your heart is pumping about 5 liters of blood every minute, and 1 liter of that volume is routed through the kidneys for cleaning! And if you happen to have only one kidney (either because you were kind enough to donate one of yours to a person in need, or you simply were born with only one functioning kidney), you are still able to live a healthy, normal life.

Osmoregulation in Humans

Land animals have an extra challenge of maintaining proper osmotic levels in a dry environment. The kidneys need to quickly excrete (or reabsorb) water when needed. If water remains in the nephron's tubule, it will be excreted as urine; if the water is moved through the nephron back into the capillaries, it will remain in the circulatory system.

Any process of moving water across a semi-permeable membrane is facilitated by these amazing little transport proteins called **aquaporins**. Aquaporins allow water to move across membranes at astonishing rates. The nephron can adjust the amount of water reabsorbed from the urine back into the bloodstream by temporarily adjusting the number of aquaporin proteins perforating the collecting ducts to which the nephrons are attached. This allows the body to maintain proper blood osmolarity. The signal that tells the kidneys to adjust aquaporin numbers is a hormone called ADH (antidiuretic hormone).

IRL

Alcohol and caffeine lower ADH secretion, thus decreasing the amount of water reabsorbed . . . and you pee more. Alcohol and coffee are, therefore, diuretics (things that make you pee), and the antidiuretic hormones counteract losing too much water in your urine.

The brain's **hypothalamus** maintains proper blood osmolarity by monitoring the dilution of the blood. Imagine, if you will, that you ate an entire bag of salty potato chips while sitting in the sun and sweating profusely. The high amounts of salt and lower volume of water mean the blood has a high osmolarity. The hypothalamus responds by releasing ADH, causing a temporary increase in aquaporins. The nephron's water permeability is increased, leading to water being reabsorbed into the bloodstream, creating concentrated urine. This is a wonderful example of negative feedback. If you drink a ton of water, your blood osmolarity is now low. Your brain realizes you need to get rid of some of that excess water, so not much ADH is released. The kidneys happily let the water flow away as dilute urine, reestablishing homeostasis once again.

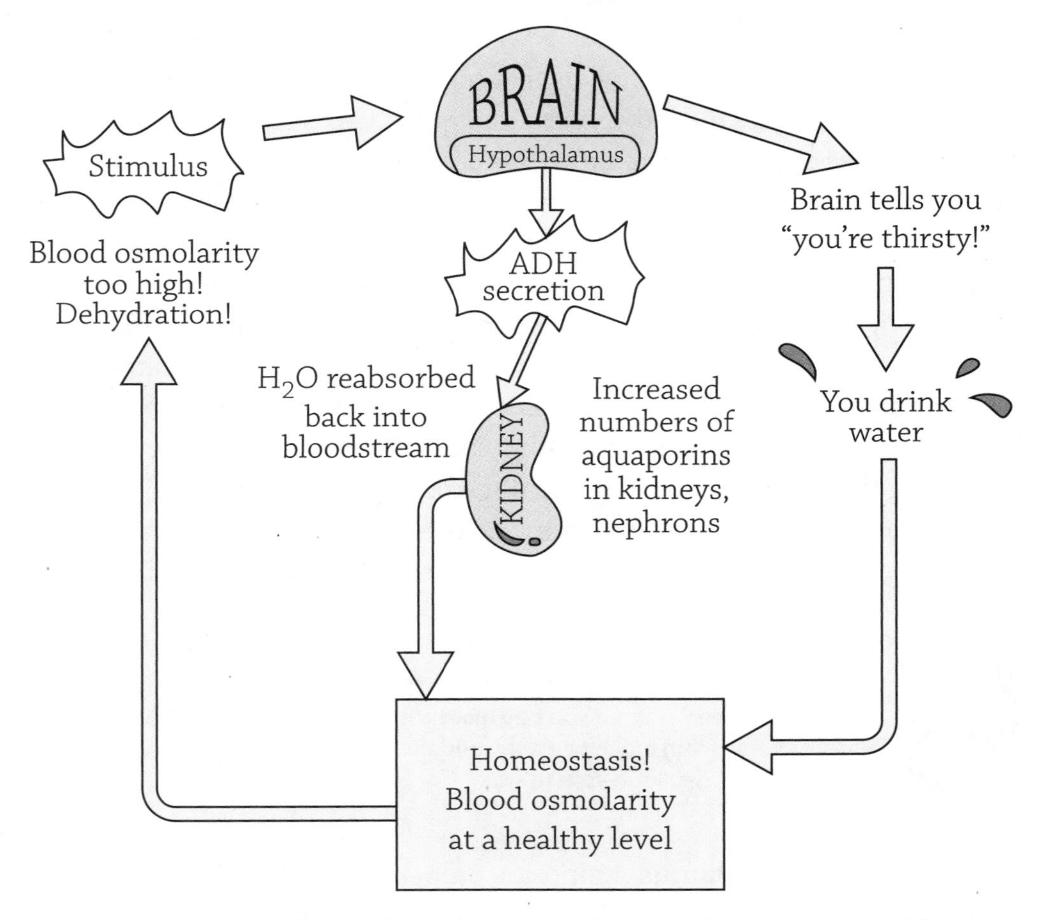

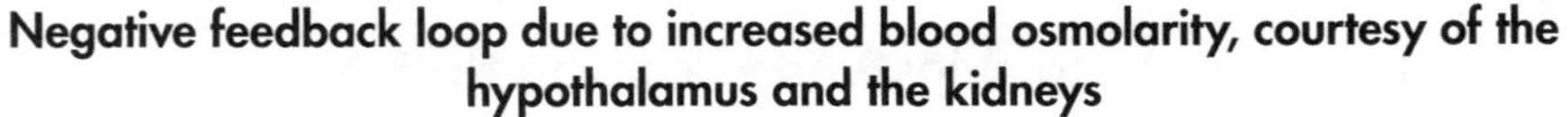
Negative feedback loop due to increased blood osmolarity, courtesy of the hypothalamus and the kidneys

REVIEW QUESTIONS

1. Why is maintaining homeostasis so important for organisms?
2. How does a positive feedback loop differ from a negative feedback loop?
3. Fill in the following table:

Form of nitrogenous waste	Notable quality	Type of critter
Ammonia		
	Won't toxify us when stored in our bodies	Mammals
		Bird

4. The process of maintaining proper levels of water and solutes in the blood is referred to as ____________________. The millions of small ____________________ in the kidneys are able to adjust the amount of water reabsorbed back into the bloodstream by changing the number of ____________________ proteins that line the collecting ducts. The chemical signal that leads to an increase in the numbers of these transport proteins is ____________________.
5. Fill in the blanks and choose the correct term from the given pair: When maintaining homeostatic levels of glucose in the bloodstream, the hormone ____________________ is responsible for **raising/lowering** blood glucose levels. One way it removes free glucose from the bloodstream is by stimulating the ____________________ to take in glucose and form the storage polymer, ____________.
6. Why is it advantageous for birds and insects to produce uric acid as their form of nitrogenous waste?

7. If the stimulus was an increase in blood osmolarity, what two responses would occur to reestablish homeostasis?

8. Choose the correct term from the following pair and then fill in the blank: Land animals have kidneys that work more like **marine/freshwater** fish because land animals are also battling water loss. The kidneys save water by reabsorbing it back into the ____________ instead of allowing too much to be excreted in the urine.

9. Which of the following is a correct homeostatic response?
   a. When blood glucose levels increase, glucagon is released.
   b. If salt levels in the blood get too high, ADH secretion is inhibited.
   c. Aquaporins help slow water movement through cell membranes.
   d. Ocean fish produce concentrated, salty urine.

24 Animal Digestion

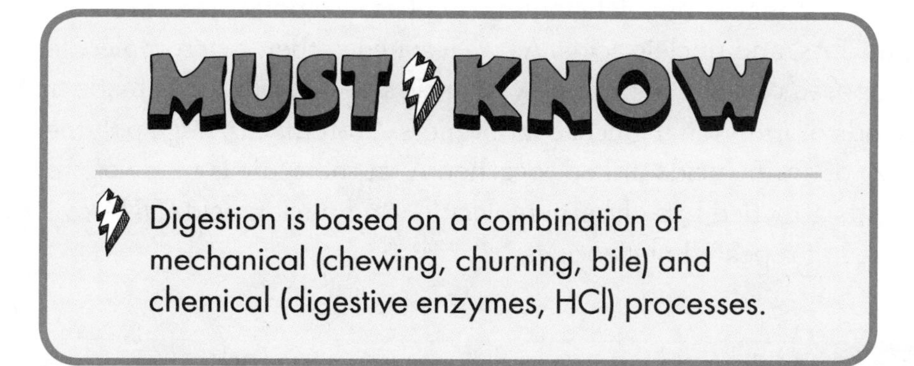

MUST KNOW

- Digestion is based on a combination of mechanical (chewing, churning, bile) and chemical (digestive enzymes, HCl) processes.

Unlike the autotrophs who make their own food, heterotrophs have to eat. Once the food goes into your mouth (**ingestion**), the process of **digestion** takes over. In its simplest form, digestion is taking large bits of food and breaking them up into smaller bits. The catabolic process of digestion is then followed by **absorption**, when those small food bits are brought into the circulatory system, where the organic molecules and essential nutrients will be used for building and energy. Finally, any indigestible material is passed out of the body (**elimination**).

Your diet consists of a delicious array of macromolecules: carbohydrates, proteins, fats, and nucleic acids. As we learned earlier, macromolecules are composed of small subunits. A gigantic starch molecule is made up of thousands of individual glucose monomers. Proteins are huge polymers made up of amino acids linked together. A fat molecule is made of three fatty acids attached to a glycerol molecule. And nucleic acid (DNA or RNA) is composed of repeated nucleotides.

Weird concept, right? You actually eat quite a bit of DNA. Think about how much of what you eat is actually made of cells. Meat is muscle (cells), salad is plant tissues (cells), bread is made of ground wheat (cells). On average, you ingest about 0.5 grams of DNA a day . . . that works out to be tens of millions of miles of the macromolecule!

Digestion is the process of dismantling these big macromolecules into their smaller components. When the macromolecules were synthesized in the first place (i.e., starch created by a plant basking in the sun), it was most likely through a dehydration reaction.

Don't forget how important dehydration reactions are in the creation of the four macromolecules (carbohydrates, lipids, proteins, and nucleic acids):

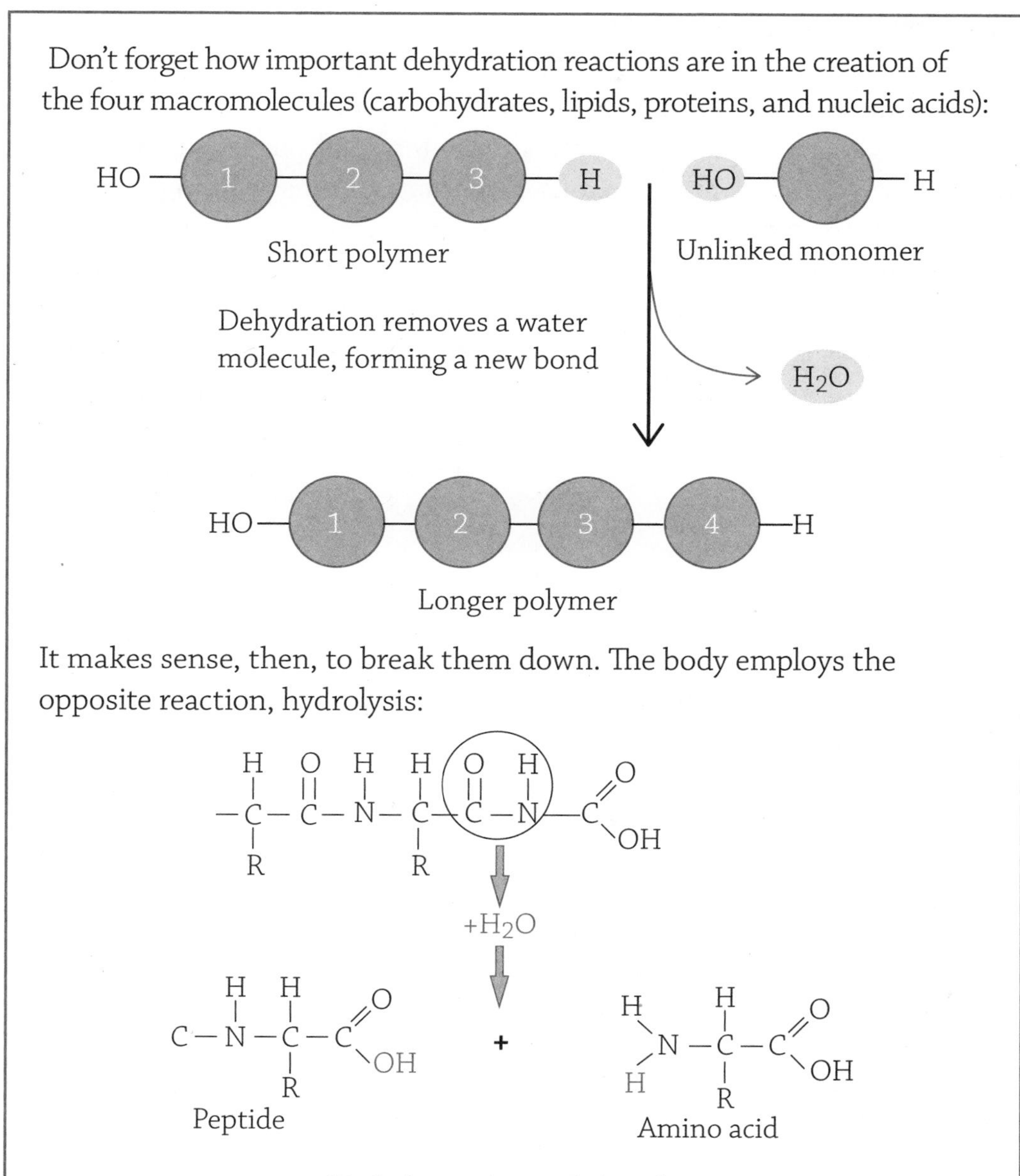

It makes sense, then, to break them down. The body employs the opposite reaction, hydrolysis:

Hydrolysis of peptide bond

A hydrolysis reaction will use a water molecule to break apart a covalent bond linking subunits together in a larger macromolecule. As we go through the stages of digestion, we will often see a hydrolysis reaction do the work of dismantling your lunch.

The Steps of Digestion

Most animals have an **alimentary canal**, a one-way tube that food passes into, through, and out of the body. Food is oftentimes pushed along by **peristalsis**, rhythmical contractions of the smooth muscle lining the canal. Along the way, food is broken up in two manners (and here is our **must know** concept for this chapter): mechanical digestion and chemical digestion. **Mechanical digestion** is the inelegant process of physically breaking up big chunks into smaller chunks, thus increasing the surface area for chemical digestion. **Chemical digestion** is when specific enzymes get to work catalyzing the hydrolysis reactions to liberate the monomers.

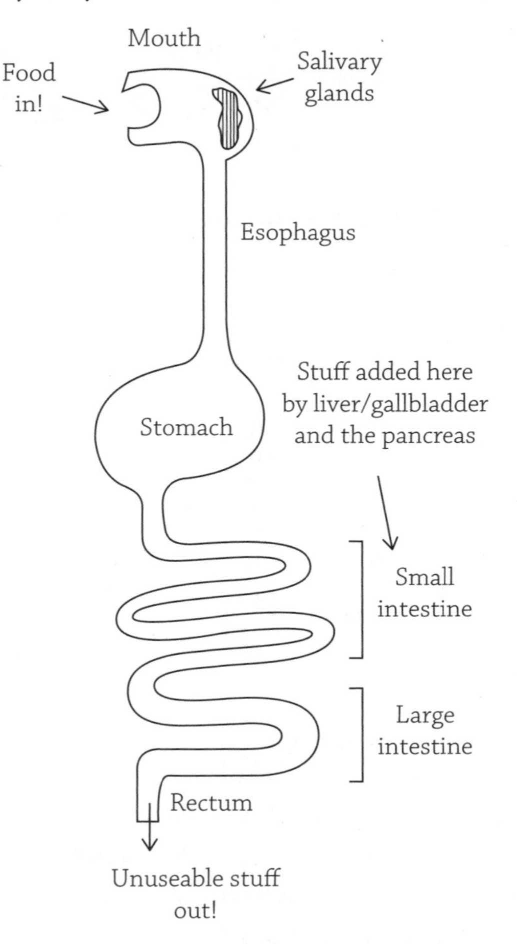

Simplified drawing of mammalian alimentary canal

Step 1 Mouth

Here it begins, that spoonful of cereal you just guided into your mouth will now meet the mechanical destruction of your teeth, grinding up big pieces into smaller pieces. At the same time, salivary glands secrete an enzyme called amylase that begins the chemical breakdown of starch and glycogen (both polymers of glucose). The saliva also contains a buffer to help neutralize acids (protects your tooth enamel) and lysozyme, an antimicrobial compound (because, I hate to tell you this, but your food may have a fair amount of bacteria contaminating it).

Meanwhile, your tongue is helping to shape this saliva-softened glob of food into a **bolus**, the official term for a saliva-laden glob of chewed up food. When you swallow, the food is guided into the pharynx (back of the throat), where it will pass down the esophagus, into the stomach. To make sure the food bolus goes down the correct esophageal path, there is a flap of cartilage that kindly covers your trachea, preventing food from entering your lungs.

Step 2 Stomach

After the food bolus is pushed down the esophagus by peristaltic waves, it passes through a **sphincter** and into the stomach. The sphincter is the connection point between the food tube and the stomach, and it helps keep gastric juice from bubbling up and burning the tender esophagus (acid reflux, or "heartburn"). The stomach is responsible for both mechanical (churning) digestion and chemical (HCl and pepsin) digestion.

Once food passes through the sphincter, the stomach revs up and begins muscular contractions that mix the stomach soup (technically called **chyme**) once every 20 seconds. Digestive fluids are dumped into the mix. The gastric juice contains hydrochloric acid (HCl) and pepsin, an enzyme that digests proteins, along with mucus to help protect the lining of the stomach. The parts of the stomach responsible for making the gastric juice are called gastric glands, and they are found all along the stomach lining. Interestingly, each gastric gland is composed of three types of cells:

Cell of the gastric gland	Product of the cell	Function
Mucous cell	Mucous	Protects lining of stomach
Chief cell	Pepsinogen	Once pepsinogen is activated by HCl, it turns into the protein-digesting enzyme pepsin
Parietal cell	HCl	Creates acidic conditions

All of these cells secrete their products to create the gastric juices. The mucous protects the lining of the stomach from self-digestion. In fact, conditions are so harsh, the stomach replaces the eroded epithelial layer once every 3 days.

Some unfortunate folks may develop an ulcer, an open sore in the lining of the stomach. For a long time, people thought ulcers were caused by stress, or eating spicy food, or maybe drinking too much coffee. We now know that ulcers are caused by these acidic-loving bacteria called *Helicobacter pylori*. Happily, we now know to treat an ulcer with antibiotics, not antacids and bland foods.

The hydrochloric acid is provided by parietal cells, and it drops the pH of the stomach to a 2. This is helpful because it kills most bacteria (unfortunately not those nasty *H. pylori*) and denatures proteins, exposing the peptide bonds for pepsin to attack.

In a clever attempt to limit self-digestion, the chief cells secrete an inactive form of the protein-digesting enzyme, called pepsinogen. Once pepsinogen hits the acidic stomach soup, however, it is converted into the active form pepsin. Not only does it begin to cleave apart proteins, the pepsin also activates more pepsinogen by converting it into pepsin. Yes, one of those relatively rare positive feedback loops!

Anywhere between 2 and 6 hours later, the chyme is slowly released through the lower sphincter that guards the passage from the stomach into the small intestine.

Step 3 Small intestine

Most of the digestion occurs here, in the approximately 20 feet of small intestine currently snuggled up in your abdominal cavity. Most of the digestion is through enzymatic hydrolysis, though there is a bit of mechanical digestion, compliments of bile. Bile qualifies as mechanical digestion because it breaks up big blobs of fat into smaller blobs of fat (**emulsification**). It is not chemical digestion because no covalent bonds are enzymatically broken.

The small intestine is made up of three sections: the first part is the **duodenum**, where most of the digestion is completed. The second and third parts are the **jejunum** and **ileum**, the major sites of nutrient absorption. The chyme from the stomach is slowly squirted into the duodenum, where it is mixed with digestive juices from the pancreas, the liver, and the gallbladder. Although we know the pancreas as the master of glucose metabolism, it also plays a large role in digestion. The pancreas helps neutralize the scalding acidity of the chyme by adding bicarbonate solution. It also secretes trypsin and chymotrypsin, a protein-digesting protease pair. The liver makes bile that is piped into the gallbladder for storage. Bile emulsifies fats, meaning it breaks up large fat globules into smaller ones.

Digesting the food is only half the battle. In order for the nutrients to do any good, they must get into the circulatory system. As the digested food travels down the inner space of the digestive tract (the lumen), the nutrients will pass through the walls of the small intestine and into the capillaries. This process is facilitated by the highly folded inner lining of the small intestine. There are large folds, each covered by smaller folds called **villi**. On top of that, each villus is covered with epithelial cells that have their own tiny hair-like projections called **microvilli** (also called the **brush border**).

BTW

A highly folded tissue is an excellent example of form fits function. If nutrients need to pass through the intestinal wall, natural selection would favor a high surface area. A highly folded structure provides a high surface area. When you flatten out the folds—all the villi and microvilli—you get a surface area the size of a tennis court!

■ Step 4 Large intestine

This is the final path along our digestion journey. Once all the useable nutrients are absorbed by the small intestine, the stuff that's left over moves into the large intestine. The large intestine includes the colon, cecum, and the rectum. The cecum is a dead-end pouch where the large intestine connects with the small intestine, and it is the happy home for bacteria that ferment cellulose. It makes sense, therefore, that the more cellulose and plant matter an animal eats, the larger the cecum. Once again, natural selection has sculpted a form that best fits its function.

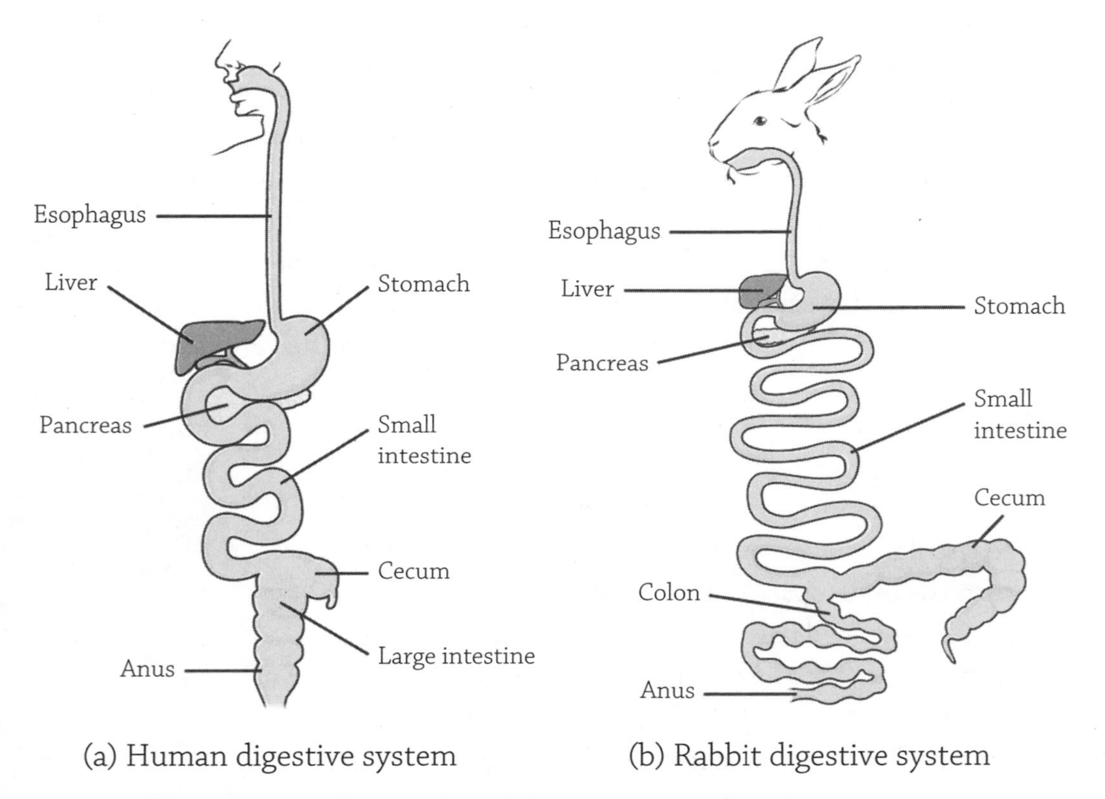

The relative lengths of the cecum in carnivore versus herbivore

Authors: Charles Molnar and Jane Gair. https://opentextbc.ca/biology/wp-content/uploads/sites/96/2015/03/Figure_34_01_05ab.jpg

The major role of the large intestine is to reclaim water before it passes out of the body. About 90% of the water that enters the digestive tract is reabsorbed by the small and large intestines! Problems occur if the feces move along too quickly and not enough water is absorbed (diarrhea) or they move too slowly and too much water is removed (constipation). Once it's all said and done, feces will pass out of the body through the rectum. Fecal matter consists of undigested material, such as cellulose fiber. About a third of poop is composed of the harmless bacteria that live inside your colon. The brown color is due to the bile that is dumped into the mix earlier in the small intestine. Bile contains a pigment called bilirubin that is produced when the hemoglobin in red blood cells is broken down. As bilirubin is digested, it produces the brown color of your feces.

REVIEW QUESTIONS

1. The first stage of digestion is ____________, when you take a bite of food. That food is broken up during the process of ____________, when large molecules are broken up into smaller components. The small components are more easily taken up into the bloodstream through the process of ____________. Finally, any indigestible material is passed out of the body through ____________.
2. What type of enzymatic reaction is responsible for chemical digestion of food?
3. What structural detail of the small intestine helps increase its ability to absorb nutrients?
4. Fill in the following table about gastric cell function:

Cell of the gastric gland	Product of the cell	Function
Mucous cell		
	Pepsinogen	Once pepsinogen is activated by ____________, it turns into the protein-digesting enzyme ____________
Parietal cell		Creates ____________ conditions

5. Explain the difference between mechanical digestion and chemical digestion.

6. Match the following terms with their correct definition:

a. alimentary canal	1. Location of protein digestion by an enzyme that only functions in acidic conditions
b. esophagus	2. Stores bile
c. liver	3. First site of mechanical digestion
d. large intestine	4. Secretes a large number of digestive enzymes (plus insulin!)
e. mouth	5. Undigested food is passed through here
f. pancreas	6. Source for salivary amylase
g. salivary glands	7. The majority of nutrient absorption occurs here
h. small intestine	8. Term for entire length of digestive tract
i. rectum	9. Produces bile
j. gall bladder	10. Major role is reabsorption of water
k. stomach	11. Food tube that doesn't help in digestion

7. If you compared the lengths of the ceca (plural of cecum) of a tiger versus a rabbit, what difference would you expect and why?

8. Interesting poop trivia! About a third of your poop is composed of ____________, and poop is brown because of pigments produced from the breakdown of ____________.

9. Which macromolecule undergoes enzymatic digestion in both the mouth and the small intestine?
   a. Protein
   b. Lipids
   c. Starch
   d. Glucose

25 Animal Circulation and Respiration

MUST KNOW

- The circulatory and respiratory systems work together to move oxygen and carbon dioxide between the tissues and the outside environment.
- A high surface-area-to-volume ratio helps facilitate diffusion.
- Gases diffuse from high partial pressure to low partial pressure.
- Oxygenated and deoxygenated blood does not mix in the most advanced four-chambered heart.

Circulatory System

All the cells of your body need to obtain oxygen and get rid of carbon dioxide. When these gases (CO_2 and O_2) are dissolved in water, they are more than happy to diffuse into (and out of) body cells. However, movement by diffusion is fast only over very short distances. It is necessary for each and every cell in your body to have a source of oxygen nearby. The circulatory system takes care of oxygen delivery. I want you to keep the **must know** concept in mind as you continue reading. When oxygen (or any dissolved gas) moves from one area to another, it must do so down its concentration gradient. This means the gas will move from an area of high concentration (where there is a lot of the gas dissolved in the fluid) to an area of low concentration (where there is less of the gas dissolved in the fluid).

Some critters (such as flatworms and sea jellies) lack circulatory systems and instead rely only on diffusion to move oxygen into their body tissues. As pointed out by our **must know** concept, oxygen will move down its concentration gradient from the surrounding fluid into the oxygen-starved tissues. In order to make this an effective means of transfer, the critters' body plans have evolved to maximize tissue exposure to the environment. For example, a flatworm is, well, *flat*, and a sea jelly has a hollow cavity called a gastrovascular cavity:

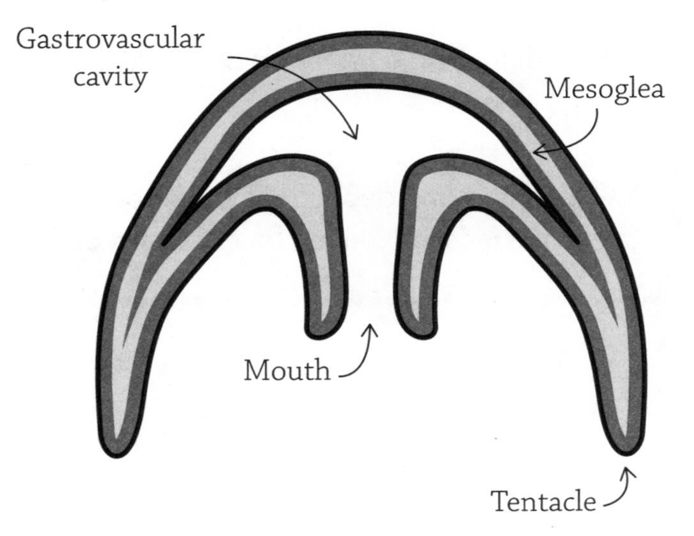

Cross section of a sea jelly

What these guys have in common (besides both being invertebrates) is that their body plans have maximal surface-area-to-volume ratios (SA:VOL), meaning there is a lot of surface area for the given volume of the body.

BTW

The higher the SA:VOL ratio, the better the diffusion rates. This means, for example, a tiny cell can deal with substances moving in and out much better than a large cell:

Small prokaryotic cell

Diameter	=	**2.0 μm**
Surface area	=	**12.6**
Volume	=	**4.19**
SA:VOL	=	**3.0**

Large eukaryotic cell

Diameter	=	**50 μm**
Surface area	=	**7850**
Volume	=	**65,449**
SA:VOL	=	**0.12**

Surface area of sphere $= 4\,\pi r^2$

Volume of a sphere $= \frac{4}{3}\pi r^3$

This is significant from an evolutionary perspective! The first cells to evolve were the small prokaryotic cells. Diffusion was sufficient for these early life-forms because prokaryotic cells have a high SA:VOL ratio. As cells evolved into the more complex eukaryotic cells, they became bigger. Their SA:VOL ratios decreased to the point that they could no longer rely on simple diffusion. In order to compensate, internal compartments helped by subdividing the contents and increasing the membrane's surface area. These "compartments" are the eukaryotic organelles!

By far, most animals must rely on a circulatory system in order to reach all those oxygen-hungry cells. There are two options: an open circulatory system or a closed circulatory system.

Open Circulatory System

An open circulatory system is "open" because the circulating fluid (called "hemolymph") freely washes through open cavities in the body, bathing organs with oxygen-rich fluids. Eventually, through the pumping of the simple heart and contractions of the muscles, the fluid is drawn through pores and funneled back into the heart. This works for arthropods (such as insects and crustaceans) and some mollusks (such as clams and snails). It is not very efficient, however, because the oxygen-rich and oxygen-poor blood can mix in the body. This just isn't going to cut it for more active and physiologically complex critters . . . which is why natural selection has resulted in the second option: the closed circulatory system.

Closed Circulatory System

Consider for a moment the glorious octopus. Like the clams and the snails mentioned above, it is a member of Phylum Mollusca. But unlike all the other mollusks, members of Class Cephalopoda (octopi, squids, and cuttlefish) are graced with a closed circulatory system. Octopi and others of their class are avid hunters who need to quickly and nimbly chase down their prey. Such rapid movement requires a lot of energy, more than the inefficient open circulatory system their peers can provide. This evolutionary event is a perfect illustration of the benefits of a closed circulatory system—its increased efficiency leads to much greater energy production.

A closed circulatory system is composed of a heart, blood, and vessels (arteries, veins, and capillaries) through which the blood flows. Because the blood is confined to vessels, it results in higher blood pressure, necessary for effective oxygen delivery in large and active species. The heart pumps blood into vessels that decrease in size, increase in branching, and supply blood to all the tissues of the body. Along with most cephalopods, annelids and all vertebrates benefit from a closed circulatory system.

Vertebrate Circulatory System

Our circulatory system is ridiculous . . . the total length of all vessels would circle Earth. Twice. Along with the muscle behind the movement (the heart), we have arteries to move blood away from the heart, and as they branch off into smaller arterioles, they provide pathways to all the organs in our body. Eventually the branching reaches maximum branchy-ness as tiny capillaries, the site of gas transfer between the blood and the cells of the body tissues. As suggested by our **must know** concept, the movement of the gases between the cells and the capillaries is based on concentration gradients. The blood then makes its way back to the heart as capillaries fuse into venules and finally the larger veins.

There are different circulation pathways, depending where on the chain of evolutionary events you stand. Fish, for example, are the first vertebrates to evolve, and they possess the simplest option of a two-chambered heart powering a single circulation pathway.

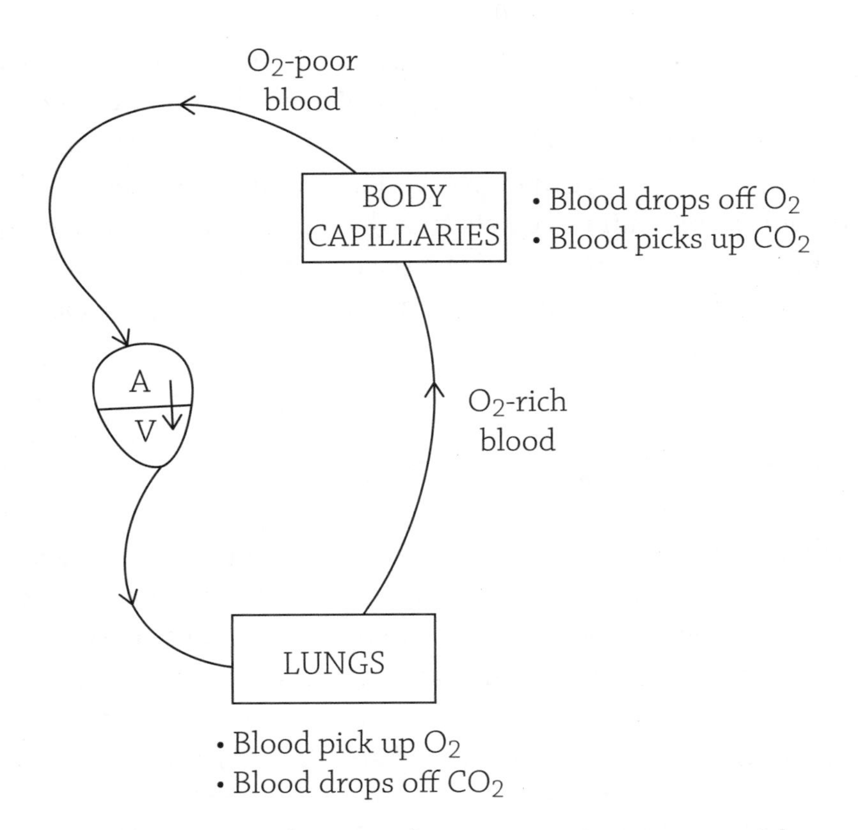

Single-circuit circulation pathway. A = atrium; V = ventricle

Blood passes through the heart only once per circuit, as it is powered out to the gills to pick up oxygen and release carbon dioxide. Then it travels on to the body tissues to drop off that oxygen (and pick up waste carbon dioxide). Blood pressure is pretty low at this point, since it traveled the entire circuit with a single pump of the heart. When the blood returns, it collects in the atrium and is pumped out through the ventricle for another loop. Once evolution took that great step from water onto land, there needed to be an option with a bit more power. Therefore, the amphibians of today (the closest vertebrate cousin to the fish) and reptiles (one more evolutionary step onto land) have a new-and-improved double-circulation pathway with a three-chambered heart. This provides better blood flow because the heart repressurizes the blood before it goes to the body tissues and again when it travels to the lungs. Yet there is still some mixing of oxygenated and deoxygenated blood in that single ventricle:

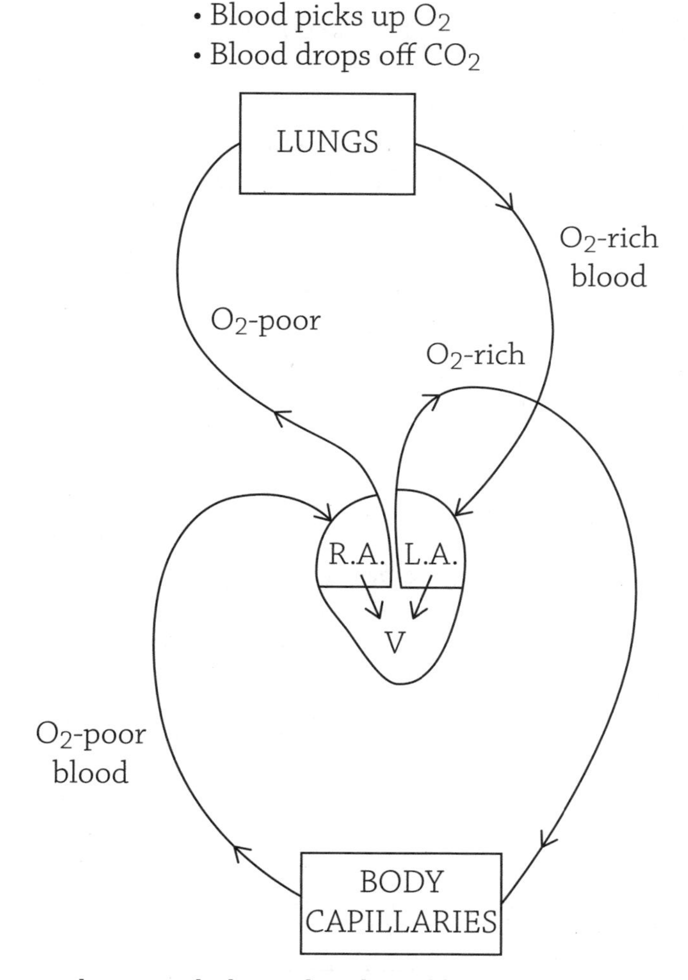

Double circulation with three-chambered heart. R.A. = right atrium; L.A. = left atrium; V = ventricle

Once we get to endotherms such as birds and mammals, we see the most efficient of systems: double circulation with a four-chambered heart. As it was for amphibians, there is a circuit that pumps blood to the lungs, and another circuit that pumps blood to the rest of the body. This keeps blood pressure strong, even when traveling to the farthest reaches of your toes. Now, however, mammals and birds have a four-chambered heart, with the right and left sides separated by a muscular wall called a **septum**. This prevents the oxygen-rich blood from the pulmonary circuit mixing with the oxygen-poor blood of the systemic circuit.

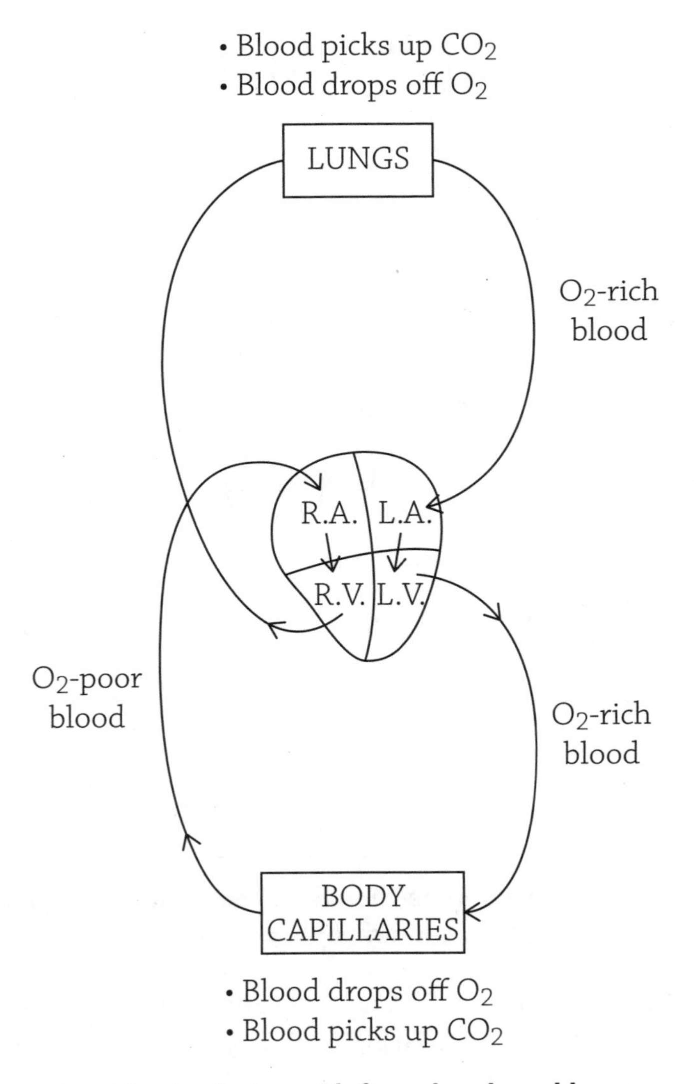

Double circulation with four-chambered heart

Since mammals and birds use about 10 times as much energy as their ectothermic cousins, it is necessary to deliver much more fuel and oxygen to the tissues.

Mammalian Details

The heart provides all the force needed to propel blood through the entire body. The atria receive the blood from the body: the left atrium gets the freshly oxygenated blood from the lungs, and the right atrium gets the spent, oxygen-poor blood from the body tissues. Once in the small atria, the blood passes through one-way valves into the larger ventricles, which then pump it away from the heart out to either the lungs (right ventricle's pulmonary circuit) or the rest of the body (left ventricle's systemic circuit). Since the blood has to travel much farther through the systemic circuit, it makes sense that the walls of the left ventricle are thicker and stronger.

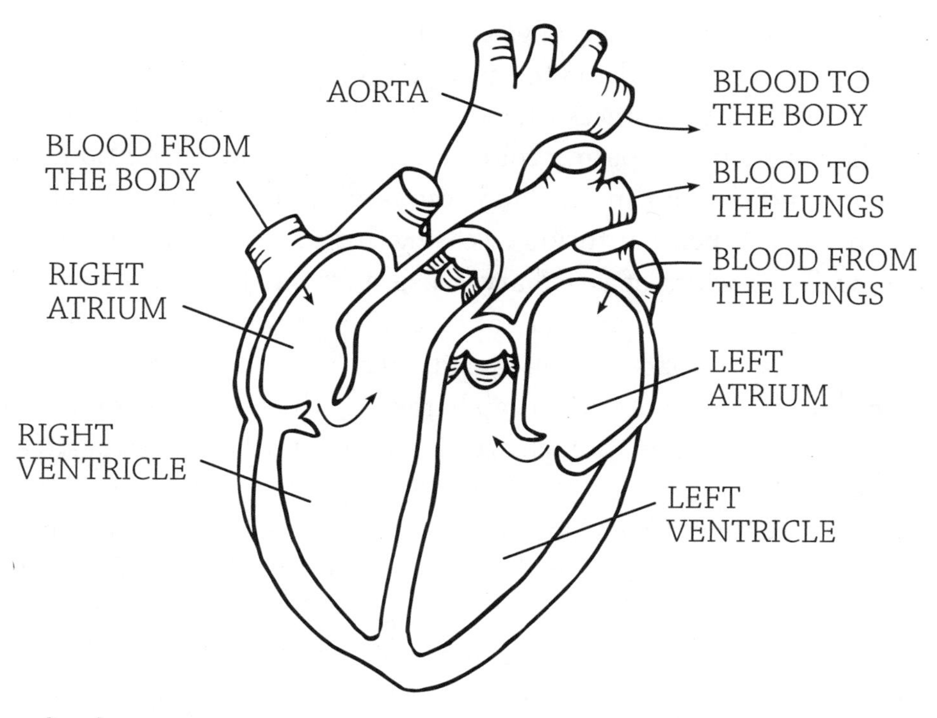

Mammalian heart anatomy

Author: Firkin. https://openclipart.org/detail/274066/heart-diagram-annotated

Blood pressure is a measurement of the force the blood exerts on the walls of the blood vessels; **systolic** pressure is the higher number (the force when the heart contracts) and **diastolic** pressure is the lower number (the pressure when the heart is at rest).

IRL Even though it's depicted as blue, deoxygenated blood is not actually that color.

Arteries take blood away from the heart. Considering the heart is the source of all blood pressure, the arteries must be strong to withstand the force of the beating heart. The walls of arteries tend to be thick and muscular. The strongest artery of the body is the one closest to the heart: the **aorta**. **Veins**, on the other hand, bring blood back to the heart after gas exchange occurs. Their walls are floppy and thin, since the pumping force of the heart is so far away. Blood pressure in some veins can be so low that there are even one-way valves to prevent backflow!

Capillaries are the location of gas exchange (both oxygen and carbon dioxide) between the blood and the surrounding environment. Gas diffusion works best over close distances when the blood is traveling at slow speed, and capillary structure is perfect for creating both of these conditions.

The walls of capillaries are thin and porous (only one endothelial cell thick!). This greatly helps diffusion by easing the passage of gases through the capillary wall. The width of a capillary is also surprisingly narrow. In some locations, red blood cells must squeeze through in single file. The width of a single red blood cell is about 7 μm; capillary diameters range from 5 to 10 μm.

A **capillary bed** is a network of capillaries and the site for gas exchange in tissues. As blood enters through arterioles, the vessel branches again and again into the tiny capillaries. This leads to a huge increase in total cross-sectional area through which blood flows. The increase in area results in a drop in blood pressure; the blood flow slows down significantly, allowing time for diffusion to occur. This is a perfect example of form helping function!

Oxygen is ferried around the circulatory system by red blood cells. These cells are chock full of a chemical pigment whose sole job is to hang onto the oxygen until it is needed in the body. The pigment molecule responsible for grabbing onto oxygen is **hemoglobin**. It is composed of four subunits, each able to snag an oxygen molecule from the alveolar air. The cool thing about hemoglobin, it relates back to a concept we learned way back in Chapter 1. Remember allosteric enzymes? They had multiple subunits, and they could change their shapes into an active form or an inactive form. Well, hemoglobin is an example of a specific kind of allosteric regulation called **cooperativity**. Here, the binding of an oxygen molecule at one subunit will cause a shape change in the other subunits, making them even *better* at grabbing oxygen. That means if there's oxygen waiting to be picked up, it's going to happen quickly. And the opposite is true, too. When the blood reaches an area of the body that is low in oxygen, one molecule will break free and diffuse out of the blood into the oxygen-starved tissues, and all the other subunits will revert back to the shape that isn't as good at holding onto oxygen. That way, when it is time to unload oxygen, it happens all at once.

BTW

Keep in mind the big picture. Why, exactly, do all the cells in your body need oxygen? Because cells need oxygen to produce ATP (the process of cellular respiration):

$C_6H_{12}O_6 + 6CO_2 \rightarrow 6CO_2 + 6H_2O +$ energy (ATP)

Respiratory System

At its most basic, the respiratory system supplies oxygen to the cells of the body. The specifics of how the respiratory system managed to funnel oxygen throughout the body is different, depending on the critter. Those flatworms and sea jellies, with their high surface areas, can manage without a specific respiratory system and rely on diffusion directly from the environment. Fish have gills, highly vascular tissues that absorb oxygen from the surrounding water. Insects have these funky internal tubes that run throughout their bodies and manage to supply every cell with air; the tubes themselves are connected to openings spaced along the surface of the bug's body. Spiders have book lungs, fascinating structures that are so highly folded, it looks like the pages from a book. Yet as we did in the circulatory system, we will focus on the mammalian model (though all of these examples are fascinating!).

Mammalian Respiratory System

Gas exchange occurs across respiratory surfaces, where the circulatory system crosses paths with the respiratory system. Our **must know** concept is the same for the respiratory system as it was for the circulatory system: movement of respiratory gases is based on concentration gradients. In order to help diffusion of gases (oxygen and carbon dioxide) occur, the respiratory surfaces must be moist and have a high surface area. In mammals, this occurs at the dead-end passages of the lungs, tiny air sacs called **alveoli**:

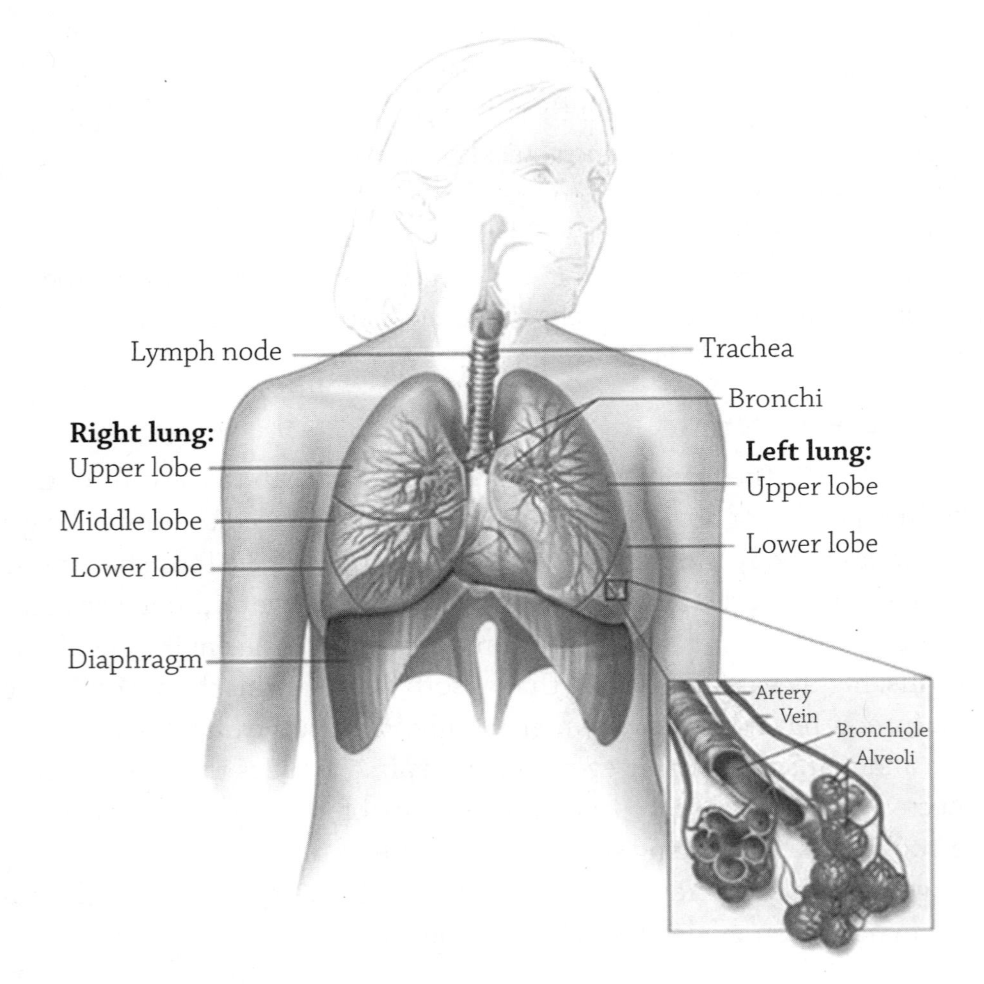

Alveoli of the respiratory tract

Author: National Cancer Institute (NCI). https://visualsonline.cancer.gov/details.cfm?imageid=7235

Human lungs contain literally millions of alveoli, and the total surface area is about 50 times that of our skin (or, more disturbingly, the size of a tennis court). Oxygen from the air entering the alveoli quickly diffuses across the epithelia into the capillaries surrounding each alveolus, where it's whisked away to body tissues in need of oxygen. Meanwhile, the waste carbon dioxide moves from the capillaries and into the alveoli airspace, where it will be exhaled and removed from the body. The way to determine the direction gases will diffuse is through something called **partial pressure of gases**. The more gas in a mixture of gases, the higher its partial pressure. A gas will diffuse from a region of high partial pressure to low partial pressure (once again, our **must know** concept).

BTW

This sounds a lot like other diffusion scenarios. Diffusion, in general, is movement of something from a region of high concentration to low concentration. Osmosis (diffusion of water) occurs from a region of high water potential to low water potential. Diffusion of gases occurs from a region of high partial pressure to low partial pressure.

The carbon dioxide and oxygen partial pressures in the body vary, depending on the location. For example, blood that just dropped off oxygen to body cells and picked up the cells' waste carbon dioxide will have a higher CO_2 partial pressure and a lower O_2 partial pressure. The opposite partial pressures would occur in the blood that just entered the tissues of the body, carrying with it a fresh supply of oxygen; here it would have a high O_2 partial pressure and a low CO_2 partial pressure. The pressures of each gas determine the directions they will move. Here are two examples:

- **Scenario 1** A capillary bed supplying body tissues with oxygen

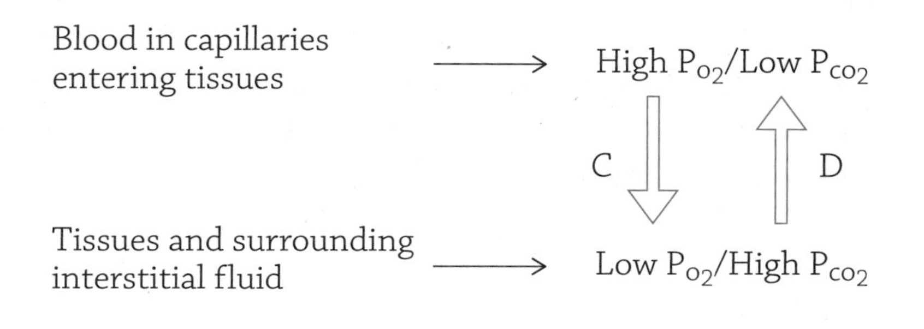

This is a model of partial pressures of gases in the capillaries supplying body tissues with oxygen-rich blood. The tissues, meanwhile, have been producing carbon dioxide as a waste product of cellular respiration:

A. The partial pressure of oxygen (PO_2) is higher in the freshly oxygenated blood compared to the PO_2 in the tissues; oxygen will diffuse from the bloodstream into the surrounding tissues.

B. The blood PCO_2 is low because blood just left the alveoli where it dumped of excess carbon dioxide to be exhaled; the carbon dioxide will now diffuse from the tissues into the bloodstream, so it can travel back to the lungs.

- **Scenario 2** Blood supply arriving at the lungs to pick up oxygen and dump off carbon dioxide:

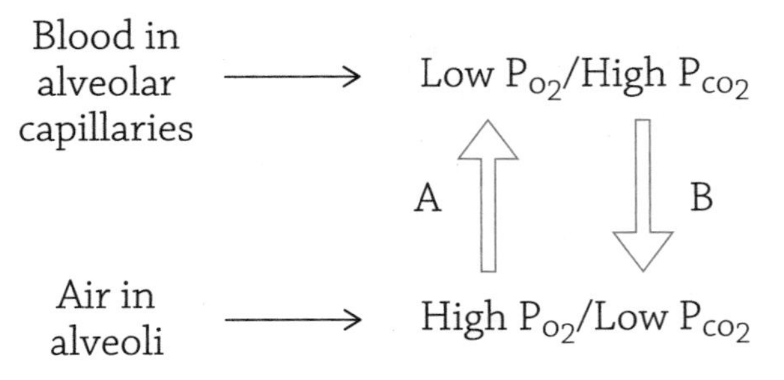

This shows what happens when a fresh inhalation of air settles into the alveoli.

A. The partial pressure of oxygen (PO_2) is lower in the blood compared to the PO_2 in the air; oxygen will diffuse from the air into the bloodstream.

B. The blood PCO_2 is high because it just returned from picking up waste CO_2 from body tissues; the carbon dioxide will diffuse from the bloodstream into the alveoli.

REVIEW QUESTIONS

1. In order for an organism to rely on direct diffusion for oxygen transport (without the aid of a circulatory system), what quality must its body plan have?

2. Why is a four-chambered heart more efficient than a three-chambered heart?

3. The place of gas exchange between the respiratory system and the circulatory system occurs in the ____________ of the respiratory system and the surrounding ____________ of the circulatory system.

4. Where would the partial pressure of oxygen be higher than the partial pressure of carbon dioxide?

5. How did surface area play a role in eukaryotic cell evolution?

6. The mammalian ____________ ventricle has a thicker wall than the ____________ ventricle. The ____________ ventricle needs to be stronger because it powers the blood through the entire body (the ____________ circuit). The ____________ ventricle only needs to power the blood out to the ____________ (the ____________ circuit).

7. How are the circulatory system and the respiratory system related in their functions?

8. Choose the correct term from each of the following pairs: Blood coming from the lungs and entering the tissues has a high **oxygen/carbon dioxide** partial pressure. The tissues have a high **oxygen/carbon dioxide** partial pressure because the cells have been producing a lot of **oxygen/carbon dioxide** through the process of cellular respiration. These partial pressures facilitate the movement of oxygen from the **tissues/blood** into the **tissues/blood**.

9. Which of the following helps facilitate the movement and transport of gases in animal bodies?
   a. When a cell has a high surface area to volume ratio
   b. Having equal concentrations of a gas on both sides of a cell membrane
   c. Arteries have thick and muscular walls
   d. A three-chambered heart allows mixing of oxygenated and deoxygenated blood

Animal Neurons and Signal Transduction

MUST KNOW

- Neuron function is based on chemicals diffusing down their concentration gradients.
- A signal passes down a single neuron as a positively charged wave.
- A signal passes between adjacent neurons as a chemical called a neurotransmitter.

Our nervous system is based on specialized cells called **neurons**. These cells are awaiting an input from another neuron or from a specific stimulus (such as stepping on a wall tack you just dropped on the floor and couldn't locate . . . until now). As soon as the pointy end of the offending office product jabbed into your heel, you jerked your leg up and pulled it out. This coordinated series of actions was made possible by neurons. The cool thing is, the number of neurons needed to move a signal around your body is staggeringly large; each individual cell is between 4 μm (micrometer) and 100 μm long. The brain alone contains something along the lines of 100 billion of these cells. The path a signal must take between the area of input (your foot), up to your brain, and then back to the area of response (get that tack outta my foot!) is huge. To make the journey, nervous signals must travel along an individual neuron, and then jump between adjacent neurons.

Just for your information, a micrometer is one-millionth of a meter. A 4 μm neuron is 0.004 millimeters; a human hair is huge in comparison . . . it's 75 μm wide.

The method by which a single neuron moves a signal down its own cell body is different from how the signal is then shuttled between adjacent neurons. Either way, as our **must know** concept states, the signals are nothing more than chemicals diffusing down their concentration gradients. First, let's talk about how a single neuron responds to a stimulus—action potentials!

Action Potential

This is the general structure of a neuron:

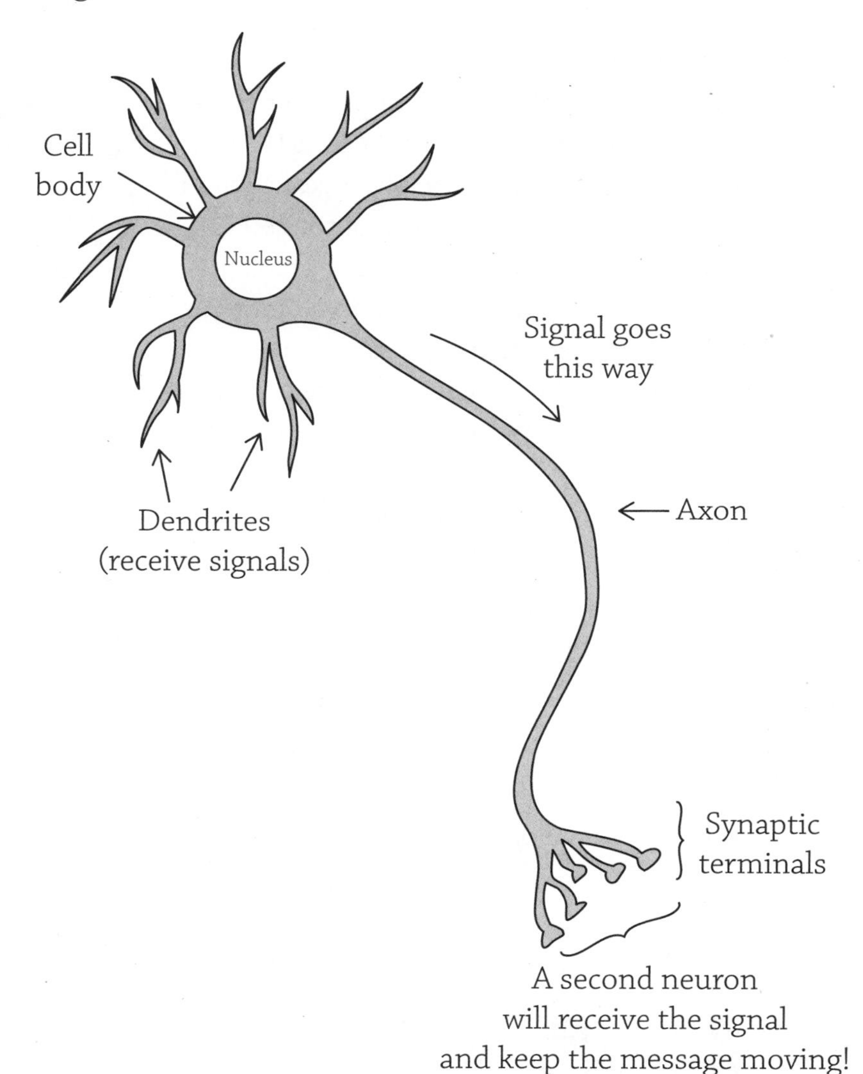

Neurons send signals in one direction: from the dendrites, down the axon, ending at the synaptic terminals. Within a single neuron, this signal takes the form of a ~whoosh~ of positive ions diffusing along the axon of the cell. A cell that is "resting" is not actively sending a signal; there is no positive wave occurring. Therefore, the inside of a neuron at rest has a more negative charge compared to its surroundings. This difference in charge is created by the cell itself, using a special membrane pump called the **sodium-potassium pump**.

Whenever there is a mention of a "pump," it should make you think of active transport. A pump requires energy and moves something against its concentration gradient (meaning it creates a high concentration of something on one side of the cell's membrane).

In the neuron, the sodium-potassium pump moves three sodium ions (Na^+) out for every two potassium ions (K^+) it brings into the cell (part A in the figure below). Because more positively charged ions are being pumped out of the cell, it results in a net positive charge on the outside of the cell. Furthermore, there are these big negatively charged proteins stuck inside the cell, further contributing to the charge differential (more positive on the outside and more negative on the inside). Because of this difference in charge, the resting neuron is now ready to fire!

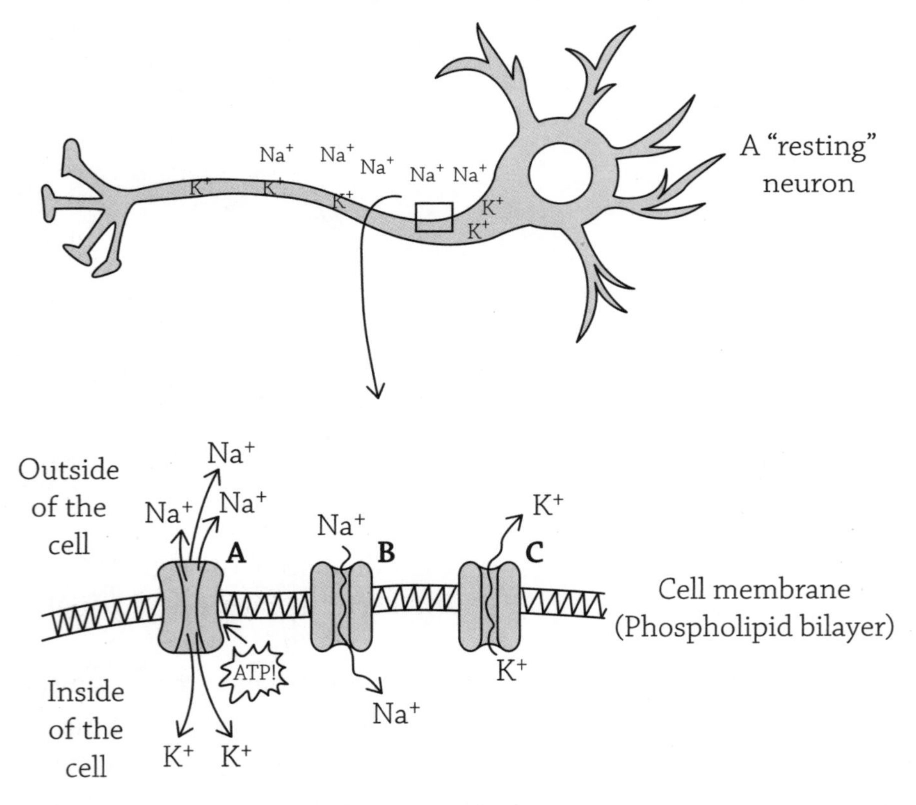

A resting neuron

The ability of a neuron to send a message all depends on this charge differential; it is called a **membrane potential**. When a neuron fires (sending the signal along its axon), it does so by opening up doorways and allowing the sodium ions to diffuse down their concentration gradient, back into the interior of the cell (step A in the figure below). Remember, our **must know** states that a neuron's function is based on diffusion of chemicals!

The special doorways that offer a passive pathway through which the sodium ions can diffuse are called **sodium channels**. This is the beginning of the positive wave and our neuron's **action potential**. The signal is moved down the length of the neuron as a sort of chain reaction; a positively charged interior at one section of the axon (B) causes more sodium channels (C) a bit farther down to open. Sodium ions enter (E) and turn that section positive, causing more sodium channels to open, and so on. This creates the ~whoosh~ of positive charge that's moving down the axon toward the synaptic terminals.

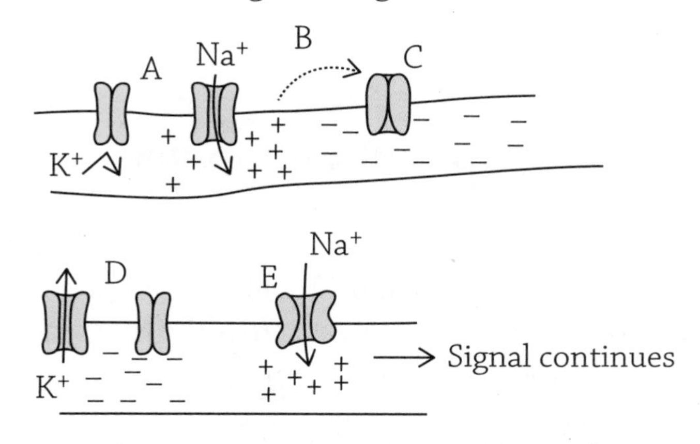

A neuron creating an action potential

IRL Puffer fish is a delicacy in Japan (*fugu*), though it is dangerous to eat unless prepared by specially trained chefs. The fish contains a toxin (tetrodotoxin) that interferes with sodium channel function if ingested. Tetrodotoxin binds to neurons' sodium channels and blocks the diffusion of sodium ions. If sodium ions cannot diffuse into the cell, the neuron cannot fire!

In the wake of this positive charge, a second type of doorway—the potassium channel—opens, allowing K^+ ions to flow *out* of the cell (part D in the previous figure). The potassium ions happily diffuse out of the cell, down their concentration gradient. In so doing, positive charges are removed from the inside of the cell, returning the inside to its "normal" resting negative charge.

There's one final thing the cell needs to do in order to fire again. Even though the charge differential is correct at this point (more negative on the inside; more positive on the outside), things aren't quite right . . . the sodium and potassium ions are on the wrong sides of the membrane! This brings us back to the start of the story: the sodium-potassium pump takes over, and with the help of energy (ATP), it pushes the sodium ions back out and pulls the potassium ions back in. The resting conditions are reestablished and the neuron is once again ready to send a message.

But what happens when the positive wave hits the end of the synaptic terminal? How do neurons communicate with each other? Let's move on to neurotransmitters and synapses.

Neurotransmitters and Synapses

When a single neuron fires and sends its message, it is in the form of a positive charge moving down the axon. This electrical charge, however, cannot jump from one neuron to the next across the tiny space (**synapse**) separating adjacent neurons. Instead, a chemical needs to diffuse across the synapse to move the signal from one neuron to the next . . . and it reinforces the importance of our **must know** for this chapter. The chemical responsible for moving the signal from one neuron (presynaptic cell) to the next (the postsynaptic cell) is called a **neurotransmitter**.

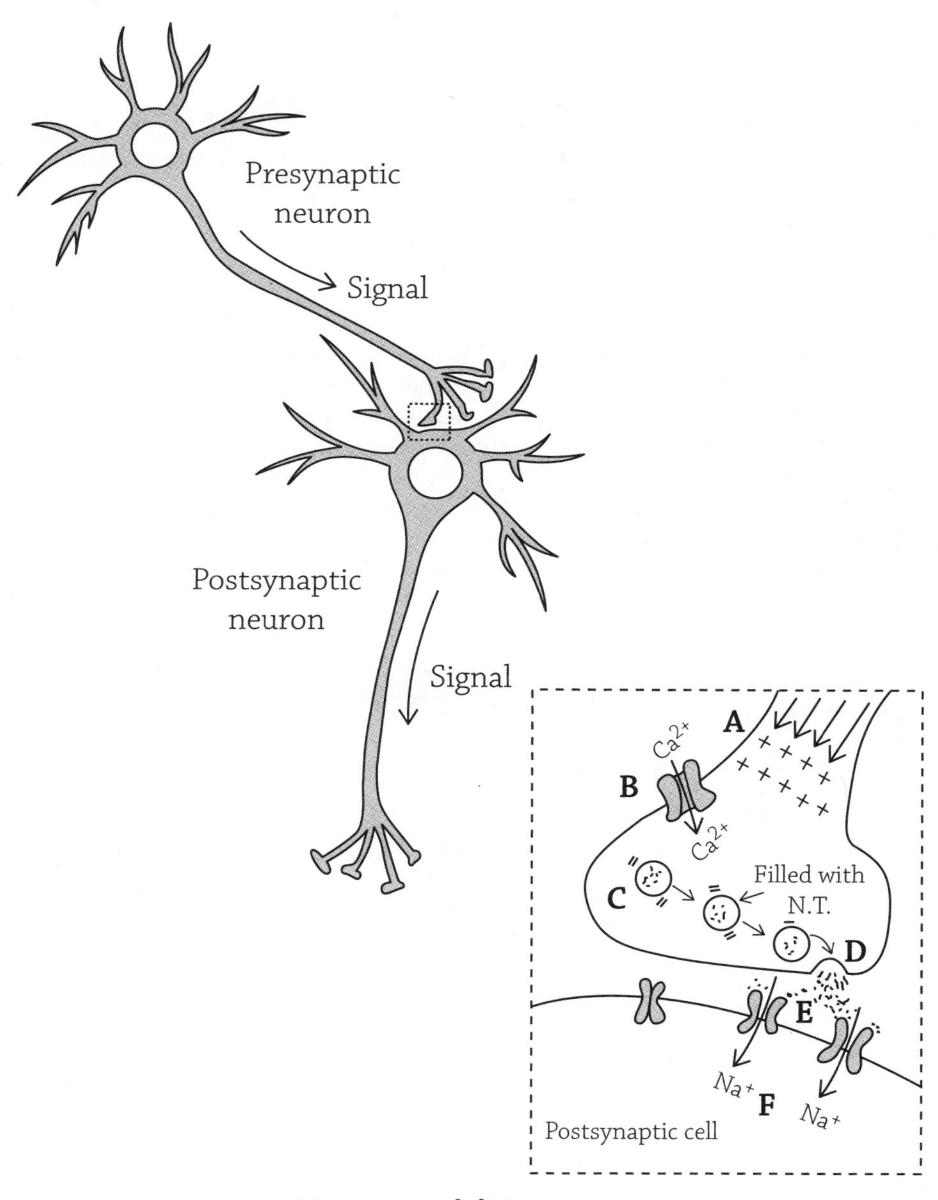

Neurons and their synapse

When the action potential hits the end of the first neuron (A), it signals for calcium channels to open (B), allowing calcium ions to flood inward. The high levels of Ca^{2+} ions cause vesicles filled with neurotransmitters (C) to fuse with the presynaptic neuron's membrane, dumping the neurotransmitter into the synapse (D). The chemical neurotransmitters are able to diffuse across the space between the two neurons until they bind to another sodium channel protein (E) in the postsynaptic neuron. The postsynaptic cell's sodium channel swings open, allowing sodium to diffuse through (F), starting another action potential in the second neuron.

IRL

The bacterium *Clostridium botulinum* is responsible for producing one of the most potent natural toxins on the planet: botulinum toxin. The botulinum toxin inhibits presynaptic release of a specific neurotransmitter called acetylcholine. If a muscle doesn't receive acetylcholine's signal, the muscle will not respond properly by contracting. Botulism is a deadly food poisoning caused when someone accidentally ingests food containing botulinum toxin; BOTOX is an anti-wrinkle treatment where someone intentionally injects tiny amounts of botulinum toxin into their facial muscles.

EXTRA HELP

Back in Chapter 8 you learned about signal transduction. Here is a perfect example of how a signal molecule (the neurotransmitter) can elicit a response in the receiving cell without itself passing into the cell. The neuortransmitter binds to the postsynaptic neuron and the sodium channels open, allowing sodium ions to rush into the postsynaptic neuron (the response).

REVIEW QUESTIONS

1. Choose the correct answer from each of the following pairs: A resting neuron has a higher concentration of **K^+/Na^+** on the inside of the cell and a higher concentration of **K^+/Na^+** on the outside of the cell. The net charge on the inside of the cell is **positive/negative**.

2. A neuron transmits a signal by generating an ____________, which is a wave of ____________ charge generated when ____________ ions rush into the cell.

3. Put the following events of a signal crossing a synapse in the correct order:
   a. Neurotransmitter binds to the postsynaptic cell's membrane
   b. Calcium channels open
   c. Neurotransmitter is dumped into the synapse
   d. Vesicles containing neurotransmitter fuse to the presynaptic neuron's membrane
   e. Action potential hits the end of the neuron
   f. Sodium channels on postsynaptic cell open

4. Explain how a resting neuron maintains a net negative charge on the inside of the cell.

5. Put the following events of a neuron firing in the correct order (this is a general overview of the process; not all the steps are included in this list):
   a. K^+ channels open
   b. Sodium ions diffuse into the cell
   c. Sodium-potassium pumps move ions into their starting positions
   d. The positive charge inside the neuron causes adjacent sodium channels to open
   e. Potassium ions diffuse out of the cell
   f. Na^+ channels open
   g. A neuron receives a signal to fire and send a message

6. When an action potential reaches the end of the __________ neuron, it must rely on the chemicals called __________ to cross the space between adjacent neurons (the __________). Once the chemicals bind to the __________ neuron, sodium channels will open and the action potential will continue.

7. If a bacterial toxin interfered with vesicle movement and fusion within a neuron's synaptic terminal, which function would be inhibited?
   a. Opening of sodium gates
   b. Operation of sodium-potassium pumps
   c. Action potential production
   d. Neurotransmitter release

The Animal Immune System

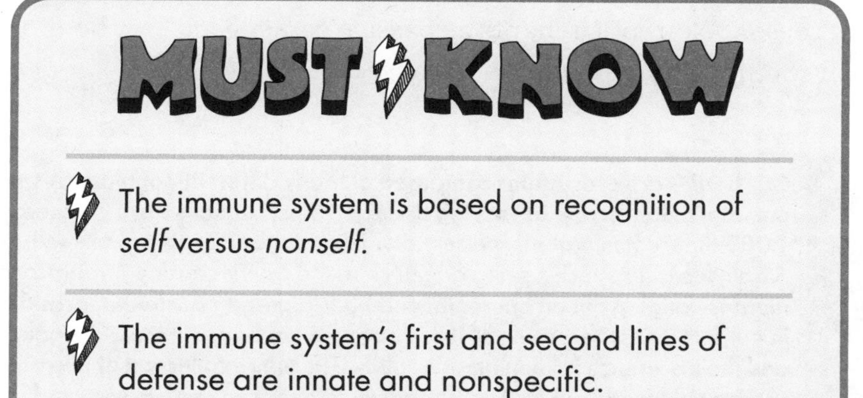

We are constantly waging a battle against foreign invaders. If your body has the upper hand, you don't even know the battle is being fought; if you are losing the war, you'll know: fever, aches, runny nose, and maybe even some swelling. Your symptoms during sickness are by-products of your immune system revving up to full speed in order to destroy the invading pathogens. Your **must know** is to understand the immune system wages a you-versus-not-you battle, immunologically speaking.

You are a happy collection of cells: obviously your own cells, but also all the bacterial cells that are colonizing the inside and outside of you. Do not discount these prokaryotic tenants as passive passengers—without our microbial community, we could not survive.

IRL

Our bodies are ecosystems composed of many different populations spread out upon a landscape of varying tissue terrains. Humans have on average 1,000 different microbial species colonizing our bodies. There are salt-tolerant *Staphylococcus* species living on our skin, the cavity-causing *Streptococcus mutans* hanging out on our teeth, and fiber-digesting *Bacteroides* residing in our intestines. Yes, some of these bacteria are bad, but most are good, and their presence is keeping us healthy. The entire collection of microbial species colonizing our bodies is referred to as the microbiome, and research is continuing to elucidate how important our microbiomes are to our general health. Some bacteria, for example, provide a direct benefit, such as *Lactobacillus* providing B vitamins in our large intestine.

Oftentimes, the benefits aren't as obvious (but are just as important). When we have these good bacteria taking up residence on our bodies, they are effectively preventing bad bacteria from moving into the neighborhood. As you will soon learn, ecological niches can only be inhabited by a single species; we want friendly species inhabiting our bodies' niches.

Unfortunately, you often don't realize the benefit until it's gone. If you have ever taken antibiotics in order to fight an infection, you might have suffered some side effects such as diarrhea and cramping. This occurred because the antibiotics also wiped out some of your good gut bacteria. The collateral damage of removing a happy inhabitant opened up a niche that can now be occupied by the not-so-welcomed. An extreme example is an infection of *Clostridium difficile*, a bacterial species that is an opportunistic invader of the colon, and can cause severe diarrhea or even life-threatening inflammation.

Your body has multiple layers of defense, essentially protecting your insides from the outside. We will go over each of these, as if a foreign invader is trying to get into your body. First, it will push against a sweaty, acidic wall (your outside layer), then it will break through to be surrounded by a swarm of hungry eating machines (second line of defense), and finally it will face a tailor-made army of specific cells designed for this particular pathogen (third line of defense).

The Immune System's Three Lines of Defense

Innate Immunity		Adaptive Immunity
First line of defense: external	Second line of defense: internal	Third line of defense: internal
Skin and secretions block invaders from entering	Non-picky phagocytic cells eat anything "Nonself" that broke through the first line of defense	Specific cells are called to battle the "Nonself" invasion • Humoral response (B cells) • Cells mediated response (T cells)

First Line of Defense: Outer Barriers

The first line of defense is the very barrier that separates you from your surroundings: your skin! The epidermal layer is a watertight wall that prevents things from easily passing through. Not only does it physically block things from entering, it also creates a hostile chemical environment by secreting acidic and salty sweat. Even the bits that seem like easy ways to sneak past the wall (such as your eyes and mouth) have their own defensive abilities: both tears and saliva contain the enzyme **lysozyme**, which destroys bacterial cell walls. This is a passive and nonspecific response to invasion because it need not be triggered for each specific infection—it's ready to do its job, regardless of the type of pathogen.

Second Line of Defense: Phagocytes and Inflammation

Imagine you accidentally cut your hand against a tree branch. After a bit of light bleeding and mild pain, the wound becomes swollen, red, and tender. Welcome to the second line of defense! Something got past your skin (easy to do if there is a cut), and the invader is met by a team of specialized cells called **phagocytes**. A phagocytic cell is a cell of the immune system that ingests and breaks down bacteria and other foreign substances. There are specific types of phagocytic cells called **macrophages** and **neutrophils**. The pain/heat/swelling of an injury is all supposed to help these phagocytes swoop into the injury site. When tissues are damaged (the cut), a chemical called **histamine** is released, causing the capillaries in the vicinity to dilate (increased blood flow causes the heat) and leak fluids into the surrounding tissue (the swelling). This brings in the phagocytes and allows them to sneak through the capillary walls into the tissues, where the infection lies. As it was in the first line of defense, these phagocytes act innately and quickly, regardless of the infectious agent. For this chapter, our second **must know** refers back to what we just learned: the first and second lines of defense are innate and nonspecific.

Third Line of Defense: Adaptive Immunity

Now things get interesting. Mammals have this amazing thing called **adaptive immunity**, where the body mounts an immune response when challenged by outside invaders. Keeping our **must know** concepts in mind, if the first and second lines of defense are both innate (work immediately) and nonspecific (doesn't matter the pathogen), the third line of defense is specific (the type of pathogen matters) and needs to first learn about the pathogen before it can mount a full offensive response (it is *not* innate and immediate). Before I get into the details, let's first learn some helpful vocabulary. Our immune systems are based on this idea of self versus nonself, right? We call any potentially pathogenic nonself invader an **antigen**. To be even more specific, an antigen does not need to be an entire, intact bacterium or virus; it can be a *part* of a pathogen, such as a specific viral protein or piece of bacterial cell membrane. This idea will be important later, when we talk about vaccinations. If an antigen is the bad guy, your **lymphocytes** are the good guys. Lymphocytes are special immune system cells (commonly called **white blood cells**), of which there are two types: **B cells** and **T cells**. The B cells are responsible for the immune system's **humoral response** (targeting antigens that are freely floating around in the bloodstream), and the T cells are in charge of the immune system's **cell-mediated response** (targeting invasions that directly involve your own body cells).

Before we look at the humoral and cell-mediated responses in more detail, I want to first provide a big-picture overview for you. Yeah, okay, the immune response is *specific* and has *memory*. What exactly does that mean? Well, the two concepts actually go hand in hand. When a specific pathogen enters your body, only one circulating B or T cell is equipped to deal with it. Seriously, it's mind-blowing. When each of your lymphocytes are created, there is a slight (intentional) difference in their cell membrane surface receptors. This means each lymphocyte can latch on to a unique antigen:

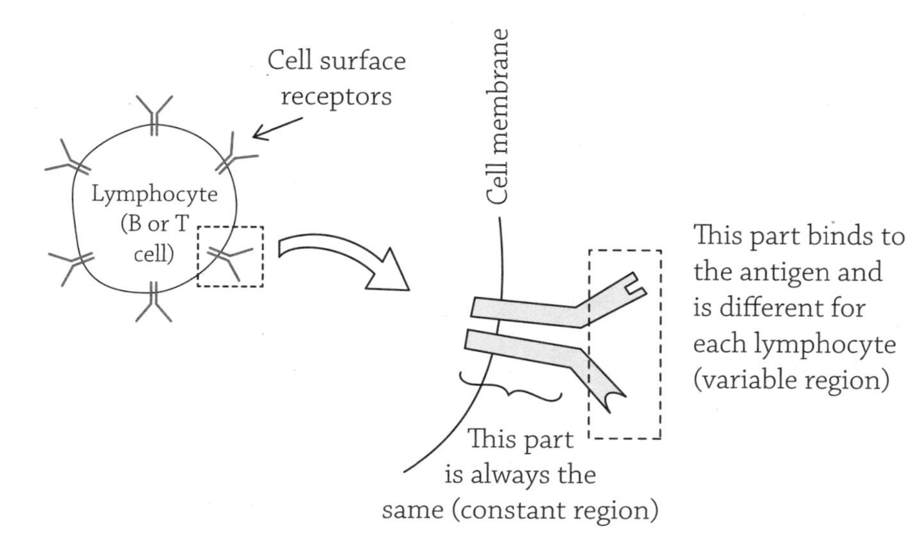

Variable and constant regions of lymphocyte cell surface receptors

That's the "specific" part of your immune system. There is a countless number of variations in lymphocytes in order to match the countless number of different antigens that may assault your body. But clearly, a single lymphocyte isn't enough to counteract an invasion. That one special lymphocyte, however, immediately begins to copy itself once it latches on to the pathogen; the antigen-lymphocyte contact is like a call to arms. The "chosen" lymphocyte now knows there is an invasion occurring, and it is the only one able to stop it! By **cloning** itself (copying itself over and over again), it creates an army of cells that all match the invading antigen.

So here is where the *memory* part comes into play: once the invading pathogen is destroyed and the threat is removed, the numbers of lymphocytes specific for that particular pathogen *never completely go away*. There remains a troupe of **memory cells**, just in case that one pathogen tries its luck again in a second invasion. If it does invade a second time, it won't have a chance . . . those memory cells are waiting, clone themselves immediately, and wipe out the nonself invader before you even have a chance to feel sick!

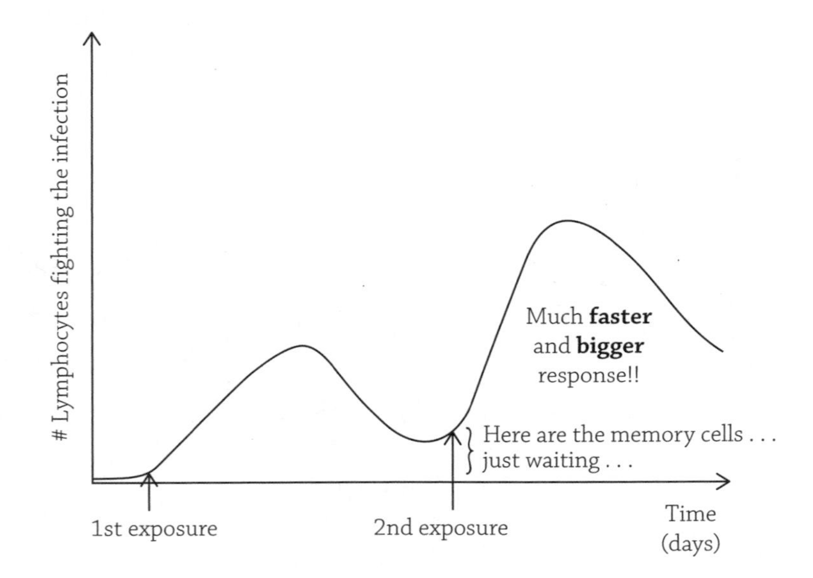

Primary and (faster) secondary immune responses to the same pathogen

You can see that the heightened secondary response is due to those memory cells, just waiting for a second chance to fight the infection. Keep this in mind when we consider the science behind vaccinations, later on in the chapter. Also keep in mind our third **must know**: the third line of defense is specific and has memory.

B Cells and the Humoral Response

As I mentioned earlier, the humoral response and the cell-mediated response target slightly different enemies. First, let's talk about the humoral response. This is the immune system's reaction to extracellular pathogens (*extracellular* means *outside of a cell*). As we learned, a lucky B cell will be activated because its cell membrane receptor perfectly matches the offending antigen. Once the B cell latches on to the antigen, it will begin to clone itself, and each B cell will begin to secrete these tiny proteins called **antibodies** that look just like the B cell's membrane receptor. These little antibodies take down the infection by binding to the virus (or toxin, or bacterium), causing them to clump up into easy-to-eat aggregates. This entices the phagocytic cell (remember that guy? From the second line of defense?) to gobble up the pathogen. The B cells that secreted all those helpful antibodies are called **plasma cells**; there are other B cells that will abide their time and wait for a second exposure to that same pathogen. They comprise the memory part of our immune system and are called **memory B cells**.

BTW

Take note of the fact that these B cells make direct contact with the antigen (the little black triangle in the illustration on the next page). This is possible because the humoral response focuses on extracellular pathogens, floating around freely in the body fluids, not trying to hide in your own body cells. The sneaky hidden pathogens require a different approach. Call in the cell-mediated immune response!

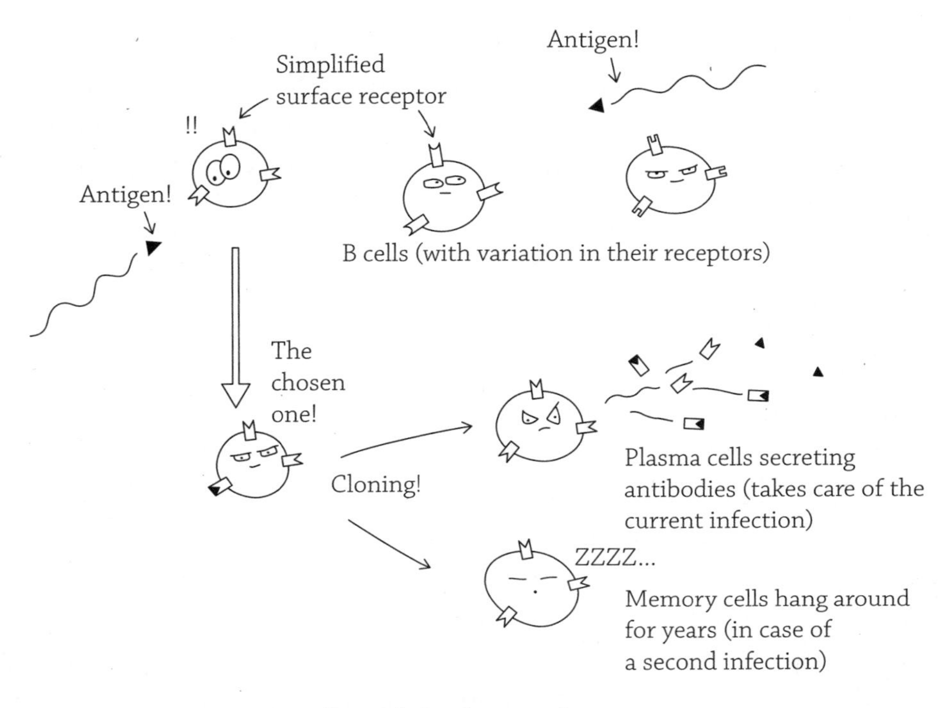

B cells and the humoral response

T Cells and the Cell-Mediated Response

Oftentimes, a viral infection is "hidden" in your body. Viruses are obligate intracellular parasites, meaning they cannot reproduce unless they invade and commandeer one of your body cells (how rude). Luckily, T cells are able to deal with this sneaky sort of invasion. It requires the help of the infected cell in which the antigen is concealed. A T cell (specifically called a **cytotoxic T cell**, or **T_c**), is called to arms if an infected body cell raises an alarm. Specifically, the infected body cell will take a fragment of the virus and combine it with a special protein that identifies "self," called **MHC**.

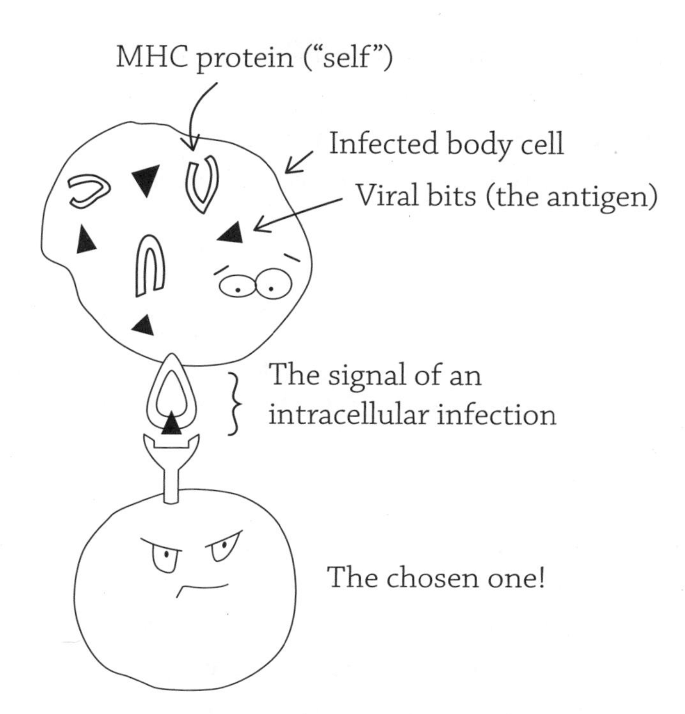

T cells and the cell-mediated response

BTW

Everybody has a unique chemical marker on the surface of their body cells. This marker—called a major histocompatibility complex (or MHC for short)—is how our immune system can identify self from nonself. Interestingly enough, everyone's unique MHC also contributes to their body odor (and may play a subconscious role in mate preference).

When a roving T_c comes into contact with a body cell holding out the combination of self (MHC) and nonself (the antigen), it knows there's an invasion happening. The T_c cell will clone itself (just like our B cell did) and start to immediately deal with the infection.

Just like in our B cell scenario, the one cytotoxic T cell that matches the antigen will clone itself to create two populations: some memory T cells, in case the infection happens again, and a population of active cytotoxic T cells that will take care of the current

EXTRA HELP

The T cell's *cell-mediated response* focuses on intracellular pathogens. That's why the infected cell needs to present the problem to a T cell, capable of handling these "hidden infections." The infected cell in this scenario is appropriately called an Antigen Presenting Cell (APC).

The signal that there is an infection must come from a reliable source (your own body cells); therefore, the APC combines a piece of the invader (the antigen) with the special protein that is unique to your body cells (MHC).

infection. How does the immune system deal with a viral invasion? Its best approach is to get ahead of the infection by destroying the already-infected cells in order to stop the virus from reproducing. Interestingly enough, T_c cells also attack another form of compromised body cells: cancer.

IRL There's yet one other scenario where your protective cytotoxic T cells attack other cells that seem abnormal: organ transplants. It makes perfect sense! These are body cells that appear a bit *off* (T cells' expertise). Therefore, when a person gets an organ transplant, one of the medications they take is to calm down their immune response, so their body doesn't reject the new organ.

Active Immunity, Passive Immunity, and Vaccinations

If you catch the flu from your friend, your immune system will rev up and deal with the infection. This is an example of **active immunity**, because your immune system is reacting to the introduction of an antigen and mounting a response by finding that lone B or T cell that best matches the invader, cloning it, and dealing with the immediate threat (plus creating a population of memory cells, in case the threat comes back). Here is another scenario: you are out hiking and a venomous rattlesnake bites you. Are you willing to wait for your immune system to mount a response to the venom? Sit idly by as the correct B cell clones itself and creates a sufficient concentration of antibodies to inactivate the venom that is sweeping through your body? Um, no. That is why hospitals in rattlesnake regions have something called antivenin (or antivenom), a medication that contains antibodies against specific poisons, such as those in the venom of snakes, spiders, and scorpions.

IRL The terms *venomous* and *toxic* are easy to use incorrectly. Something is considered venomous if its poison is transferred to you through a bite; something is toxic if the poison is ingested (eaten, like tetrodotoxin from the puffer fish we learned about earlier).

Antivenin is cool because you are injecting the already-made antibodies directly into the bloodstream, without waiting for your own body to create them. This also occurs in newborns, who have yet to develop their own immune systems. They are given antibodies through the placenta and, once born, in their mother's milk. Chances are, the mother is exposed to the same environmental antigens as their baby, so why not share some antibodies? Once the newborn has a functioning immune system (around three months old), they no longer have to rely on the passive transference of antibodies from their mother. That is the key term: passive. If antibodies are passively given to you, you (and your immune system) did not go through the work to create them. Active immunity, however, is when your immune system directly responds to the infection and creates its own antibodies.

Regardless of how your body acquires the antibodies, our **must know** concept still applies: both are a matter of self versus nonself. But what about vaccinations? A vaccine is a medicine created from parts of a pathogen, in order to intentionally challenge a person's immune response to that particular pathogen. If this active or passive immunity? Vaccinations are an example of active immunity, regardless of it being a purposeful administration instead of a natural exposure. A vaccine not only refers back to our first **must know** (self versus nonself) but is also an example of the third **must know**: the immune system's third line of defense is specific and has memory.

BTW

What is one good thing and one bad thing about the passive immune response? It's good because it's fast (snake bite!) and antibodies are kindly given to you. But if you are bitten again by a rattlesnake (maybe a sign that you need to rethink your hiking locations), is your immune system now ready to protect you? Nope. The bad thing about the passive immune response is that there is no memory. Since your own body did not go through the process of cloning the correct lymphocytes (B or T cell), there is no population of memory cells hanging about and providing protection in case of a second attack.

IRL

Speaking of memory, a vaccination's memory cells can last years, even decades, depending on the disease it is protecting against. If the memory cells begin to dwindle, that is when you get a booster shot (a second vaccination, intended to remind your immune system about that disease you were vaccinated against years ago).

The importance of vaccinations cannot be underestimated. Here is an easy means to protect you and your loved ones against potentially deadly diseases. Some vaccines are composed of parts of the actual pathogen (or, in some cases, an *attenuated*—weakened—version of the pathogen). Others, such as the vaccine against COVID-19, instead carry tiny snippets of viral mRNA. Your own body cells will then do the work of translating the mRNA into tiny bits of viral proteins, and your body then mounts an immune response against these foreign proteins. You may experience mild symptoms due to your immune system doing its job, but this is still much, much preferable to *actually contracting the disease*. Furthermore, any ickiness you may feel after a vaccination means your immune system is responding to the pathogen, which is a GOOD THING. You want those B and T cells to create a population of memory cells, willing to hang around and protect you against the flu, or measles, or cervical cancer, or tetanus, and so on. Not only are vaccinations important for your health, they are also important for the health of others that are themselves unable to vaccinate due to illness or infancy. There is a phenomenon called herd immunity that helps to protect those at risk by stopping the spread of a disease by ensuring most of us are vaccinated. It's a social contract to help not only ourselves but also others that are unfortunately unable to protect themselves.

REVIEW QUESTIONS

1. What do the first and second lines of defense have in common (not shared by the third line of defense)?

2. Match the word with the appropriate definition:

1. antigen	A. also called white blood cells
2. B cells	B. large phagocytic (eating) cell
3. lysozyme	C. secretes antibodies for extracellular invasions
4. histamine	D. response of second line of defense that increases blood flow and capillary leakage
5. inflammation	E. nonself invader
6. lymphocytes	F. targets invasions involving body cells
7. macrophages	G. chemical that causes capillary dilation
8. T cells	H. enzyme that destroys bacterial cell walls

3. Choose the correct term from each pair: The cell-mediated response targets **intracellular/extracellular** pathogens such as **viruses/pollen** and relies on **B cells/T cells**.

4. Choose the correct term from each pair: How does the immune system "remember" an infection and protect you against a second invasion? How is it related to vaccine function?

5. Choose the correct term from each pair: What is the first line of defense when an antigen is attempting to invade your body?

6. Choose the correct term from each pair: The humoral response targets **intracellular/extracellular** pathogens and relies on **B cells/T cells** to produce **antigens/antibodies** to bind to and help remove the invader.

7. How does passive immunity differ from active immunity? When is this sort of immunity important?

8. Every year, you need to be revaccinated against the seasonal flu virus. Why is this necessary, unlike other vaccines that offer protection for years?

9. Which of the following is incorrect?
   a. Antivenin provides active immunity against snake venom.
   b. B cells are the key players in the humoral response.
   c. Histamine is part of the second line of defense.
   d. T cells target intracellular pathogens.

PART SIX

Ecology

Imagine yourself outside, relaxing under a tree, reading your biology book. Without even making an effort, you are interacting with your ecological surroundings. Ecology is an important subject because it takes a large-scale view of biology. Instead of focusing on the systems contained within a single organism, ecology considers how the organism interacts with its surroundings. Nothing lives in true solitude! You are constantly sharing your environment with other life-forms, whether they're obvious (other humans, swaying trees, that pesky mosquito that is buzzing in your ear) or not so obvious (mutualistic bacteria colonizing your digestive tract, or that annoying athlete's foot infection).

Populations

The living world is filled with different populations of organisms coexisting and interacting with one another. Furthermore, as our **must know** points out, these populations are not static and unchanging. Through either natural events or man-made circumstances, a population's numbers may change drastically (or even disappear altogether).

When you refer to a **species**, you are talking about organisms that can reproduce together and create fertile offspring. A group of the same species living in the same area is called a **population**, and a bunch of different populations occupying the same area is referred to as a **community**. But if you look around you, there are certainly things in your environment beyond living organisms (referred to as **biotic factors**). As you can probably guess, **abiotic factors** refer to the nonliving properties of an environment, such as the weather, water, oxygen, temperature, and sunlight. All of these—the communities and abiotic factors with which they interact—create that inclusive term **ecosystem**.

For example, consider the meerkat. These little mongooses live in the Kalahari Desert of southern Africa. They form groups of between 20–50 individuals called gangs or mobs, usually consisting of extended family members. Meerkats are omnivores (eating both plants and animals), but prefer delicious insects over anything else. The species is well adapted to life in the desert and has evolved dark patches of fur around their eyes to reduce sun glare. This is important, because meerkats need to keep a sharp lookout for predators, such as cheetahs and eagles.

Meerkats

Author: George Hodan. https://www.publicdomainpictures.net/en/view-image.php?image=39399&picture=suricate-or-meerkat-sitting

A meerkat's ecosystem is in the Kalahari Desert.A population of meerkats live amongst other populations, such as cobras, palm trees, beetles, and cheetahs. This community (collection of different populations) shares the abiotic factors of the desert, including temperature and water availability. This sum of abiotic and biotic factors defines the ecosystem of the Kalahari Desert.

Organism	
Population	

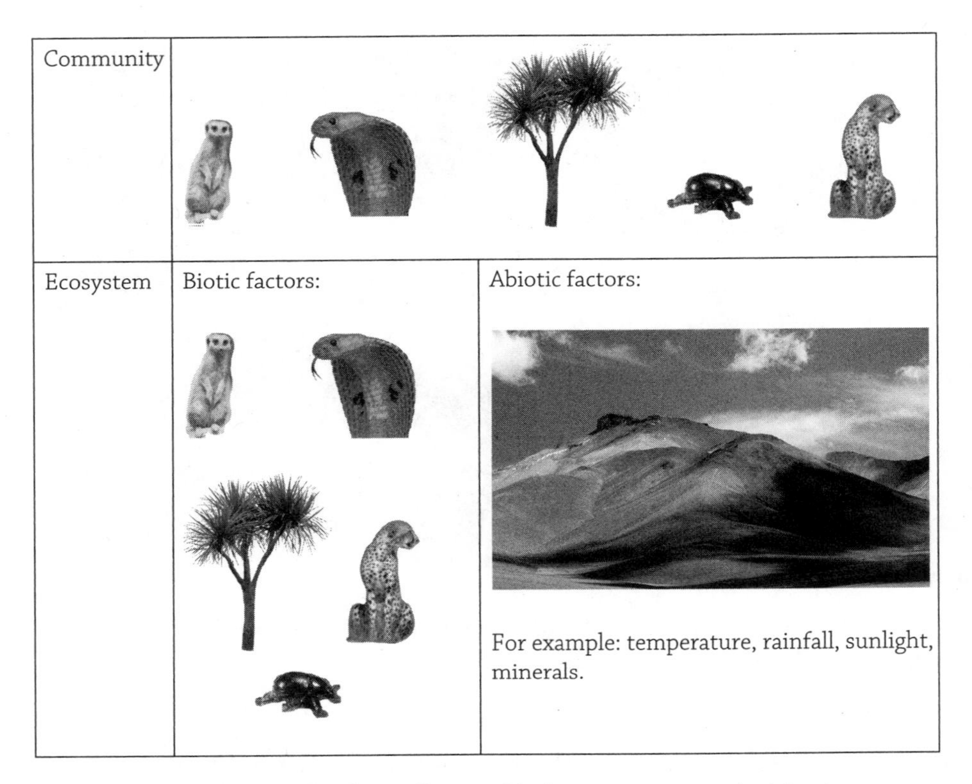

Meerkat image author: George Hodan. https://www.publicdomainpictures.net/en/view-image.php?image=39399&picture=suricate-or-meerkat-sitting

Cobra image author: Kamalnv. https://commons.wikimedia.org/wiki/File:Indiancobra.jpg

Palm tree image author: Karen Arnold. https://www.publicdomainpictures.net/en/view-image.php?image=38462&picture=palm-tree

Cheetah image publisher: U.S. Fish and Wildlife Service. https://digitalmedia.fws.gov/digital/collection/natdiglib/id/4289

Beetle image author: Heiti Paves. https://commons.wikimedia.org/wiki/File:Metsasitikas.jpg?fastcci_from=10362&c1=10362&d1=15&s=200&a=fqv

Desert image author: Luca Galuzzi. https://commons.wikimedia.org/wiki/File:Colors_of_Altiplano_Boliviano_4340m_Bolivia_Luca_Galuzzi_2006.jpg?fastcci_from=48485&c1=48485&d1=15&s=200&a=fqv

When you consider the impact a population has on the ecosystem, or if you are worried about an endangered species' fate, you must consider the size of the population. If there are too many individuals for a given area, it may have a negative impact on their habitat. Conversely, if there are too few young individuals in a population, it may be in danger of dying out completely. Changes in population size—our **must know** concept—are natural occurrences and are often determined by two factors: the birthrate and the death rate.

Birthrate

Suppose you have a population of warblers (a type of small, vocal, perching bird). There are some old warblers, some middle-aged warblers, and some young warblers. This distribution of age groups describes the **age structure** of the population and helps to determine the birthrate. Is the population of warblers having a lot of babies? That means it has a high birthrate! This is significant because a population with many individuals in their reproductive years will have current (and future) growth due to a strong birthrate.

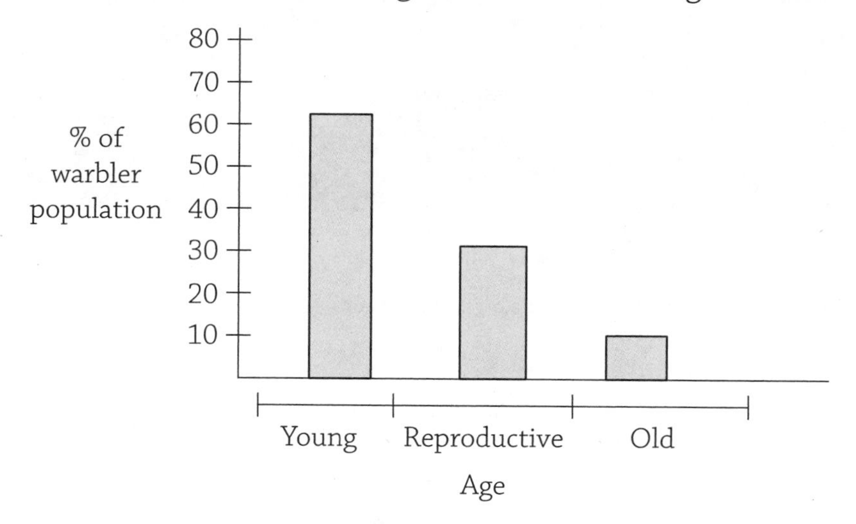

Age structure of a growing warbler population

In the previous figure, the highest percentage of members are young individuals whose reproductive years stretch ahead of them. There is definite potential for future reproduction and growth.

If, however, a population is mainly composed of older individuals, the population at best will remain stable (at worse, it will decline). There are many sad scenarios where habitat loss has doomed an endangered species because the population cannot thrive and reproduce. One example is the kakapo of New Zealand:

Kakapo

Author: Mnolf. https://commons.wikimedia.org/wiki/Strigops_habroptila#/media/File:Strigops_habroptilus_1.jpg

The kakapo (also called "night parrot") is a large, round, flightless, nocturnal parrot of New Zealand. This adorable bird is a heavy and rotund little guy who does not fly because it evolved on an island without predators. But when humans moved in and introduced predators (such as cats and ferrets), the kakapo population crashed. Even though conservation efforts began as early as the 1890s, nothing really helped until the 1980's Kakapo Recovery Plan. The current surviving population is kept safe on three predator-free islands. Even so, the adorable kakapo is critically endangered. As of 2018, fewer than 150 adults exist.

Age structure of a declining kakapo population

These percentages are hypothetical and should not be considered actual data. Yet the dire situation depicted in this graph is unfortunately true.

As the graph shows, there are not many young individuals, and the older kakapos are no longer able to have babies. Who will create the next generation? This is an extreme (and sad) example of our **must know** concept: populations change, and the kakapo population may even be entirely removed from the island ecosystem.

Death Rate

The death rate is obviously working against the birthrate. If there's a high death rate, many members of a population are dying and the overall numbers are dropping (unless the birthrate is high enough to offset the deaths). And the death rate can vary for different populations, depending on when the majority of deaths occur. For example, a fish may lay hundreds of eggs, but only a small percentage of these individuals reach adulthood (there is a high death rate early in the lives of the fish). Large mammals, however, have relatively few offspring, but most survive and reach old age (most of the deaths occur late in life). A **survivorship curve** is a diagram comparing numbers of survivors to the percentage of their maximum life span.

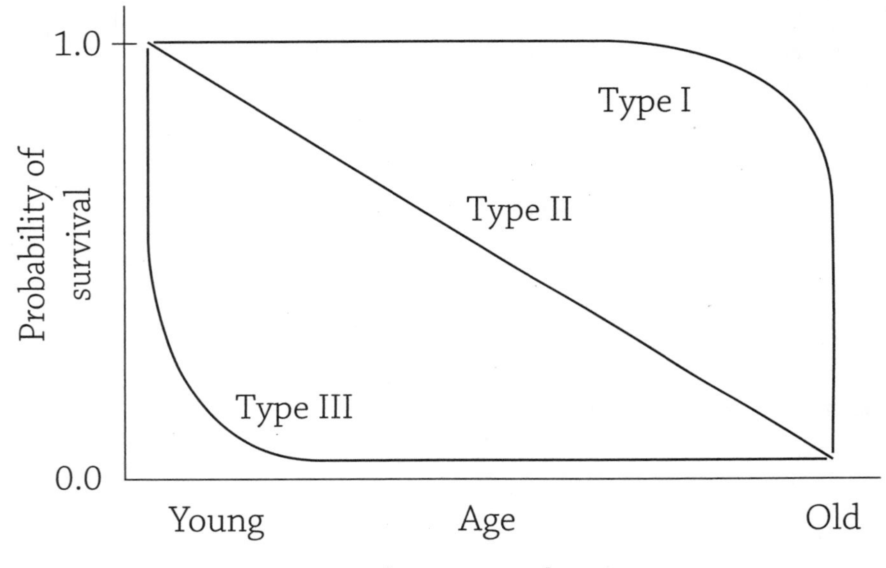

Type I, II, and III survivorship curves

Type I survivorship shows a low level of infant mortality, probably because the parents care for their young; this curve is for the large mammals, including primates. Type III survivorship has a rapid drop-off early on, with relatively few survivors. Species showing this survivorship curve (plants, invertebrates, and fish) need to have a large number of offspring to compensate for this loss, and animals that suffer a high predation rate also tend to produce large numbers of offspring. The Type II survivorship curve is common in birds, some reptiles, and smaller mammals, because they tend to have an equal chance of dying throughout their lifetimes.

Limiting Population Growth

Clearly, there are many environmental factors that keep down a population's size. The steep initial drop-off of a Type III survivorship curve may be due to predation. A population's growth eventually slows due to limited resources such as food and space. Anything that limits the population size is called a **limiting factor**, and can be categorized as either density-dependent or density-independent. A **density-dependent limiting factor** gets worse as the population increases and crowding gets worse. Because this type of limitation is due to interactions with living organisms, most density-dependent limits are biotic.

Disease, for example, is caused when a pathogenic organism (bacteria or parasitic worms, for example) is passed from host to host. A disease can more easily spread and wreak havoc if the members of the population are crammed in together without much space. *Competition* is worse if there are a huge number of individuals fighting for the same resources. If there are a ton of prey milling about, *predation* is going to be significant. In contrast, **a density-independent limiting factor** will impact a population regardless of its density. *Weather* events such as floods and droughts can severely impact an ecosystem's resources, and *natural disasters* such as earthquakes, fires, and tsunamis will kill a great number of organisms in the area, regardless of the population density. And, unfortunately, *human activity* has significant effects on populations. If we destroy entire ecosystems by clearing forests or draining wetlands, there is little chance of the native population's survival.

EASY MISTAKE!

Labeling something as *density-dependent* or *density-independent* can be a little misleading. For example, let's say a flood kills a large population of crickets. Yes, if there are more crickets in the flood zone, there will be an increased number of dead. Yet a flood is still a density *independent* factor because regardless of the number of crickets present, the flood will remain as deadly (say, by wiping out 50% of the population). On the other hand, disease—a density *dependent* factor—will kill a higher proportion if it has a highly-packed population to spread through.

Limiting Factors

Density-dependent limiting factor	Density-independent limiting factor
Competition ▪ Space ▪ Food ▪ Mates	Natural disasters ▪ Fire ▪ Earthquake ▪ Flood
Disease ▪ Bacteria ▪ Viruses	Industrial accidents ▪ Oil spill
Predation	Habitat destruction

Even though fluctuations occur, the number of individuals in a population (number of individuals/area) is defined as the **population density**. The limiting factors we just learned about can have a significant impact on a population's density. For example, a change in population density can be due to resource availability. If there are unlimited resources, then a population will grow very, very fast. This isn't a realistic scenario, however, because an environment will eventually run out of nutrients, space will become limited, and there will be an increase in predation. This is why **exponential growth** doesn't often happen (probably a good thing . . . could you imagine if the chipmunk population grew exponentially? You'd find them in your sock drawer.).

The one time you would see exponential growth is when a species moves into an uninhabited area. No predators! Plenty of food and space! Rampant reproduction! It is way more common, however, to see a **logistic growth** pattern. This is when a population increases slowly at the beginning, enjoys a short burst of exponential growth, and then eventually levels off at a stable (and sustainable) population size. This maintainable population density is referred to as an ecosystem's **carrying capacity**.

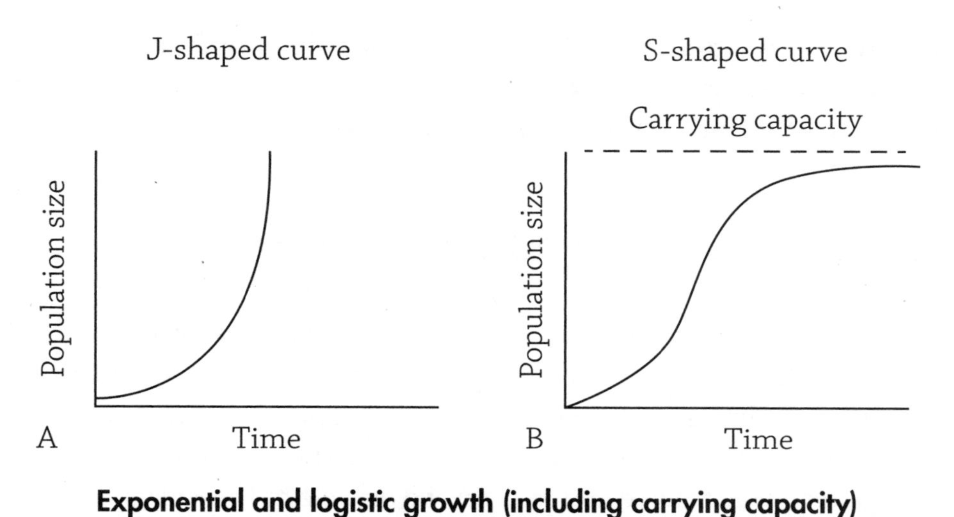

Exponential and logistic growth (including carrying capacity)

Interspecies Interactions

Predation and competition are two harsh examples of how species can interact. Consider the tardigrade, a tiny animal who enjoys sucking the juice from moss and algae (more on this little guy in the next chapter). The tardigrade is the **predator**, and it is feeding upon another organism, in this case, an algal blob. Now, if another tardigrade decides to take up residence in the same moss clump, there will now be **competition** for the same limited resources (delicious algal blobs). Competition can occur between members of the same species, or entirely different species. If a resource is limited, there's gonna be a battle.

> **EASY MISTAKE!**
> You may accidentally use the term symbiosis to mean a relationship between individuals that is beneficial to both parties. This is not necessarily true! A symbiotic relationship can be really, really bad for one participant. You probably mean "mutualism."

When individuals of different species live in direct contact with one another, this is referred to as a **symbiotic** relationship. There are three types of symbiosis: mutualism, parasitism, and commensalism. Each of these refer to an ecological relationship between different species.

Symbiotic Relationship: Parasitism

In a parasitic relationship, one organism benefits while the other is harmed. This is unlike predation because the parasite reaps the most benefit by keeping its host alive as long as possible (unlike a predator, who immediately kills and eats its prey). Some parasites live inside the host (endoparasites), whereas other parasites live and feed on the outside of the unsuspecting host (like ticks and lice and . . . shudder . . . leeches). Not only are parasites creepy, but they can have a significant impact on a host population. Parasitic interactions are, in fact, perfect examples of a density-dependent limiting factor. Usually, the parasites that come to mind are the obvious worms and other creepy-crawlies, but plants and fungi can be parasites, too. There are so many awesome examples of cool/horrific parasitic relationships it's hard to choose. Here are two:

Cool Parasite 1

Mistletoe brings to mind wintery seasons and happy holiday decor. But a far more sinister truth lies beneath that beautiful sprig of glossy green leaves and beautiful white berries . . . mistletoe is a plant parasite that sucks the life-juice out of its photosynthetic victims with nutrient-absorbing projections called **haustoria**. A growing mistletoe will wrap itself around a host plant and use its haustoria to puncture through to the host-tree's circulatory system (xylem and phloem) to steal its water and minerals.

A plant sinking its haustoria into its prey plant

Author: Chrissicc. https://commons.wikimedia.org/w/index.php?search=haustoria&title=Special%3ASearch&go=Go#/media/File:Cassytha_pubescens_haustoria_.jpg

Cool Parasite 2

Toxoplasma gondii is a unicellular, parasitic protist that can cause a disease called toxoplasmosis. Cats are a common host for *T. gondii*, which is why pregnant women shouldn't clean out cat litter boxes (the parasite can be passed through cat feces). Cats themselves usually aren't affected by the infection, but their prey—rats—are another matter. When a rat is infected by *T. gondii*, it causes unusual changes in neural activity and the rat loses its fear of cats. Even worse, the rat becomes attracted to the smell of cat urine! This may help the parasite spread farther, because there's a good chance that bold rat is going to end up as some cat's dinner.

Furthermore, recent research suggests that *T. gondii* could contribute to mental disorders in humans. A study in mice showed that infection with *T. gondii* may cause brain cells to release a higher-than-normal level of the neurotransmitter dopamine; altered dopamine levels are linked to some mood disorders. Other studies have discovered a potential link between infection by *T. gondii* and schizophrenia. There is no definitive answer to the causation between infection and mood disorders, but the research suggests a significant link. It's also notable that infection by *T. gondii* is so widespread, up to *half* of all humans on Earth are infected. Luckily, most of the time there are no symptoms and you don't even know you have it. If you're curious, your doctor can perform a simple blood test to screen for antibodies against the protist to see if you have ever had the infection!

Symbiotic Relationship: Mutualism

A mutualistic relationship is a peaceful and happy symbiosis where both parties benefit—everyone's a winner! This sort of relationship is often the result of coevolution, with changes in one species affecting the survival of the other species (and vice versa); it's like a friendly arms race. For some fascinating examples, read further, friend.

Cool Mutualistic Relationship 1

If you have ever gone out for a walk and looked closely at the trunks of some trees, you may have noticed a dry, flat, greenish growth on the bark. Or maybe when flipping over large rocks to see what scurries out from underneath, saw patches of green flakey stuff that looked too dry and dead to be a thriving life-form. If so, you have just noticed some lichen.

Author's photos of lichen

Lichen is not a single organism. Instead, it is a permanent mutualistic relationship between a non-photosynthetic fungus and a photosynthetic algae or cyanobacteria.

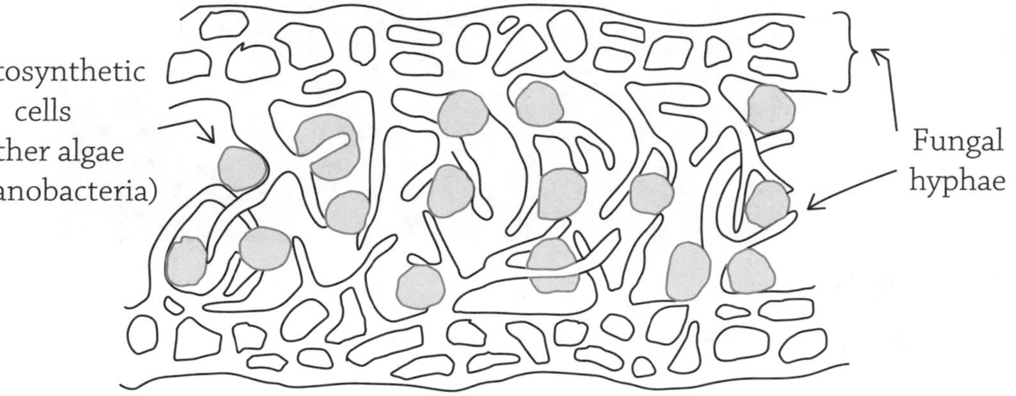

Diagram of the mutualistic relationship that makes up lichen

Filaments of fungus (also called hyphae) surround and cradle their photosynthetic partners. And "partners" is a perfect description. Fungi, by definition, are unable to photosynthesize. Instead, they reap the benefit of their partner alga (or cyanobacteria) producing too much glucose—excess sugars and vitamins are passed from the alga onto the fungus. The alga is happy because it is protected by the fungal hyphae, reducing the chances of it drying out in the hot sun. This allows lichen to colonize many different habitats and they are remarkably resistant to drought.

Cool Mutualistic Relationship 2

There is this clever little leafhopper called a glassy-winged sharpshooter that feeds on the sap of woody plants (including grapevines, to the chagrin of vineyard owners in California).

A leafhopper

Image source: https://pixabay.com/images/search/leafhopper_macro_insect_biology/

Sap, however, is not a very healthy diet . . . it's lacking essential amino acids and vitamins that are needed in the insect's diet. Cleverly enough, the insect has mutualistic relationships with not one but TWO different microbes. These microbes produce the nutrients not supplied by the leafhopper's primary diet: one species of bacterium uses sap-derived carbon to make amino acids (but cannot make any of the required vitamins). The other species of bacteria uses the carbon to make vitamins (but can't make any of the needed amino acids). The insect is supplying the two bacterial species with a steady source of carbon, and the bacteria are, in turn, providing either the amino acids or the vitamins needed by the leafhopper. Furthermore, the next-door bacterial species are supplying each other with chemicals needed to make their own nutrients.

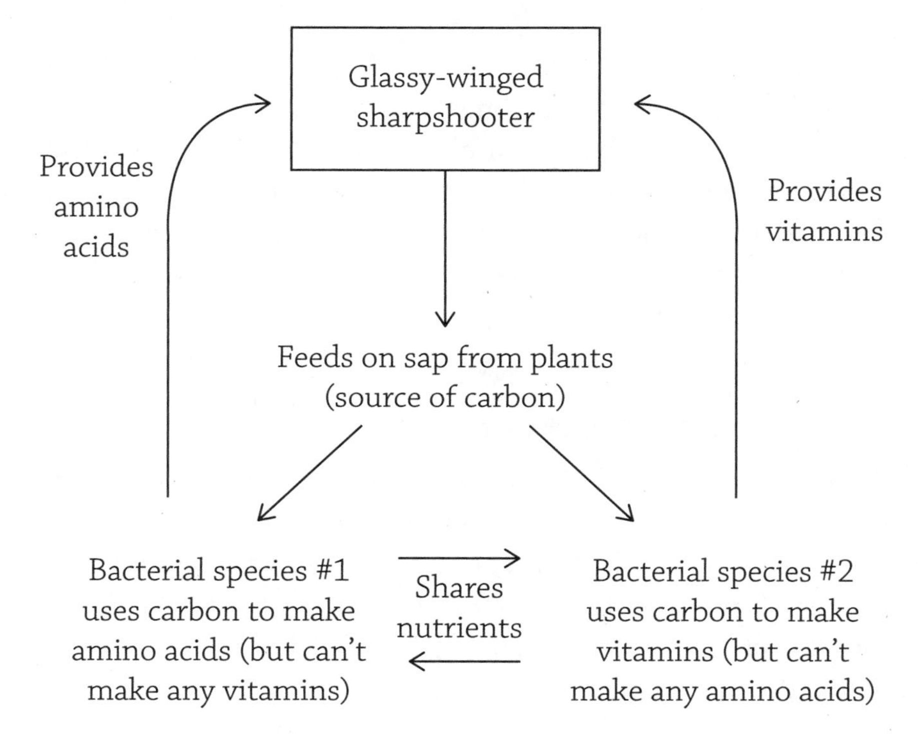

Multiple mutualistic relationships within the glassy-winged sharpshooter

These mutualistic relationships are so important to all three participants that the insect evolved to have specific housing for its two resident microbes. These houses are called **bacteriomes** and they reside on each side of the insect's abdomen.

Symbiotic Relationship: Commensalism

Commensalism is a symbiotic relationship between two organisms where one benefits and the other one doesn't really care either way (neither harmed nor helped). I personally think this is the most difficult one to identify because so often it turns out to be a mutualistic relationship instead. For example, a common example of commensalism is when a species of bird is "hitchhiking" on the back of a buffalo. When the buffalo walks through the grass it stirs up bugs, which the bird then eats. If this was pure commensalism, the bird would benefit (tasty bug meal), and the buffalo would neither benefit nor be harmed. But this could be a mutualistic example, since the buffalo is benefiting from a reduction of annoying bugs in its personal space. I will give you some more clear-cut examples of commensalism.

Cool Commensal Relationship 1

The soil community is a lively and crowded one, populated by myriad bacterial and fungal species. There can be one species of bacteria that releases a chemical that is happily gobbled up by another species of bacteria. The bacteria that produces the chemical really doesn't care what happens to the by-product, but the second species is thrilled to receive an essential nutrient (discarded from its microbial neighbor).

Cool Commensal Relationship 2

A species of coral reef jellyfish will have a juvenile fish swimming along within the protective safety of its tentacles. The jellyfish is "meh" about this ride-along, but the young fish is enjoying an umbrella of protection from predators. It's important to note that the jellyfish does not dine on its ride-along friend (otherwise, this would quickly turn into an example of predation).

Symbiotic Relationship: It's Complicated

It's Complicated 1

Cymothoa exigua is called the "tongue-eating louse," and for good reason. It is a crustacean that lives in the Gulf of California (among other places) and parasitizes at least eight different species of marine fish. The male *Cymothoa exigua* takes up residence by attaching to the fish's gills. The females, however, are a bit more creative in where they live in the unsuspecting fish. She causes degeneration of the fish's tongue until there's nothing but a stub left. She then attaches to the stub with seven pairs of hooked legs and *replaces the fish's tongue*! The parasite fits so perfectly it serves the same mechanical function as the tongue when the fish feeds, and the fish can still eat normally. Because of this, some people consider this an example of a commensal relationship, instead of the more obvious parasitic comparison.

Cymothoa exigua, or the tongue-eating louse

Author: Marco Vinci. https://commons.wikimedia.org/wiki/File:Cymothoa_exigua_parassita_Lithognathus_mormyrus.JPG

It's Complicated 2

Wolbachia is a common and widespread group of bacteria that hang out in the ovaries and testes of many different insects, as well as some spiders and nematodes. This bacterial infection is actually passed down generation to generation (also called "vertical transmission"), because it's inherited in the cytoplasm of the egg. Since the bacteria don't hang out in mature sperm, only infected females (and their *Wolbachia*-laden eggs) pass the infection on to their offspring.

A phenomenon called cytoplasmic incompatibility helps *Wolbachia* spread like wildfire through the host population. Males that are infected with *Wolbachia* are unable to mate with uninfected females. The uninfected egg "phenotype" is selected against, because uninfected eggs will not become fertilized. In order to further help the spread, *Wolbachia* causes infected males to turn into infected females (feminization). In some species of insects, *Wolbachia*-infected females can even reproduce without the help of males (a phenomenon called parthenogenesis)! It must be mentioned, however, that *Wolbachia* isn't a clear-cut parasite. Some infections appear to be a bit mutualistic. Elimination of the infection from some nematodes, for example, leads to the worms' sterility or even death. Infection of *Wolbachia* in fruit flies helps them to better resist viral infections. But overall, any infection that messes around with a species' ability to reproduce is considered parasitic.

What if we could harness *Wolbachia* and use it to control an insect population that spreads more death and misery than any other insect in the world? Yes, the evil mosquito. Studies have suggested that intentionally infecting a mosquito population with *Wolbachia* may reduce the mosquitos' ability to spread horrible viruses such as Zika and dengue. Let's go, science!

REVIEW QUESTIONS

1. Match the following term with the correct example:

a. abiotic factors	1. blue-ringed octopi can reproduce together and produce fertile offspring
b. biotic factors	2. a sun-warmed nutrient-filled lake populated with a variety of fish and amphibians
c. community	3. all the red-tailed hawks that live on campus
d. ecosystem	4. temperature, rainfall, wind, and nutrients
e. population	5. all the chipmunks, squirrels, and raccoons that live in a forest
f. species	6. plants, fungi, bacteria, animals, and protists

2. Which type of survivorship curve necessitates having a large number of offspring? Give an example of an organism with this type of curve.

3. Of the two growth curves, the S-shaped logistic model is more realistic. Explain why, and draw/label an example curve.

4. The glassy-winged sharpshooter houses two species of bacteria that, in turn, provide essential nutrients to the leafhopper. What kind of symbiosis is this an example of?

5. Choose the correct word from each pair: A population with a **high/low** death rate and a **high/low** birthrate would experience an overall increase in the population numbers. This population would most likely have an age structure with mostly **young/old** individuals.

6. Draw the three types of survivorship curves, and explain which one best describes large mammals.

7. For each of the following examples, indicate whether it is a density-dependent liming factor (DD) or a density-independent limiting factor (DI):
    a. A viral disease
    b. Competition for mates
    c. Volcanic eruption
    d. Predation
    e. Wildfire
    f. Oil tanker spill in ocean
    g. Clearcutting forest

8. The female "tongue-eating louse" removes and replaces a fish's tongue. In order for this to be an example of commensalism, what characteristic of the relationship must be true?

9. Which of the following would cause an increase in a population's growth?
    a. The population's age structure shows a higher percentage of old individuals.
    b. The death rate is higher than the birth rate.
    c. There is a decrease in density-dependent limiting factors.
    d. A population reaches its carrying capacity.

29 Ecosystems and Their Interconnectivity

MUST KNOW

- The interconnectivity of ecosystems means that different populations will have an impact on each other.

As mentioned earlier, an ecosystem is a collection of organisms and nonliving things. The number of interactions that occur within an ecosystem is vast and varied. A critter, for example, must call some specific location its home, or **habitat**. A habitat includes both biotic and abiotic factors. A tardigrade, for example, may enjoy spending its days in a freshwater droplet held within a lush bed of moss.

The glorious tardigrade—aka "water bear," aka "moss piglet"—is one of the most amazing animals on Earth. They are obviously adorable little critters, but don't mistakenly think you can put it on a leash and take it for a walk—they don't get any bigger than 1 mm. They have eight legs with tiny little hands and tiny little claws.

Tardigrade

Authors: Schokraie E, Warnken U, Hotz-Wagenblatt A, Grohme MA, Hengherr S, et al. https://commons.wikimedia.org/wiki/File:SEM_image_of_Milnesium_tardigradum_in_active_state_-_journal.pone.0045682.g001-2.png

Their superpower lies in their ability to withstand ridiculous environmental conditions. Tardigrades can survive temperatures from minus 300°F to over 300°F. Vacuum of space? No problem. Heavy radiation? Sure. Their survival abilities are rooted in their powers of cryptobiosis—going into a death-like state that enables them to slow their metabolism to 0.01% levels of normal. They pull in their tiny little legs and turn into a teeny dehydrated ball called a tun. Just add water and . . . voila!—rehydrated tardigrade, ready to go and find a nice, new patch of moss to call home.

Well, back to our tardigrade habitat scenario. The temperature is 76°F and there are plenty of plant cells for it to eat. This is the tardigrade's habitat, or *where* it lives. Now, *how* the tardigrade lives is a bit more complex. This is referred to as a species' **niche**, and along with all the abiotic and biotic factors of its habitat, other factors are considered. When does the tardigrade eat? What might decide to prey upon the tardigrade? Will the tardigrade decide to seek out a partner and reproduce sexually, or undergo parthenogenesis (a form of asexual reproduction via the development of an embryo from an unfertilized egg)?

A bunch of different organisms occupy the same habitat, but they have their own distinct niches. When so many different species live so closely together, there will undoubtedly be interspecies interactions. In fact, two or more species cannot occupy the same niche. The species who is a bit better at acquiring resources will have a higher survival rate and reproduce more rapidly. The inferior species will die out. This is called the **competitive exclusion principle**, and it is an excellent example of our **must know** concept: the interconnectivity of ecosystems means different populations can have an impact on each other.

There is a very well-known example of the competitive exclusion involving two different species of *Paramecium* participating in a tiny, microscopic cage fight . . . two species go in, but only one will survive. This specific investigation was conducted by the Soviet biologist G. F. Gause in 1934, using the two *Paramecium* species *P. caudatum* and *P. aurelia*. He introduced the two species into the same ecological niche and then measured their population densities over time. Imagine if you wanted to conduct a similar experiment with Paramecium A and Paramecium B. Your hypothesis might be something like this: due to the competitive exclusion principle, one species will win out over the other, and there will be a decline in the loser species' numbers because the dominant species will use up all the resources. In order for you to really know if any die-off was due to competition, you need a control experiment against which you can compare your experimental data. Your controls are the growth curves of each species living independently.

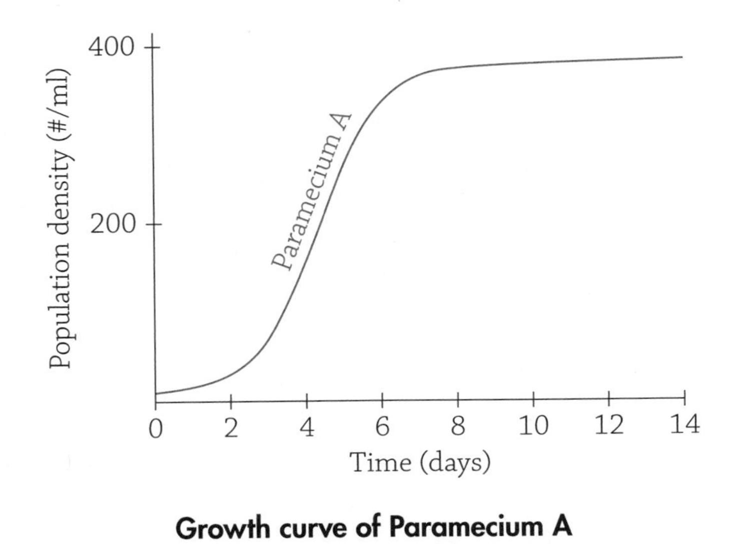

Growth curve of Paramecium A

The first graph shows you the growth curve of Paramecium A population if it lives alone in its very own bacteria-filled flask (your paramecia love to eat bacteria). It follows a healthy S-shaped growth curve of a population enjoying an initially brisk boom, followed by a leveling off once it hits the habitat's carrying capacity (at about 380 critters per milliliter).

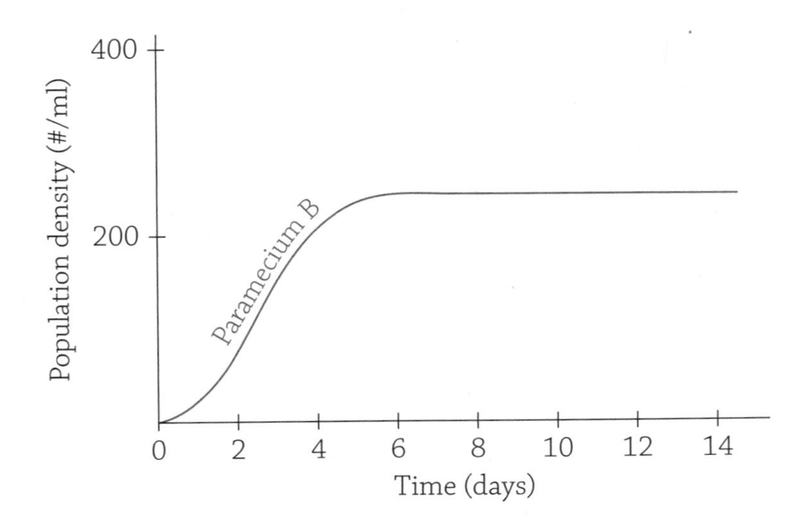

Growth curve of Paramecium B

Paramecium B also grows well in its own flask, and though its initial growth spurt is faster than its cousin's, it has a smaller carrying capacity (240 critters per milliliter).

In the actual experiment run by Dr. Gauss, Paramecium B was a larger species than Paramecium A. It makes sense, then, that the carrying capacity for the larger species was a smaller number—bigger paramecia need bigger bacterial meals!

Finally, you are ready to run the competitive exclusion experiment, and you happily combine both species A and B into a single food-filled flask.

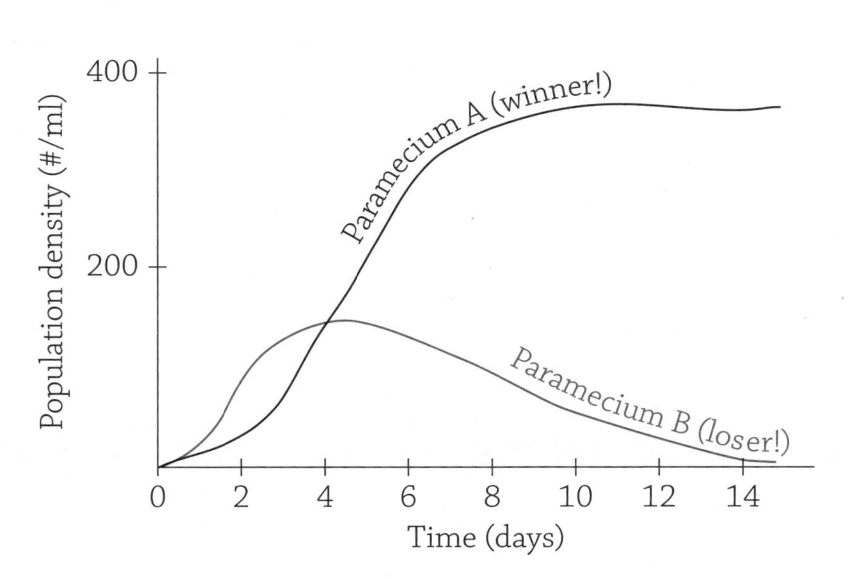

Growth curve when species A and B are combined

For the first 2 days, everything seems normal. Paramecium B's population increases faster than Paramecium A's. But then around the fourth day, something happens . . . Paramecium B numbers start to decline! Meanwhile, the Paramecium A population is growing like crazy, and continues to do so as their cousins dwindle in numbers until they disappear. Your hypothesis was correct, and species A outcompeted species B.

Matter and Energy Flow in an Ecosystem

It would be remiss not to address the issue of energy and matter flow within an ecosystem. You can learn about a specific ecosystem with its distinct critters and habitats, and it's easy to think of it as a closed system that can exist on its own. Nothing could be further from the truth. There is a constant flow of energy into an ecosystem from the sun, and after the light energy is converted to a chemical form by autotrophs (photosynthetic organisms), it is then transferred throughout the ecosystem via a web of intricate and detailed food chains and complex food webs. Though this interconnectivity is not as obvious as different species competing with each other for niche domination, it is still a perfect example of our **must know** concept. Please recall the laws of thermodynamics we talked about in Chapter 1.

A **food chain** is a model of how energy is transformed as it is moved from one organism to another (through predation). Every chain must begin with a

EXTRA HELP

Don't forget those two laws of thermodynamics:

- Energy cannot be created, nor can it be destroyed. It can, however, be converted from one form to another.
- Any time energy is converted from one form to another (meaning it is being used by the critters of Earth), some of the potential energy is lost as heat. It is a thermodynamic rule that when energy is transformed, the process is never ever 100% efficient (and most of the time, the lost energy escapes as heat).

critter that captures the sun's energy and transforms it into a chemical form that is used by all subsequent members of the food chain. An **autotroph** (also called a **producer**) is a photosynthetic organism at the base of every chain. Most of the time, a plant comes to mind, but many bacteria and protists are able to photosynthesize as well. Each step in a food chain is referred to as a **trophic level**.

If the first trophic level is a producer, then the second trophic level must be composed of critters that eat the producers, either an **herbivore** (that only eats plants) or an **omnivore** (that eats both plants and other animals). A plant-eating critter that is the second member of a food chain is referred to as a "primary consumer." A food chain cannot have too many steps because energy is lost along the way (recall the second law of thermodynamics). Eventually, there is not enough energy left to support another trophic level.

In summary, a chain goes like the following:

> Producer → primary consumer → secondary consumer → tertiary consumer

In an actual ecosystem, however, there are no distinct and isolated food chains (once again referring back to our **must know** concept); instead, many chains are interlinked into complex **food webs**.

IRL Carl Sagan said, "We're made of star stuff." It's beautiful and true. The atoms on Earth today are the same atoms from 4.6 billion years ago, constantly recycled into new shapes, new forms, and new purposes.

Every move from one trophic level to the next only transfers about 10% of the energy locked within the tissues of the prey. When a leafhopper eats a plant, it receives only 10% of the plant's energy (the rest is lost as heat). When a bluebird eats the grasshopper, it gains only 10% of the energy locked up in the grasshopper's tissues. A model that quantifies the amount of energy available at each trophic level is called an energy pyramid, and the pyramidal shape is derived from the significant loss of energy at each subsequent trophic level.

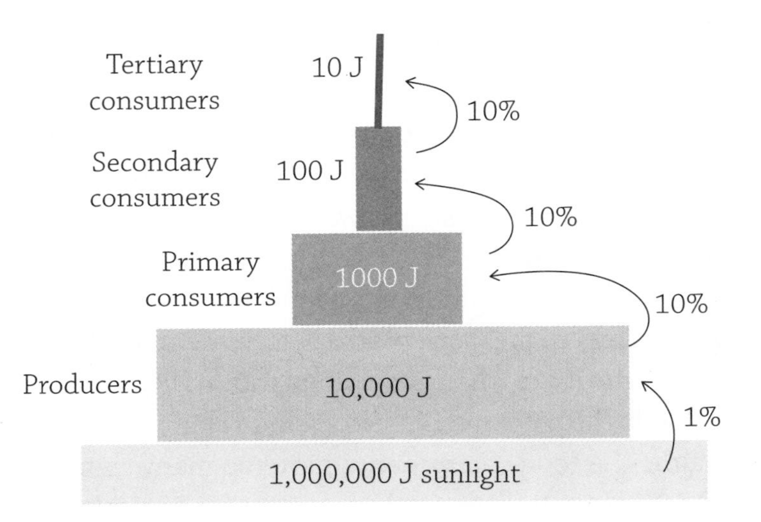

An energy pyramid

Notice that each trophic level receives only 10% of the energy from the level below it.

A biomass pyramid has the same shape because it compares the amount of biomass at different trophic levels within an ecosystem. **Biomass** is a term used to describe the total dried weight of a group of organisms in a particular ecosystem. It makes sense that if each trophic level only receives a fraction of the energy stored within the previous trophic levels (because of energy and heat loss), there are fewer organisms. Each level shows the mass needed to support all the organisms above it! Luckily, the sun keeps replenishing the base.

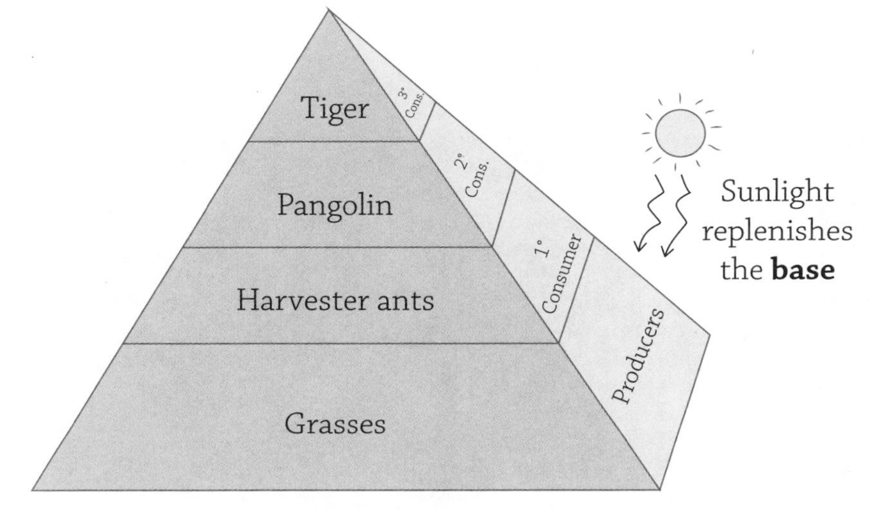

African biomass pyramid

IRL

Look at the secondary consumer in the above biomass pyramid . . . do you know what that is? A pangolin is *the* most adorable critter, and it looks sort of like a cross between an anteater and a pinecone.

Indian Pangolin (*Manis crassicaudata*) in Kandy, Sri Lanka
Author: Dushy Ranetunge. https://commons.wikimedia.org/wiki/File:Scaly_ant_eater_by_Dushy_Ranetunge_2.jpg

IRL

A pangolin in defensive posture, Horniman Museum, London
Author: Stephencdickson. https://en.wikipedia.org/wiki/File:A_pangolin_in_defensive_posture,_Horniman_Museum,_London.jpg.

It waddles along on its hind legs as if its center of gravity is a bit too far back, and when it's scared, it rolls up into a scaly sphere and lies still until the threat goes away. It does no one any harm (well, except to those delicious ants it ladles into its mouth with its long, sticky tongue). It breaks my heart that the pangolin may disappear from Earth because humans poach them for their scales, which are believed to have medicinal properties. Luckily, there are programs like the African Wildlife Foundation that are working to save these beautiful creatures. For more information, please check this out: https://www.awf.org/wildlife-conservation/pangolin.

All of these trophic levels are showing organisms being eaten by other organisms, and they themselves being devoured . . . and so on and so forth. The interconnectivity is stunning; we are all linked. Bacteria, belugas, brown bears, or bamboo, we are all tethered to one another through cycled energy and matter.

Earth is an *open system* in regards to energy but a *closed system* in regards to matter. A **closed system** is like a vacuumed-sealed environment that doesn't have anything new coming in (nor is anything leaving). Imagine one of those cool glass sphere ecosystems filled with water and plants and a tiny invertebrate or two. It is filled with water and, unless you drop it on the floor, there won't be any water lost from it through evaporation. Also, you cannot add water to it. The water is an example of a closed system.

Inside, there are plants and animals composed largely of carbon, and any of this organic matter contained within the sphere is certainly going to stay in there (another closed system). Inside of the closed sphere there are food chains and energy transformations as the plants are eaten by the tiny herbivores, and once these critters die, decomposers unlock the carbon stored within their tissues. The carbon will be taken up by the plants through photosynthesis, renewing the bottom trophic level and providing food for the next generation of herbivores.

But recall that pesky second law of thermodynamics—every energy transformation has an inherent loss of energy, heat that escapes this tiny ecosystem and is lost forever. How, therefore, can these cycles keep going, if so much energy is lost at every trophic level step along the way? The answer is because this living sphere may be a closed system in regards to matter, but it is an **open system** in regards to energy. Sunlight is constantly streaming down on this sphere, renewing the energy cycling throughout the ecosystem. The producers grab the sun energy and transform it into glucose. Consumers eat the plants and convert the stored glucose into energy through the process of cellular respiration. The waste produce of carbon dioxide is freed into the water and captured by the plants, continuing the cycle. Meanwhile, a constant stream of sunlight balances the constant loss of energy (heat). If you place your glass ecosystem in the dark, it will eventually die after all the enclosed energy is spent, without any source to renew it. Open systems require a constant input (in this case, of sunlight).

This glass sphere is a model of Earth. Our planet has a closed system for the cycling of matter (carbon cycle and the water cycle, for example) but an open system for energy, thanks to the sun. Next up, we will talk about how carbon and nitrogen are cycled within our closed Earth system.

Carbon Cycle

Carbon is arguably the most important element for life on Earth, and the study of organic chemistry is based on the study of carbon-based chemicals. Every one of the organic compounds critical to life contains carbon. Clearly, the cycling of carbon is really important to an ecosystem's health.

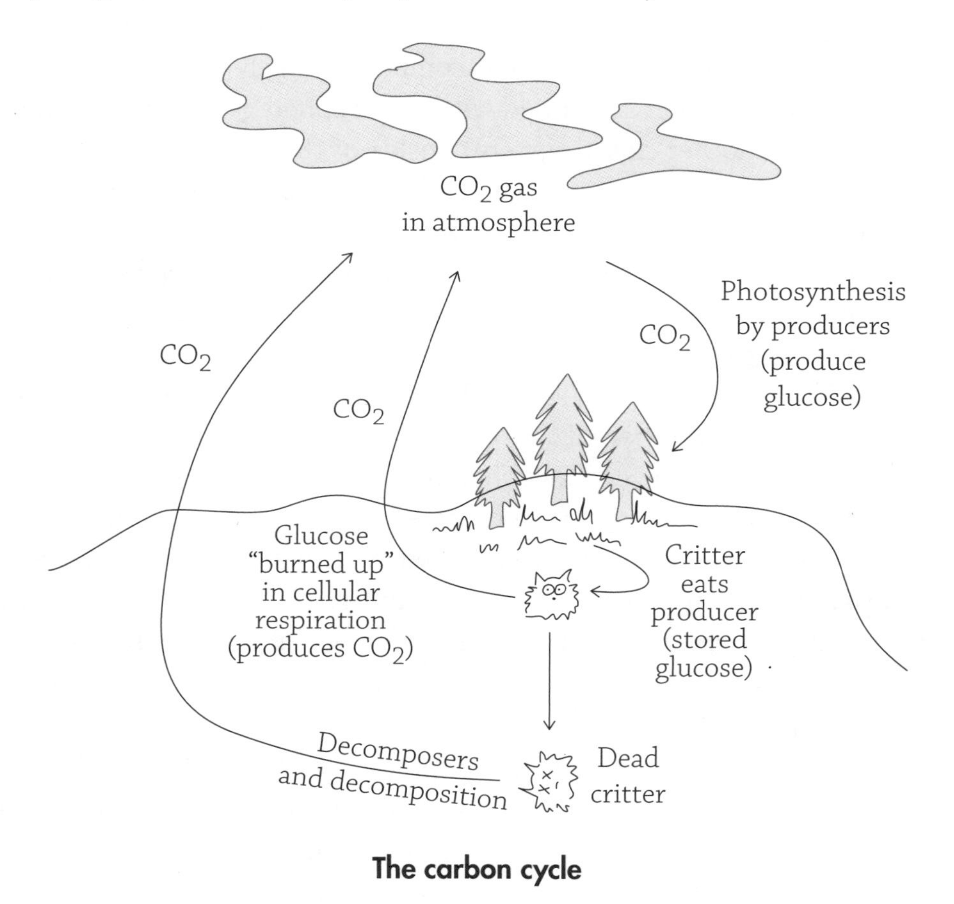

The carbon cycle

Carbon is found in the atmosphere as the gas carbon dioxide (CO_2). Autotrophs (i.e., producers) capture this gas in the process of photosynthesis, fixing it into the organic compound glucose, thus making it accessible to all heterotrophs. The carbon moves along the food chain as each trophic level devours the tissues of its previous trophic level, ingesting the carbon locked in its meal. Along the way, carbon is released back into the atmosphere as each form of life converts the glucose back into carbon dioxide through the process of cellular respiration. Furthermore, when an organism dies, the decomposers that consume the tissues also unlock the carbon through cellular respiration. Finally, large amounts of carbon dioxide are also exhausted back into the atmosphere through the burning of woods and fossil fuels.

It's important to notice the number of sources *adding* CO_2 to the atmosphere, versus the number of sources *removing* CO_2 from the atmosphere. Even though there are many pathways releasing carbon dioxide into the air, the only metabolic pathway removing it is photosynthesis. An unfortunate unbalance. Along these lines, why is it really bad when we clear-cut and burn huge expanses of forest and grasslands? By doing so, not only do we remove a CO_2 sink (the plants), we release a ton of CO_2 into the atmosphere by burning all the plant matter.

IRL **We can never underestimate the impact of plants' removal of atmospheric carbon dioxide on the global climate. When the huge fern forests of the Carboniferous era (359–299 million years ago) first formed, they removed so much carbon dioxide from the atmosphere, it caused global cooling and glacier formation! It's unfortunate that this is now occurring in reverse. Increasing levels of atmospheric CO_2 are increasing global temperatures, and the blame lies with us (humans) instead of it being a natural event.**

Nitrogen Cycle

I love the nitrogen cycle because bacteria play such an important role in ensuring this essential element reaches all corners of an ecosystem. Nitrogen is important because it is a building block of proteins and nucleic acids. The main reservoir (storage area) for nitrogen is in the atmosphere, and it is in the form of nitrogen gas (N_2). Since the gaseous form is literally floating around the air, it needs to first be brought down to Earth and "stuck" in a form that plants can use. This process is called **nitrogen fixation**—converting gaseous nitrogen into forms that cells can use—and it is how nitrogen enters an ecosystem.

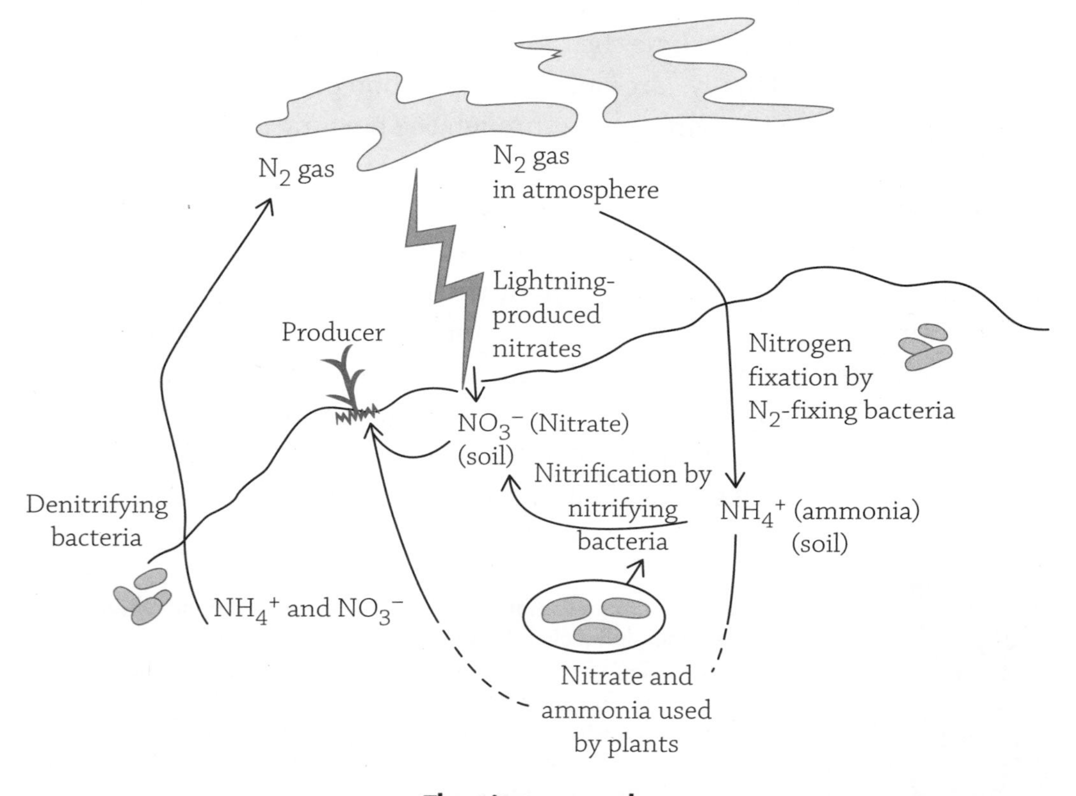

The nitrogen cycle

Nitrogen fixation can happen when lightning strikes! The enormous energy in a bolt of lightning shatters the N_2 molecule, allowing it to combine with atmospheric oxygen to form nitrate (NO_3^-), which then dissolves in rainwater and falls to Earth. More commonly, we rely on bacteria to grab this nitrogen gas and turn it into a form that cells can work with (cells can't grab nitrogen gas). There are many different forms of nitrogen-containing compounds, and they can be used by different life-forms.

Cell type	Nitrogen compound
Plants	NH_4^+ (ammonium)
	NO_3^- (nitrate)
Bacteria	NH_4^+ (ammonium)
	NO_3^- (nitrate)
	NO_2^- (nitrite)

In case you're wondering why animals aren't listed above, it's because we get our nitrogen by eating plants and animals, and their tissues contain nitrogen as it moves through the food chain. I am only going to describe how nitrogen is initially fixed and converted into different forms through terrestrial cycling.

Bacteria that are able to grab gaseous nitrogen from the atmosphere and turn it into a usable form are called (unsurprisingly) nitrogen-fixing bacteria. Probably one of the most well-studied N_2 fixers is a bacterium of the genus *Rhizobium*. Along with being a perfect example in the nitrogen cycle, it is also an excellent example of mutualism!

Rhizobium forms a close relationship with plants of the legume family, such as clover, alfalfa, peas, and beans. The plant gives the bacteria a protected, nice place to live and a steady source of nutrients. The bacteria, in turn, fix atmospheric nitrogen and produce ammonia for the plant. How this beneficial relationship grows (literally) in the first place is pretty cool. The bacteria hang out in the soil during the winter and infect a plant when it germinates in the spring. The bacteria are attracted to certain chemicals produced by the plant, and they slowly make their way over to the roots. The bacterial cells move into the root tissue, and the plant cells in the invaded region are stimulated to grow and divide, forming a small spherical "house" (a **nodule**) in which the *Rhizobium* reside. Each nodule is chock full of *Rhizobium* bacteria that are now nitrogen-fixing machines, creating a slew of ammonia for the plant to use.

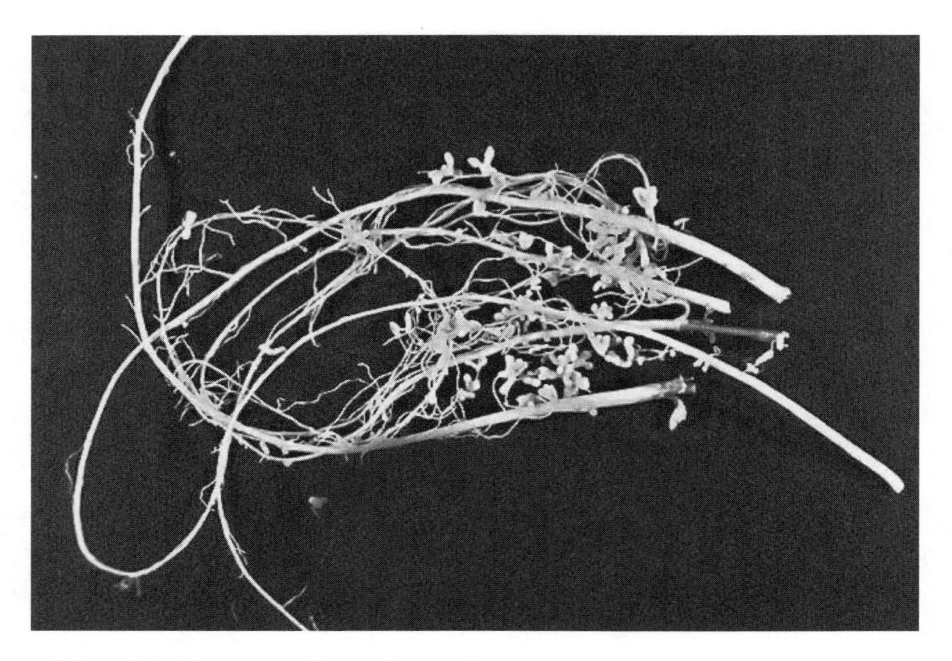

Root nodules along the roots of a soybean plant

Author: United States Department of Agriculture. https://commons.wikimedia.org/wiki/File:Soybean-root-nodules.jpg

Once nitrogen gas is turned into ammonia (NH_4^+) and deposited in the soil, it is available for plants to use. Another type of bacteria (nitrifying bacteria) also find this soil ammonia and convert it into another chemical form called nitrate (NO_3^-), which plants also love to use. Plants take in the nitrogen—an essential nutrient—and incorporate it into their tissues. When critters eat these plants, their own tissues take up the nitrogen . . . and when they are eaten by *other* critters, the nitrogen is then used in their own tissues, and so on and so forth. Once an animal dies and decomposers help to liberate the stored organic molecules, nitrogen is once again set free into the cycle. Considering that bacteria are driving this cycle, there's one final group of microbes—the denitrifiers—that grabs on to the nitrate in the soil and converts it back into nitrogen gas, removing it from the ecosystem and releasing it back into that huge reservoir in the sky.

REVIEW QUESTIONS

1. What is the difference between an organism's habitat and its niche?

2. Explain the significance of the following graph. Both species are occupying the same hypothetical niche.

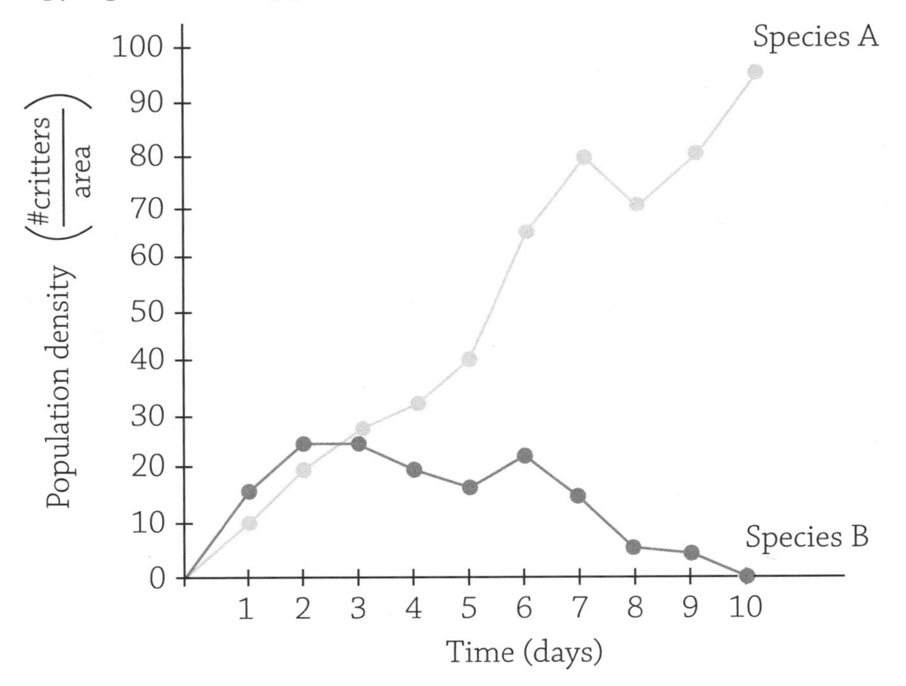

3. Why are food chains limited in steps (trophic levels)? Relate your answer back to the laws of thermodynamics.

4. What process removes carbon dioxide from the atmosphere? What processes release carbon dioxide back into the atmosphere?

5. Fill in the blanks: The competitive exclusion principle states that two different ______________________ cannot occupy the same ecological ______________________. The survivor is the one who was able to ______________________ the other.

6. The first trophic level in a food chain must be a ________________. Explain your answer.

7. Choose the correct term from each pair: Earth is a(n) **open/closed** system in regard to matter, but a(n) **open/closed** system in regards to energy.

8. The first step of the nitrogen cycle is ________________, when nitrogen gas is fixed into organic molecules. The two means of nitrogen fixation are ________________ and ________________.

9. Which of the following would increase the amount of nitrogenous compounds found in the soil?
   a. A decrease in the population of denitrifying bacteria.
   b. A decrease in the population of nitrogen-fixing bacteria.
   c. An increase in the population of legumes in a soil ecosystem.
   d. A reduction of lightning activity in a given area.

30 Human Impact on Ecosystems

MUST KNOW

- Humans depend on ecosystems for resources, and our needs sometimes cause damage.

healthy ecosystem is a complex web of many different populations. In general, the higher the biodiversity (different number of species), the more stable the ecosystem. Yet no matter how stable an ecosystem may be, it is still subject to change. Anything that disrupts the communities in an ecosystem is referred to as a **disturbance**. Disturbances to ecosystems can be natural or human-caused. Our **must know** for this chapter will focus on the human-induced stress: we need resources, and sometimes ecosystems suffer for it.

Natural Disturbances and Ecological Succession

Even without us humans around, ecological disturbances occur. Natural disasters can wipe out entire communities, such as when a tsunami or hurricane hits the coast. A tornado has the same devastating effect, but its swath of destruction is only along the path it travels across the landscape. Volcanoes, lightning-sparked forest fires, and floods are all examples of natural disasters. Once the disturbance is over, the ecosystem will work on repairing itself by reestablishing the populations that once inhabited the area; this is called **ecological succession**. There are two types of ecological succession, depending on the severity of the damage: primary succession or secondary succession.

Primary succession is starting from scratch . . . there is literally nothing but rock and not the slightest spark of life is present. The types of natural disturbance capable of creating these harsh conditions include volcanic eruptions (lava!) and landslides. The first forms of life to colonize the previously uninhabited area are called **pioneer species** and are usually moss and lichen blown in by the wind or brought in by migrating animals. These unassuming life-forms are able to live on the rocks and slowly help their disintegration into smaller particles, contributing to the formation of new soil.

Moss is a primitive type of plant, but lichen is not a plant. Lichen is also not a fungus . . . nor a bacterium . . . nor a protist. Strangely enough, lichen isn't actually a single type of life: lichen is a permanent mutualistic pairing of a photosynthetic bacterium (cyanobacteria) or protist, plus its buddy, a fungal filament that offers protection and housing. We learned about this symbiotic relationship in a previous chapter!

As the rock disintegrates and microorganisms begin to colonize newly formed niches, soil forms. Happily, grasses can now grow. After a few decades, a young forest will emerge, and eventually transition to an older forest with larger trees. A **climax community** is achieved once the ecosystem is back in a healthy equilibrium after years of succession.

Secondary succession is an easier ecological "healing" process. In this case, a natural disruption such as a fire or flood leaves the soil intact. The process of recolonization of plants is much quicker, because a healthy soil is already available. Primary succession takes a long, long time because the process of forming soil from rock is the most time-consuming step; secondary succession regrowth is much quicker.

For example, consider a building construction site that has been cleared of its vegetation. In no time at all, new plants around the area begin to grow and fill in the patches of empty soil—a perfect example of secondary succession. The process of regrowth continues until a climax community forms.

Human-Caused Disturbances

Our **must know** focuses on the stress the human population exerts on natural ecosystems. Consider the startling exponential growth curve of human population growth.

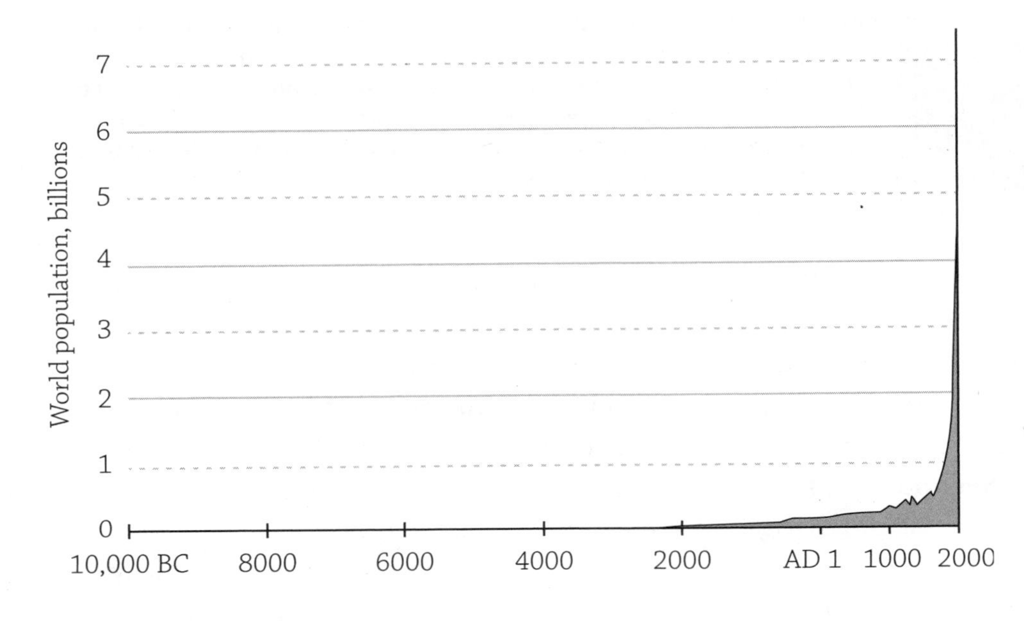

World population increase

Author: El T. https://commons.wikimedia.org/wiki/File:Population_curve.svg

There have been amazing advances that allowed the human population to increase exponentially, such as industrialized farming, better sanitation, and advances in medicine. But recall from an earlier chapter the significance of a population undergoing exponential growth. It is almost impossible to maintain such growth, because it required unlimited resources and space . . . certainly not sustainable when you consider Earth as a whole.

This, unfortunately, puts a strain on the planet's natural resources. Some resources are renewable, meaning they can be replaced at the same rate by which they are consumed. **Renewable** resources include solar energy, wind energy, hydroelectric, and geothermal. Others are **nonrenewable** because they are used faster than they are able to form. Fossil fuels (such as coal, natural gas, and petroleum) are nonrenewable resources.

IRL **Unfortunately, the means by which we acquire these nonrenewable resources are often damaging to ecosystems: hydraulic fracturing (fracking), drilling, and mining.**

CO_2 Emissions and Climate Change

Ecosystems innately strive for homeostasis and balance. Because matter on Earth remains in a closed system, an ecosystem will reach a happy equilibrium as an element (such as carbon) is cycled throughout the atmosphere and fixed back into organic matter as the biotic members of the ecosystem live, breathe, and eat. Sadly, the carbon cycle can be thrown out of whack by an "outside" introduction of atmospheric carbon dioxide. If this imbalance becomes too big, it can lead to climate change. **Climate change** refers to a global phenomenon brought about by burning fossils fuels and increasing the amounts of carbon dioxide in the atmosphere.

Carbon dioxide is considered a **greenhouse gas** because it increases the amount of heat trapped in the atmosphere. Solar (sun) radiation enters the atmosphere, heats Earth's surface, and the energy is "bounced back" as infrared (IR) radiation. This IR radiation is absorbed by atmospheric gases and ends up warming the air . . . the greenhouse effect in action. Human-generated greenhouse gases contribute greatly to this effect. Other greenhouse gases include methane, nitrous oxide, and chlorofluorocarbons (CFs).

By studying carbon dioxide levels locked in ice samples from hundreds of thousands of years ago (and more recent readings collected directly from the atmosphere), data can be studied to show a statistically significant increase in atmospheric carbon dioxide, beyond that of natural geological cycling.

EASY MISTAKE

Do not make the mistake of thinking the greenhouse effect is a bad thing. There would not be life on Earth if it wasn't for the (natural) greenhouse effect. Without it, the atmosphere would not be warm enough to sustain life!

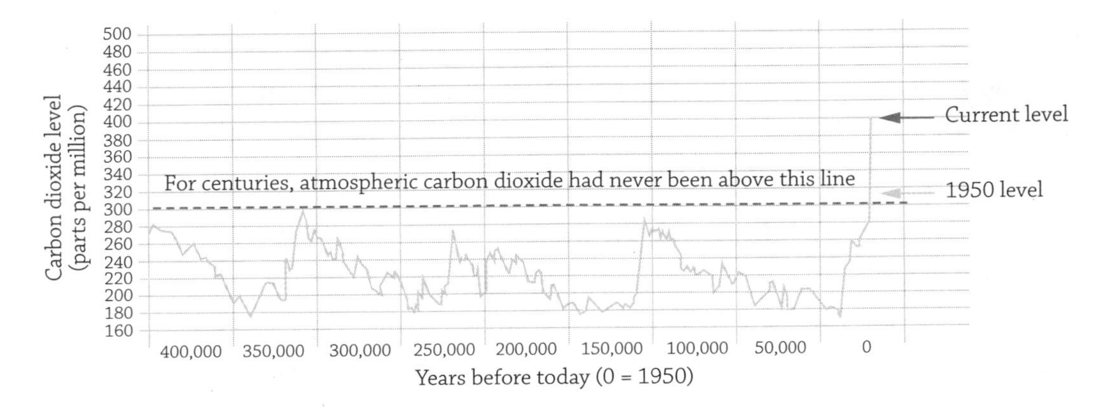

Atmospheric CO_2 levels during the last 400,000 years

Author: Vostok ice core data/J.R. Petit et al. https://climate.nasa.gov/evidence.

Pollution

Pollution is defined as adding something to the environment that has a negative impact on the environment. Pollution can be as obvious as trash floating along the edges of a stream, or more insidious such as acid rain. We will discuss two general types of pollution, based on whether it originates in the air or in the water.

Air Pollution

Most of our air pollution comes from transportation emissions. There are a lot of automobiles out there, and most of them rely on fossil fuels. Furthermore, power plants, refineries, and industrial plants also rely on fossil fuels and add to the atmospheric pollution. Air pollutants can be categorized as primary pollutants and secondary pollutants. Any nasty chemical released directly into the atmosphere is a **primary pollutant**. If that chemical reacts with the atmosphere to create a new (and also nasty) second chemical, that is referred to as a **secondary pollutant**. On the following page is a table summarizing some of these pollutants.

Primary air pollutant	Secondary air pollutant
Nitrogen oxides	Smog
Sulfur dioxide	Acid rain
Carbon monoxide	
Particulate matter (microscopic bits of soot and metals)	

Smog, a secondary air pollutant, is a hazy, low-hanging layer that obscures the view of the horizon. It is composed of particulate matter and something called ground-level ozone. **Ozone** (O_3) is normally found in a layer in Earth's upper atmosphere, where it plays an important role of absorbing dangerous UV radiation that would otherwise damage cells. Ground-level ozone is not natural, and it is not beneficial. The nitrogen oxides released from burning fossil fuels react with atmospheric oxygen and sunlight to form ground-level ozone. Breathing even relatively low levels of ozone can reduce lung function and cause a variety of health problems.

Acid rain is another secondary pollutant, formed when water molecules react with atmospheric gases.

Atmospheric gas	Resulting acid pollutant
Carbon dioxide	Carbonic acid
Nitrogen oxide	Nitric acid
Sulfur oxide	Sulfuric acid

Rainwater is normally slightly acidic (neutral pH is 7; anything lower than 7 is considered acidic), and naturally occurring carbonic acid alone can acidify rain to a pH of 5.6. Add to that mix human-generated sulfuric and nitric acid, and the pH can reach levels as low as 3! Acid rain can have devastating effects on aquatic environments, and the animal and plant inhabitants of such acidified ecosystems suffer greatly.

Water Pollution

Water pollution can mean the obvious trash contamination of water, or a more subtle chemical imbalance that causes unhealthy events, such as algal blooms. An **algal bloom** is just what it sounds like—a sudden increase in the amount of algae growing in either a freshwater or saltwater ecosystem. In truth, the term *algal bloom* can also refer to an overgrowth in cyanobacteria (these photosynthetic prokaryotes are often mistaken for algae, which is a protist). Either way, the hallmark sign of an algal bloom is unusually green water (though an algal bloom can also color the water red or brown). Why is this bad, you ask? Green means lots of happy photosynthetic life, and how could that be bad? If the process of photosynthesis releases oxygen as a by-product, why would that hurt an aquatic ecosystem? But I'm getting ahead of myself. Let me explain why this overgrowth happens in the first place.

A **limiting factor** refers to a nutrient needed for something to grow, and its scarcity limits the growth of the organism. Phosphorus is a limiting factor in the soil for plant growth, which is why it is a major component of fertilizer. Sometimes excess fertilizer leaches from the soil into water runoff, ending up in nearby lakes, ponds, and other bodies of water. Phosphorous also acts as a limiting factor in the water, keeping algal growth in check. Once runoff enters the aquatic ecosystem, the algae happily begin to grow like crazy (which is why the water turns green!).

The problem occurs when the algae begins to die (they are short-lived little critters). Any dead organic matter will begin to decay and be consumed by decomposers. These busy decomposers are undergoing cellular respiration and consuming the dissolved oxygen at an alarming rate, causing hypoxic (low oxygen) conditions. Because the oxygen levels are dropping, aquatic animals and plants start to die off.

Unfortunately, fertilizer runoff is not the only source of phosphorous water pollution. Many household cleaning solutions contain phosphorous, and their overuse can also lead to algal blooms.

REVIEW QUESTIONS

1. Compare and contrast the two types of ecological succession.

2. A _______________ resource can be replaced at the same rate by which it is consumed. Give an example of this type of resource: _______________.

3. How are primary and secondary pollutants related?

4. Using the following terms, explain how phosphorus can damage an aquatic ecosystem: **algal bloom, decomposers, fertilizer, hypoxic, limiting factor, phosphorus.**

5. For each of the following examples, indicate whether it is primary succession (PS) or secondary succession (SS):
   a. Chernobyl nuclear contamination
   b. California wildfires
   c. Mount St. Helens volcanic eruption
   d. Deforestation of Amazon rainforest
   e. Retreating glaciers in Alaska

6. What important role does lichen play in primary succession?

7. How does the burning of fossil fuels contribute to the phenomenon of climate change?

8. Which of the following is a nonrenewable fuel resource?
   a. Natural gas
   b. Solar energy
   c. Geothermal energy
   d. Hydroelectric power

Answer Key

1

Chemical Bonds and Reactions

1. This is an example of an intermolecular bond because it is occurring between two separate molecules (in this case, two water molecules) and is due to the slightly positive side of one molecule attracting the slightly negative end of a different molecule.
2. Both laws of thermodynamics apply to this example. Breaking up glucose and converting it into another form of energy describes the first law (energy cannot be created, but it can be converted from one form to another). The second law (every energy transformation includes some energy lost as heat) is why our metabolism provides body heat.
3. A reaction that breaks bonds is called a **catabolic** reaction because energy is **released.**
4. a. This is a noncompetitive inhibitor, meaning it binds at a site other than the active site. b. The effects of the inhibitor cannot be overcome by adding more substrate because when the non-competitive inhibitor binds to the enzyme, it changes the shape of the active site.
5. A **covalent** bond involves sharing two electrons between two atoms.
6. Reaction A requires an input of energy; it creates products that contain more energy than reactants. Reaction B has a negative delta G.

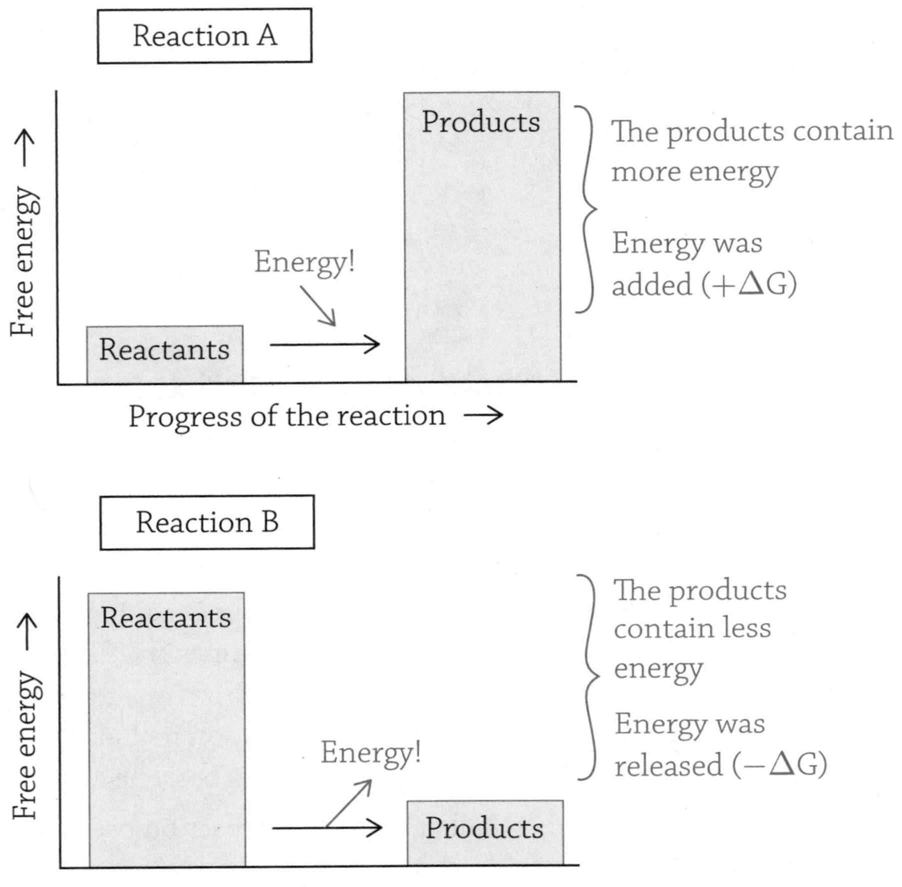

7. Enzymes are catalysts because they speed up reactions. They're biological because all enzymes are proteins that are made by cells.
8. A catabolic reaction is one where bonds are broken and energy is released. An enzyme helps this to occur because when it grabs onto the substrate, it squeezes the molecule and puts stress on the covalent bonds, making them easier to break.
9. Both allosteric enzymes and non-competitive inhibitors rely on a substance binding to the enzyme at a site other than the active site. The difference, however, is when something binds to that other site it can either decrease or increase the activity of an allosteric enzyme. A non-competitive inhibitor can only decrease the activity.

10. **d.** An anabolic reaction is one where energy is used to build something that stores the energy. In this case, energy from the sun is used to build an energetic molecule of glucose from carbon dioxide and water.

Properties of Water

1. Two water molecules involved in a hydrogen bond:

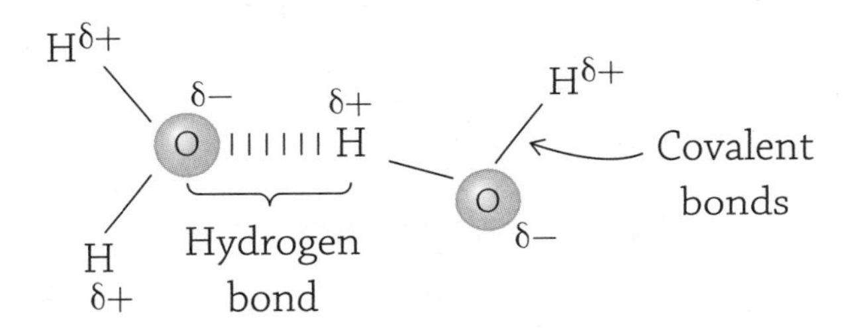

   Remember to draw your hydrogen bond as a dashed line and NOT a solid line (solid lines are reserved for covalent bonds).

2. When a water molecule forms a hydrogen bond with another water molecule, this is called **cohesion**; when a water molecule hydrogen bonds to another substance, it is called **adhesion**.
3. Water becomes less dense when it freezes (which is why your ice floats on the surface of your drink).
4. Water resists quick changes in temperature and doesn't easily evaporate (or freeze); this is due to water's **high heat capacity**.
5. It is only a partial charge (either negative or positive) because electrons are not evenly distributed throughout the molecule. If you said a full positive or negative charge, you would mean a complete loss (or gain) of an electron, which would make a water molecule an ion (which it's not!).
6. The polarity of a water molecule (as described in the previous question) allows the partially positive side of one molecule to attract the partially negative side of another molecule. It's like two magnets sticking to one another.

7. Cohesion creates surface tension because the water molecules are holding on to one another.
8. The ice floating on the surface acts as insulation to the water below, keeping it warmer. Furthermore, if ice sank, entire bodies of water would freeze solid in the winter, killing all life within it.
9. When water melts or boils, the hydrogen bonds between individual water molecules are being broken apart. The covalent bonds within individual water molecules do NOT break; otherwise, when you boil water, you would be producing oxygen and hydrogen gases (you're not!).

Macromolecules

1. A polymer is a large **macromolecule** (term for large molecule) that is composed of repeating subunits called **monomers**.
2. The creation of a peptide bond between two amino acids (via a dehydration reaction):

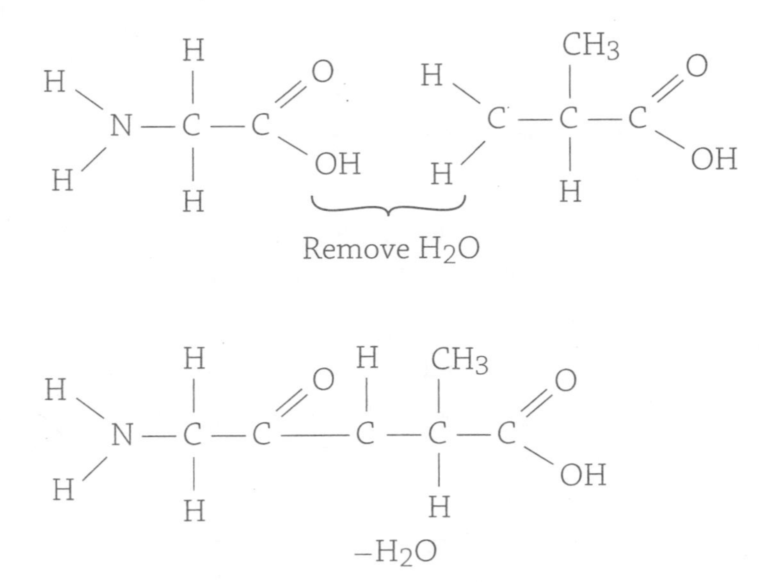

3. a – 3; b – 5; c – 1; d – 6; e – 4; f – 7; g – 2
4. DNA is double stranded; RNA is single stranded.

   DNA contains the sugar deoxyribose; RNA has ribose.

   DNA has the base thymine; RNA has uracil.

   DNA remains in the nucleus; RNA travels out to the cytoplasm.
5. First of all, to ensure the width of DNA remains constant, a pyrimidine (single-ringed base) must always pair with a purine (double-ringed base). Furthermore, both adenine and thymine like to form two hydrogen bonds, whereas guanine and cytosine like to form three hydrogen bonds.
6. a – 2; b – 5; c – 6; d – 3; e – 1; f – 4; g – 7
7. If a protein had a stretch of hydrophilic amino acids, you would expect them to **face toward** water.
8. When you digest starch, **hydrolysis** enzymes work at breaking up the starch polymer.
9. Primary structure is the sequence of individual amino acids in a protein. The statement describing that one subunit is composed of 141 amino acids and the other subunit is composed of 146 amino acids is describing the primary structure.

   Secondary structure is a regular folding pattern (alpha helix or a beta-pleated sheet) created by hydrogen bonding between the amino acids. It was not mentioned in this description.

   Tertiary structure is the 3D shape a protein makes when it starts to fold up into a particular pattern, based on the R groups of the individual amino acids. There is no direct mention of what each individual protein looks like when it's folded up.

   Quaternary structure is achieved when there are two or more separate proteins involved in the final overall structure. Hemoglobin has four total protein subunits; this is referring to its quaternary structure.
10. **c.** The monomer of protein is an amino acid. The other answers are wrong because ribose is the sugar in an RNA nucleotide, cellulose is a structural polymer, and the hydrocarbon chains compose the hydrophobic portion of a phospholipid.

Overview of the Cell

1. The main structural component of the cell membrane is the **phospholipid**.
2. Archaea prokaryotes tend to live in environments of extreme pH, salinity, and heat. This is a remnant of their ancient ancestry because the first cells to evolve on Earth had to survive in the extreme conditions present on the planet 3.5 billion years ago.
3. A. **phospholipid,** B. **protein,** C. **hydrophilic phosphate group,** D. **hydrophobic tails,** E. **phospholipid bilayer**
4. Transmembrane proteins span the entire width of the membrane (from the outside of the cell all the way to the inside of the cell), thus providing a tunnel through which things can move.
5. If a cell suddenly was exposed to much colder temperatures, the cell membrane would become too stiff. In order to counteract the stiffness, it would increase the percentage of unsaturated phospholipids. Unsaturated phospholipids have double bonds in the tails of the fatty acid chains, which make them unable to pack tightly together. Because they can't pack tightly, this increases the fluidity (and helps to counteract the stiffness caused by cold temperatures).
6. Prokaryotic cells (bacteria) are the smaller and simpler cell types that lack organelles.
7. Characteristics of life: a. made of cells, b. reproduces and creates progeny, c. has a heritable genetic code (DNA), d. uses materials and undergoes metabolic processes, e. requires energy to survive, f. strives to maintain constant internal conditions (**homeostasis**), g. responds to its environment, h. evolves
8. A single phospholipid is *amphipathic*, meaning the phosphate head group portion of the molecule is **polar** and **loves** water, whereas the two fatty acid chains are **nonpolar** and **hate** water.

9. If a cell lived in very hot conditions (such as a volcanic deep-sea vent), its phospholipids would be composed of more **saturated** fatty acids in order to prevent the cell membrane from becoming too fluid. This is because saturated fatty acids form straight chains (without kinks and bends) and can pack tightly together. If the phospholipids of the phospholipid bilayer can pack tightly together, that makes the cell membrane more solid (which is good in really hot conditions).

10. List three possible roles for cell membrane proteins, and provide a brief description of each:
    1. Transmembrane proteins can be used for transport, providing a pathway from the outside of the cell into the interior of the cell. Aquaporin is a specific example, and it provides a pathway for water molecules, increasing their movement into cells.
    2. Proteins embedded in or on the cell membrane can be enzymes and play important roles in cell metabolism.
    3. Some proteins are receptors and can latch onto a chemical signal on the outside of the cell. Once bound, it "transmits" the signal into the cell (this is called the process of signal transduction).

11.
    a. Cell membrane **Y**
    b. Nucleus **N**
    c. Flagella **Y**
    d. DNA **Y**
    e. Cell wall **Y**
    f. Ribosomes **Y**
    g. Mitochondrion (an organelle) **N**
    h. Nucleoid **Y**

12. An advantage to prokaryotic cells' simplicity is it enables them to reproduce very quickly. Furthermore, their quick reproduction helps them to evolve quicker!

13.

a. An atom
b. A molecule
c. A virus ← Nope, not alive (because a virus is not made of a cell!)
d. **An amoeba ← A single-celled amoeba is definitely alive!**
e. A tree

14. The cell membrane is how a cell can create internal conditions different from its surroundings. The semipermeability of the membrane helps to keep certain things in the cell (and keep other things out of the cell).

Eukaryotic Cells and Their Organelles

1. Organelles create compartments within the cell, which can be used for microenvironments (creating conditions different from other areas of the cell).
2. A cell with a high amount of rough endoplasmic reticulum would be specialized in exporting large amounts of proteins (the rER makes proteins destined to be exported from the cell).
3.

Organelle	Function
Chloroplast →	Converts sun energy into glucose
Golgi apparatus →	Modifies and ships proteins to their final destination
Lysosome →	Cell's "stomach"
Mitochondrion →	Converts glucose into ATP
Nucleus →	Controls all the cell's functions
Nuclear pore →	Allows mRNA to pass out of the nucleus
Rough endoplasmic reticulum →	Makes proteins for export
Smooth endoplasmic reticulum →	Important in drug detoxification

4. Both mitochondria and chloroplasts have their own DNA, their own ribosomes, and they can reproduce on their own (independently of the cell).
5. A = outer membrane; B = inner membrane space; C = inner membrane; D = matrix
6. The glucose needed for cellular respiration in an animal cell comes from **food/eating**, whereas the glucose needed for cellular respiration in a plant cell comes from **photosynthesis/the sun**.
7. Plant cells have cells walls, central vacuoles, and chloroplasts (animal cells do not have these things).
8. The lysosome forms a space in which the pH is much lower (more acidic) than other locations within the cell. It also contains a bunch of digestive enzymes that prefer to function in these acidic conditions.
9. Fill in the blanks with the correct organelles: When a protein is made for export, it is created by **ribosomes** stuck onto the surface of the **rough endoplasmic reticulum**.
10. **False!** Only plant cells have chloroplasts, but *both* plant and animal cells have mitochondria.
11. A = outer membrane; B = thylakoid; C = granum; D = stroma
12. Turgor pressure occurs when the central vacuole of a plant cell fills with water. As it swells, it presses against the surrounding cell wall. This provides structure and rigidity to the plant cell (turgor pressure!).
13. **c.** A muscle cell specializes in movement; therefore, it would need a lot of energy to do so (and would generate that energy through cellular respiration occurring in mitochondria).

Cells and Energy Transformation

1. The energy in ATP is released by cleaving the bond between: **c. The third phosphate group and the second phosphate group**

2. Cellular respiration is the process where cells convert the stored energy of **glucose** into the useable form, **ATP**.
3. Cellular respiration is a **catabolic** reaction because energy released as a molecule of glucose is broken down into smaller molecules of carbon dioxide.
4. The overall purpose of fermentation is to re-create NAD^+ from NADH in order to keep glycolysis running in the absence of oxygen. The purpose of fermentation is NOT to create lactic acid or ethanol.
5. In photosynthesis, the light reactions create **ATP** (a source of energy) and **NADPH** (a source of hydrogens).
6. The high concentration of hydrogen ions are stored in the inner membrane space of the mitochondrion, or the thylakoid space of the chloroplast.
7. If adenosine triphosphate is analogous to a battery, the fully charged form is **ATP** and the dead battery form is **ADP**.
8. In glycolysis, two ADP are converted into two **ATP**, two NAD^+ are reduced to form two **NADH**, and glucose is oxidized into two molecules of **pyruvate**.
9. The role of oxygen is to grab the low-energy electron at the end of the electron transport chain. Once the oxygen also combines with a couple hydrogen ions, it turns into water.
10. Photosynthesis is an **anabolic** reaction because energy is required in order to build a molecule of glucose from smaller molecules of carbon dioxide.
11. The Calvin cycle creates glucose from carbon dioxide. In order to do so, it needs a source of **hydrogens** (provided by NADPH) and **energy** in order to create new covalent bonds (provided by ATP). Both NADPH and ATP are provided by the **light reactions** occurring on the thylakoid membranes.

Cell Transport

1. In order for a cell to actively transport a substance across its cell membrane, there needs to be a protein pump embedded in the cell membrane and the energy (ATP) needed to make it run.
2. **Diffusion** simply means the movement of a substance from an area of high concentration to low concentration. **Osmosis** is the diffusion of water, specifically. If in order for a molecule to diffuse across a cell membrane it needs a protein to act as a tunnel, it is referred to as **facilitated diffusion**.
3. The process of endocytosis is active transport because energy is required to change the shape of the cell in order to engulf a food particle.
4. A cell living in hypotonic conditions must deal with osmosis occurring **into the cell**.
5. The paramecium would suddenly be in hypertonic conditions, which means water would diffuse out of the cell (and the cell would most likely die).
6. $\Psi_s = -iCRT$

   $i = 2$ (because NaCl will break up into Na^+ and Cl^- in solution)
   $C = 0.23$ M
   $R = 0.00831$ liter MPa/mole K
   $T =$ temperature in kelvins ($20°C + 273 = 293$)
   $\Psi_s = -iCRT = -(2)(.23M)(0.00831 \text{ liter MPa/mole K})(293)$
   $\mathbf{\Psi_s = -1.1\ MPa}$
7. Facilitated diffusion would help the passive movement of glucose molecules into the cell. Facilitated diffusion provides a tunnel through which the glucose molecules could diffuse.
8. It is bad to disrupt a cell's chemical gradients because the cell would no longer be able to do work. For example, if there were no longer hydrogen ion gradients in a mitochondrion, the cell would no longer be able to generate ATP.

9.

a. A plant cell floating in pure water: **hypotonic**.
b. A protist cell normally living in a freshwater lake transferred to a marine environment: **hypertonic**.
c. A bacterium in a solution that has the same water potential as inside the bacterial cell: **isotonic**.

10. Some cells that live in hypotonic conditions have evolved to rely on an organelle called a **contractile vacuole** to pump out the extra water that diffuses inward.

11. The overall water potential is the sum of the solute potential plus the pressure potential: $\Psi = \Psi_s + \Psi_p$

$\Psi_p = 0$ (because it is in an open container)

$\Psi_s = -2.0$ MPa

Therefore, overall $\Psi = -2.0$ MPa

12. **a.** When a cell releases carbon dioxide from cellular respiration, it diffuses from the cell because the concentration of carbon dioxide in the surrounding environment is lower than that inside the cell.

Signal Transduction and Cell Communication

1. Protein signals rely on signal transduction because the proteins themselves cannot pass through the cell membrane.
2. In order for the receptor to do its job, it needs to relay the signal into the interior of the cell. The protein does so by changing the shape of the inside portion of the receptor protein. If the receptor did not span the membrane, it could not do this.
3. c, a, d, b
4. The transduction step amplifies the message as soon as it enters the cell. This ensures that even a small signal (only a few signal molecules) is enough to elicit a huge response.

5. Reception, transduction, response.
6. When the receptor binds to the signal, the inside region of the receptor protein changes shape. This moves the message (but not the actual chemical signal) into the cell.
7. c, d, b, a
8. Neurons (the cells of the nervous system) use ligand-gated ion channels in order to transmit nerve impulses from one cell to the next.
9. Protein kinase enzymes play an important role in signal transduction because they **amplify/increase** the message by creating a **phosphorylation** cascade.
10. **d.** The yeast-mating signal pathway begins with the signal causing a protein kinase cascade, and that leads to the activation of specialized transcription factors. That in turn causes specific gene expression, which leads to directional growth of the yeast cell.

DNA and RNA: Structure of Nucleic Acids

1. From *smallest* to *largest*: **nucleotide, gene, chromosome, genome**
2. The upright portion of the DNA backbone is composed of alternating **sugar (deoxyribose)** and **phosphate** portions of the nucleotides. This backbone of the DNA is very strong because it is linked by **covalent** bonds. The two strands of the double helix hold onto one another because of **hydrogen** bonding between the **nitrogenous bases** of complementary nucleotides.
3. The width of DNA would be inconsistent because both cytosine and thymine are the smaller pyrimidine bases, whereas guanine and adenine are larger purine bases.
4. RNA nucleotides contain the sugar ribose (DNA contains deoxyribose).

   RNA is single stranded instead of double stranded, like DNA.

   RNA has the nitrogenous base uracil instead of DNA's thymine.

5. DNA sequence: ACTGACA

   Complementary RNA sequence: UGACUGU

6. First, a single-ringed pyrimidine (T and C) must pair with a double-ringed purine (A and G). Second, adenine and thymine both create two hydrogen bonds, whereas guanine and cytosine both create three hydrogen bonds.

7. What is the name of the monomer of DNA? **nucleotide.** Which part of the monomer is the basis for the genetic language? **the base (nitrogenous base).**

8. DNA sequence: GGACACTT

   Complementary DNA sequence: CCTGTGAA

9. A strand of DNA is wound around special proteins called **histones**. This mixture of DNA and protein is called **chromatin**.

10. **b.** Every correct base pairing (A-T, G-C, A-U) requires a two-ringed purine to pair with a single-ringed pyrimidine. In DNA, this makeup is especially important because it ensures that the width of the double-stranded molecule remains constant.

Replication of DNA

1. Before a cell divides, it must replicate its DNA. This ensures each resulting daughter cell has a full copy of chromosomes.

2. The double helical structure of DNA is antiparallel. This means that the 3′ end of one strand (which has a **hydroxyl/–OH** group hanging off of it) must line up with the **5′** end of the complementary strand (which has a phosphate group hanging off of it).

3. The lagging strand needs multiple RNA primers.

4. Free nucleotides are added to a growing strand of DNA through a **dehydration** reaction, which creates a new **phosphodiester** bond.

5. In order for DNA polymerase to attach a nucleotide through a dehydration reaction, there needs to be a free –OH (hydroxyl) group available; the 3′ end of a DNA molecule has a free hydroxyl group.

6.

   a. The enzyme **helicase** unwinds and unzips DNA in order for replication to occur.
   b. The enzyme **primase** must first synthesize a small piece of RNA in order for another enzyme to begin adding DNA nucleotides to the growing strand.
   c. Once DNA is unzipped, **single-stranded binding proteins** stick to the unpaired bases in order to keep the two strands apart.
   d. The enzyme **ligase** glues together fragments of newly synthesized DNA by creating new phosphodiester bonds.
   e. The bulk of the work of creating a new strand of DNA is done by **DNA polymerase**, the enzyme responsible for pairing up complementary nucleotides.

7. First, in order for DNA polymerase to be able to create two identical DNA molecules, the original DNA must be double stranded. Otherwise, if the original DNA was a single strand, the newly synthesized strand would be a complementary (not identical) copy. Second, double-stranded DNA is much more stable than a single-stranded piece.

8. **b.** One strand from the original DNA molecule and one created from new nucleotides

9. **False.** Immediately after the DNA replication (and before the cell splits into two daughter cells), another DNA polymerase will replace the RNA primers with DNA.

10. **a.** When DNA is replicated, it is crucial that the resulting daughter cells have exactly the same compliment of DNA as the parent cell.

Gene Expression and Differentiation

1. The process of **differentiation** is when an undifferentiated **stem** cell begins to selectively express certain genes and turns into a specific tissue type.

2. The promoter is a sequence in the DNA right before a gene. The enzyme RNA polymerase needs to land right before a gene to be transcribed, and the landing strip is the promoter.
3. General transcription factors are small proteins that bind to the **promoter** and help **RNA polymerase** to land properly. If the cell needs to create a lot of mRNA, however, it needs the help of specific transcription factors called **activators**. Unlike the general transcription factors, these specific transcription factors bind upstream of the promoter at sequences in the DNA called **enhancer** regions.
4. By adding methyl groups to [**DNA**/histones], gene transcription will be [increased/**decreased**]. If, instead, acetyl groups are added to [DNA/**histones**], gene transcription will be [**increased**/decreased].
5. General transcription factors bind at the **TATA box** within the promoter, whereas specific transcription factors (activators) bind at a specific combination of **distal control elements** in the **enhancer** region.
6. The *trp* operon is a repressible operon, meaning it is normally transcribing the genes for the synthesis of tryptophan. This suggests that the bacterium normally does NOT have tryptophan available in its environment, because by default, it is creating the enzymes for the synthesis of the amino acid.
7. For each of the following, indicate whether the statement is true or false:
    a. During transcription, the entire chromosome is unwound and unzipped: **False**. Only the portion of the chromosome that contains the gene being expressed is unzipped.
    b. The process of transcription involves the enzyme RNA polymerase moving down the entire chromosome and creating a piece of mRNA: **False**. Yes, the enzyme is RNA polymerase, but no, it is not transcribing the entire chromosome (refer back to part "a").
    c. For a given gene, only one side of the DNA double helix contains the correct code for protein synthesis: **True**. This is called the template strand.

8. Transcription factors are small proteins that first land on a gene's promoter, and then help RNA polymerase to align properly on the promoter.
9. A. enhancer region; B. activator proteins; C. RNA polymerase; D. promoter; E. gene; F. DNA bending protein
10. mRNA: CG **AUG** ACU AGC UGG GGG UAU UAC UUU UAG

    Protein: MET THR SER TRY GLY TYR TYR PHE [stop codon]
11. The *lac* operon is a(n) [**inducible**/repressible] operon because it is normally not transcribing the genes for lactose digestion.
12.

Operon	Purpose of structural genes	Default activity of repressor	Default expression of mRNA	Inducible or repressible?
lac	Create enzymes needed to **digest** lactose	**Active** (able to bind operator)	Transcription **not occurring**	**inducible**
trp	Create enzymes needed to **synthesize** tryptophan	**Inactive** (unable to bind operator)	Transcription **occurring**	**repressible**

13. **c.** When DNA has methyl groups attached to it, the DNA remains condensed (tightly coiled) and prevents access by the transcription machinery. If the gene cannot be accessed by RNA polymerase, gene expression is decreased.

The Cell Cycle and Mitosis

1. Mitosis is specifically focused on division of the nucleus, or more specifically, the proper division of the replicated chromosomes. Cytokinesis occurs after mitosis and is when the entire cytoplasmic contents of the cell is split in half.
2. Fill in the blanks: After the S phase, each chromosome is now composed of two genetically identical **sister chromatids** and they are attached at the **centromere**.

3. There are three checkpoints that a dividing cell must pass through: the G_1 checkpoint, the G_2 checkpoint, and the M checkpoint (during mitosis).
4. The cell is currently in the G_2 phase of the cell cycle. Each of the chromosomes have already been replicated in the S phase (as indicated by the presence of two sister chromatids attached to one another), and there are two centrosomes (organelles are copied during G_2).
5. Absolutely. There are many instances when a cell should NOT divide and instead will exit the cell cycle during the G_1 checkpoint. A cell will then be in the G_0 phase, and instead of actively dividing, will instead just do its cellular thing (whatever that may be).
6. During G_1, the cell grows in size and performs normal metabolic functions for the given cell type. During S, DNA will be replicated (synthesized) in preparation for cell division. In G_2, the cell will copy its organelles to ensure the two daughter cells have the correct assortment. Finally, the cell divides into two, first by mitosis (division of the chromosomes), and then cytokinesis (division of all the cytoplasm).
7. Any genetic differences between daughter cells after mitosis are considered to be mutations. **False**! The goal of mitosis is to create two genetically identical daughter cells.
8. If a cell fails a checkpoint but has already copied its DNA, it may undergo **apoptosis** in order to prevent a defective cell from dividing (and possibly becoming cancerous).
9. The cell is currently in metaphase. The middle dividing line (the metaphase plate) has one sister chromatid on either side. This ensures when the cell splits in two, each resulting daughter cell will receive one copy of each chromosome.
10. Growth factor
11. **b.** Apoptosis occurs when a cell is not functioning properly. Instead of dividing, the cell will destroy itself.

Meiosis

1. Meiosis creates cells with **half** the DNA of the parent cell. Specifically, each gamete contains only **one set** of chromosomes.
2. **False.** The only cells in your body that can undergo meiosis are the special cells in the ovaries and testes that produce the eggs and sperm.
3. The specialized cells in the ovaries and testes that produce the gametes start off as **diploid** cells. Once they divide by **meiosis,** they produce the **haploid** gametes (egg or sperm). When an egg and sperm fuse in sexual reproduction, it creates a **diploid** zygote cell.
4. Nothing. The sister chromatids are genetically identical, so if segments switched between the two, there would be no difference.
5. The purpose of sexual reproduction is to create new genetic combinations in the offspring. When each individual creates different variations of egg or sperm, this ensures their offspring will have their own unique genetic makeup.
6. The cell that divides by meiosis to create the gametes is **diploid** because it contains **two sets** of chromosomes (referred to as a **diploid** cell). The resulting gametes (either egg or sperm) contain **one set** of chromosomes and are called **haploid** cells.
7. Synapsis occurs when the homologous chromosomes clump together. Also, the pairs of homologous chromosomes (called tetrads) line up in the middle during metaphase I and are split up into separate cells during anaphase I.
8. When the two pairs of homologous chromosomes segregate into separate cells during anaphase I, the resulting daughter cells are haploid (because they contain only one of each homologous chromosome pair).
9. Meiosis creates gametes with different genetic combinations through the process of **crossing over** (when segments of two homologous chromosomes swap with one another) and **independent assortment** (the homologous chromosomes line up in a different pattern during metaphase I).

10. **d.** When a diploid cell begins meiosis, it has two sets of DNA (meaning it has both pairs of homologous chromosomes). Meiosis I results in the homologous chromosomes being separated into two different cells; each resulting cell—which contains only a single set of homologues—is now haploid.

Mendelian Genetics

1.

   1. e
   2. b
   3. f
   4. a
   5. g
   6. c
   7. d

2. In this example, we would use R to indicate the dominant red allele and r for the recessive ebony allele. The heterozygous red-eyed fly has a genotype of Rr, and the ebony-eyed fly must be rr.

	R	r
r	Rr	rr
r	Rr	rr

The percentage of offspring with the recessive ebony eyes (rr) would be $\frac{2}{4}$, or 50%.

3. If the male cat is heterozygous for both traits, his genotype must be BbSs. The female shows both recessive phenotypes, so her genotype must be bbss. When determining the odds that the kitten will have white, long fur (bbss), create two separate Punnett squares to determine the odds of the kitten having either white fur *or* long fur.

	B	b
b	Bb	bb
b	Bb	bb

In regards to fur color, the odds of producing a white kitten are 50%, or ½. Now set up a second Punnett square looking only at the fur length:

	S	s
S	SS	Ss
s	Ss	ss

In regards to fur length, 25% (or ¼) of the kittens would have long fur. The question is asking you about the odds of producing a white kitten with long fur, so using the rule of multiplication, multiply the odds of producing each of these traits independently:

$\frac{1}{2} \times \frac{1}{4} = \frac{1}{8}$ . . . there's a 1 in 8 chance of these two cats producing a white, long-haired kitten!

4. The man is X^CY (he has the gene for normal color vision), and you know that the woman must be X^CX^c because her father is color-blind (and the father provides the X^c chromosome to his daughter).

	X^c	Y
X^C	X^C	X^CY
X^c	X^C	X^cY

Of their four offspring, they could produce a colorblind boy (¼ chance).

5. The purple allele is dominant because it masks the white allele in the second generation (they are all purple).

6. **False.** It will always result in all the offspring having the dominant phenotype (because all the offspring are heterozygous and the dominant allele will be the "winner").

7. The white rooster (recessive phenotype) is heterozygous for a large comb, so his genotype is bbLl. The heterozygous hen is BbLl. In regards to feather color:

	b	b
B	Bb	Bb
b	bb	bb

The odds of creating a chicken with white feathers is ½.

In regards to comb size:

	L	l
L	LL	Ll
l	Ll	ll

There is a ¾ chance that the offspring will have a large comb. Therefore, the odds of creating a chick with both white feathers and a large comb is ½ × ¾ = ⅜.

8. The color-blind man has the genotype X^cY. The woman must be XX. Therefore:

	X^c	Y
X	X^cX	XY
X	X^cX	XY

In order for a daughter to be a carrier, they must have one allele for color blindness (X^cX). There is a 100% chance that their daughter will be a carrier for color blindness.

9. **d.** If you cross a homozygous dominant (e.g., AA) with a heterozygous individual (Aa), all offspring would have at least one of the dominant alleles; therefore, all offspring would exhibit the dominant phenotype.

The Theory of Natural Selection

1. **Evolution** is change over time. **Natural selection** is the mechanism that explains how it happens.

2. The current belief at the time was Earth was only a thousand years old, and thus not enough time existed to allow evolution to occur. The geologists, however, provided the insight that Earth was actually billions of years old, providing plenty of time for the slow process of natural selection. Furthermore, Earth itself could change during that huge amount of time.
3. The principle of inheritance of acquired characteristics believes that traits acquired during one's lifetime could then be passed on to the offspring. This isn't true, because a trait must be genetic for it to make it to the next generation.
4. An **adaptation** is a helpful trait found in a population that is the product of natural selection.
5. A species of woodpecker that evolved to possess brightly colored head feathers is an example of **microevolution**. The ancestral woodpecker that gave rise to many different species better adapted for their particular environment is an example of **macroevolution**.
6. a–4; b–3; c–2; d–1
7. In natural selection, the environment selects the most-fit phenotype. In artificial selection, humans select the most-fit phenotype.
8.
   a. Any beneficial genetic mutation will be passed on to the next generations. **False**. Only mutations in the gametes (eggs and sperm) are able to be passed on to the next generation.
   b. Populations, but not individuals, are able to evolve. **True**.
   c. Once a most-fit phenotype is selected through evolution, the population's adaptations are "fixed" and will no longer change. **False**. An adaptation that was considered beneficial at one point in a population's lifetime can change. If the environment changes, so do the adaptations.

9. First, there needs to be variation in a population (individuals within a population are born with genetic differences). There are too many offspring for the environment to support. Due to the limited resources, there is a lot of competition among members of a population, and any individual that has a trait that makes it better able to outcompete others will end up surviving. These most-fit individuals are the ones left to reproduce and pass on their specific genes to the next generation. This results in a larger percentage of the next generation having these "most-fit" alleles.

The Evidence for Evolution

1. The more similar the cytochrome C sequence, the closer the relation. In order from most closely related to least related: mouse, donkey, carp, corn, euglena.
2. Tail, limb buds, and pharyngeal arches
3. A butterfly wing and a bird wing are examples of **analogous** structures because they **do not** arise from a common ancestor.
4. The leaves of a cactus have been modified into spines, and the Venus flytrap's "mouth" is also a modified leaf. The spines and the "mouth" are examples of **homologous** structures in plants.
5. The **lower** the sediment layer, the older the fossilized species.
6. A previous question listed the percent similarities between different organisms and a human:

Cytochrome C protein – % similarity to humans
Mouse – 91%
Donkey – 89%
Carp – 79%
Corn – 67%
Euglena – 57%

We are mammals, so it makes sense that the two mammals on the list—a mouse and a donkey—would be our closest relatives. A carp (fish) isn't a mammal, but it is a vertebrate, so it is definitely related to us. A plant (a multicellular eukaryote like us) is next, leaving the single-celled protist (Euglena) as our most distant relative.

7. What is the relationship between homologous and vestigial structures? A vestigial structure is a homologous structure in a species that is no longer functional.
8. What is the driving force behind the development of analogous body structures? The two unrelated organisms have similar features because they live in a common environment. Their analogous body part was an adaptation for this common environment.
9. Hutton's theory of **gradualism** stated that geological structures were formed through **slow** changes over long periods of time.
10. **b.** Though every one of those traits would be evidence of common ancestry, the most reliable evidence is similarities in genetic code (since it is the "root" of the other listed traits).

Microevolution: Evolution of Populations

1. (1) No mutations allowed, (2) only random mating, (3) the population needs to be large, (4) no gene flow is allowed, and (5) no natural selection.

2. Recall that you should always start by finding the frequency of the recessive (q) allele. Tan is the recessive phenotype (bb) and 40% of the snails are tan, then bb $= q^2 = 0.4$. Take the square root of 0.4 to get $q = 0.63$. By applying the Hardy-Weinberg equation that summarizes the two alleles ($p + q = 1$), you can determine that p equals $1 - 0.63 = 0.37$. The first question asks what percentage of snails are heterozygous. You know from the second equation that 2pq equals the percentage of heterozygous critters, so by plugging in your p and q values . . . $2(0.63)(0.37) = 0.47$. The number of heterozygous snails would be 118 (rounded) snails. The number of homozygous dominant individuals would be $251 \times (p^2)$, or $251\ (0.37)^2 = 35$ (rounded) homozygous dominant snails.
3. If a population suddenly suffers a drastic reduction in numbers due to a catastrophic lava flow, the remaining population will eventually repopulate. The reestablished population has very little genetic diversity due to the **bottleneck effect**.
4. This is an example of **disruptive selection**. The most-numerous medium-smelly bugs are removed, and the two extremes (slightly smelly and extremely smelly) are now the most-fit survivors and will become the most-represented phenotypes.
5. Write the Hardy-Weinberg equation that tallies the different alleles for a given gene: $\mathbf{p + q = 1}$. Next, write the equation that summarizes all the given genotypes possible: $\mathbf{p^2 + 2pq + q^2 = 1}$. The homozygous dominant genotype is represented as p^2, the homozygous recessive genotype is q^2, and the heterozygous genotype (both pq and qp) is 2pq.
6. Non-tasters are homozygous recessive $= q^2 = 70/220 = 0.32$

$$q = 0.57$$

$$p = 1 - q = 0.43$$

7. Evolution in a population that occurs because of a chance change in allele frequency is referred to as **genetic drift**. For example, a small number of individuals may leave their original population and start their own. If these colonizing individuals happen to carry an assortment of **alleles** that are not similar to their original population, their new population will have a different **allele frequency**. This is an example of **founder effect**.

8. **Stabilizing selection** reinforces the current most-fit phenotype by removing organisms with the extreme phenotypes.
9. **c.** If an organism shows the recessive phenotype, its genotype must be homozygous recessive (q^2). You can determine the frequency of the recessive allele (q) by taking the square root of 0.35; $q = 0.59$. You also know that only two allele options are available for this particular gene (dominant and recessive), and adding them both must equal 100%:

   $p + q = 1$

   therefore

   $p = 1 - q; p = 0.41$

   The frequency of heterozygous individuals in a population is determined with 2pq:

   $2 (0.41) (0.59) = 0.48 = 48\%$

Macroevolution: Evolution of Species

1. Since the beetles' environments are different colors, there will be different phenotypes that are considered "most fit." For example, on the side with dark soil and not much grass, those beetles that happen to have a darker exoskeleton would be more camouflaged and have a better chance at surviving (not being eaten). This would lead to a change in the gene pool to have a higher percentage of dark-colored beetles. The opposite would occur on the side with more grass coverage; the beetle population would evolve to have a higher percentage of green-colored exoskeletons. If these two populations of beetles are no longer able to interbreed because brown beetles only choose to mate with other brown beetles (and green beetles mate only with green beetles), speciation has occurred!

2. One species of flower can be fertilized by another species of flower, but once the egg and sperm fuse, their differing chromosome numbers prevents the formation of a viable zygote. This is an example of **offspring viability**, a type of **post-zygotic** barrier that prevents the production of offspring between two different species.
3. About 200 years ago, an ancestral population of flies laid their eggs on tree fruits called hawthorns. Once domestic apple trees were introduced, some flies in the population chose to lay their eggs on apples, instead. The maggots that developed in the hawthorn fruit grew into adult flies who preferred to mate with other hawthorn-reared flies; maggots from apples developed into flies who preferred to mate with other apple-grown flies. This is an example of **behavioral isolation**, a type of **pre-zygotic** barrier that prevents the production of offspring between two different species.
4. Both terms refer to a genetic change in a species over time. Microevolution focuses on a change occurring in the gene pool of a specific species, whereas macroevolution looks at the bigger picture of how different species arise from an ancestral species (due to changes in their gene pools).
5. In order for speciation to occur, there needs to be some sort of barrier that divides the current population's **gene pool**, thus interrupting the gene flow between the two resulting groups. Next, the isolated gene pools need to **change/evolve** in a way that renders them unable to interbreed if brought back together.
6. One species of field cricket mates in the spring, whereas a second species of field cricket mates in the fall. This is an example of **temporal** isolation, a type of **pre-zygotic** barrier that prevents the production of offspring between two different species.
7. **c.** Offspring viability means that even though fertilization did occur and an offspring was produced, the offspring was not healthy enough to survive.

Phylogeny and Vertebrate Evolution

1. Humans and chimpanzees evolved from a common ancestor.
2.
   a. **Node 3**
   b. **Species A**
3.

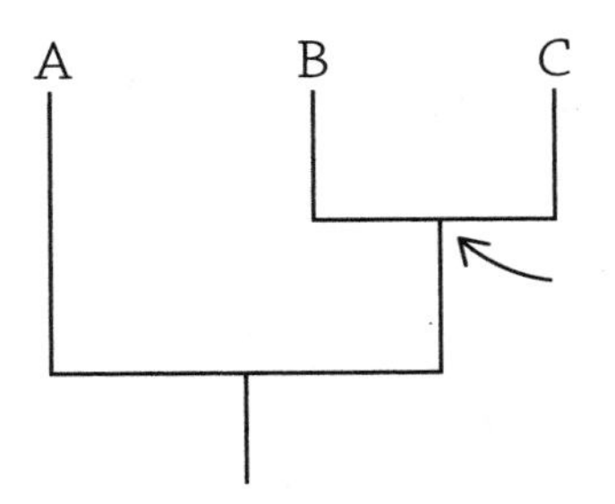

4. Species E is most closely related to species D.
5. Lungs and four limbs
6.
   a. **Salmon**
   b. **Node X**
7. **b.** Phylogenetic trees do not consider time (only common ancestry).

Plant Structure and Transport

1. Dermal tissue is the skin of the plant. It can protect the plant against water loss by secreting a waxy coating. Ground tissue composes most of the "filling" of the plant and is specialized for photosynthesis and storage. The vascular tissue is the circulatory system of the plant, responsible for moving glucose and water around the plant's body.

2. **Parenchyma** cells are the "typical" plant cell, whose main roles are photosynthesis and storage. **Collenchyma** cells provide structure without restraining growth. **Sclerenchyma** cells are specialized for structure and are no longer able to elongate.
3. The upper, sunny side of the leaf is composed of a palisade layer that is tightly packed with cells filled with chloroplasts. A leaf has openings in the bottom to allow carbon dioxide gas to enter (required for photosynthesis). The gas is able to move through the leaf tissue into the palisade layer because it can freely travel through the airspaces in the spongy layer. The vascular tissue (both xylem and phloem) forms the veins of the leaf, providing easy access to water and a place to deposit the glucose.
4. The root cells actively move ions from the soil into the root's xylem. This lowers the water potential in the root xylem (compared to the water potential in the surrounding soil). Water will move from [higher water potential] → [lower water potential], so water will passively flow from the soil into the root. This provides a bit of "push" up the body of the plant.
5. Both phloem and xylem are components of the plant's vascular tissue. Phloem moves downward, transporting photosynthetically derived glucose from the leaves down toward the roots. Xylem, on the other hand, moves upward, transporting water taken in to the roots up toward the rest of the plant.
6.
   a. Has two cell walls (S)
   b. Found in strings of celery because they provide structure without restraining growth (C)
   c. Has lignin in its cell walls (S)
   d. Primary function is photosynthesis (P)
   e. The reason a pear has a gritty feel (S)

7. The process of **transpiration** is the major force that pulls water from the roots up a plant's body. As water **evaporates** from the top of the plant, it pulls a chain of water molecules up the xylem. Each water molecule creates a chain through the process of **cohesion**, and the entire chain sticks to the sides of the hollow tracheids and vessel elements through the process of **adhesion**.
8. **b.** Stomata are the openings in the leaves for the uptake of carbon dioxide gas.

Plant Reproduction

1. It has both sperm-producing anthers and egg-producing ovaries.
2. In a flower, the **anther** produces the **pollen grains**, each of which carries **sperm**.
3.

If a sperm fertilizes the...	It forms the...	Which has a chromosome count of...
Egg	Embryo	2n (diploid)
Two polar nuclei	Endosperm	3n (triploid)

4. Once fertilization occurs, the ovary develops into **fruit**, and the ovule develops into the **seed coat** (protecting the embryo). Since the embryo is stuck inside a protective coating, it needs a packed lunch to keep it alive until the embryo grows into a photosynthesizing plant; this food source is the **endosperm**.
5. Imperfect flowers cannot self-fertilize because an imperfect flower has ONLY eggs or sperm.

6. Pollination is when a pollen grain lands on the **stigma**. Fertilization is when the egg in the **ovary** is fertilized by one of the two sperm that travels down the **style** through the tube created by the third cell in the pollen grain, called a **tube nucleus**. The second sperm will fertilize the **polar nuclei**, turning into food for the embryo, called **endosperm**.
7. [outermost] petals → ovary → ovule → egg [innermost]
8. The fruit itself does not help protect the embryo (that is the seed coat's job). The fruit, however, may be eaten by an animal that passes the indigestible seeds through its poop, thus helping to spread the seeds (and embryos) throughout the ecosystem.

Introduction to Animals

1. Animals are defined as **heterotrophic** organisms, meaning that they can only obtain their nutrients by **eating**.
2. Nematodes and arthropods
3. Flatworms' flatness creates a high surface-area-to-volume ratio. This allows oxygen to easily diffuse from the environment and permeate their tissues. Therefore, they do not need a circulatory system to transport oxygen around their bodies.
4. Based on the phylogenetic tree, the phylum that includes **echinoderms/ sea stars** is the closest relative of our phylum, the **chordates**.
5. **True.**
6. Nervous tissue and muscle tissue.
7. They have radial symmetry and, instead of a brain, they possess a nerve net.
8. Arthropods are the most numerous (it includes all insects!).

Animals Need Homeostasis

1. Maintaining a stable internal environment prevents dangerously wild swings in physiological conditions (such as temperatures, pH, salt levels, etc.).
2. A negative feedback loop always counteracts a stimulus to bring conditions back to a set point. Positive feedback, on the other hand, increases the stimulus in order to increase the response even further.
3.

Form of nitrogenous waste	Notable quality	Type of critter
Ammonia	**Very toxic and very water soluble**	**Fish**
Urea	Won't toxify us when stored in our bodies	Mammals
Uric acid	**Solid**	Bird

4. The process of maintaining proper levels of water and solutes in the blood is referred to as **osmoregulation**. The millions of small **nephrons** in the kidneys are able to adjust the amount of water reabsorbed back into the bloodstream by changing the number of **aquaporin** proteins that line the collecting ducts. The chemical signal that leads to an increase in the numbers of these transport proteins is **antidiuretic hormone (ADH).**
5. When maintaining homeostatic levels of glucose in the bloodstream, the hormone **insulin** is responsible for **lowering** blood glucose levels. One way it removes free glucose from the bloodstream is by stimulating the **liver** to take in glucose and form the storage polymer, **glycogen**.
6. Uric acid is a solid, so when it is produced by the developing embryo that is stuck in the egg, it will form a solid precipitate (instead of toxifying the fluid environment).

7. The hypothalamus would increase ADH secretion, leading to an increased number of aquaporins embedded in the kidneys' nephrons (this leads to more water being reabsorbed back into the bloodstream). The brain would also send signals of "thirsty!," increasing water intake.
8. Land animals have kidneys that work more like **marine** fish because land animals are also battling water loss. The kidneys save water by reabsorbing it back into the **bloodstream** instead of allowing too much to be excreted in the urine.
9. **d.** Ocean fish are surrounded by a hypertonic (saltier) environment, compared to their tissue osmolarity. Salt will constantly move into their tissues by diffusion, and their body will lose water by osmosis. To battle this imbalance, a saltwater fish will conserve water and rid itself of salt by excreting concentrated, salty urine.

Animal Digestion

1. The first stage of digestion is **ingestion**, when you take a bite of food. That food is broken up during the process of **digestion**, when large molecules are broken up into smaller components. The small components are more easily taken up into the bloodstream through the process of **absorption**. Finally, any indigestible material is passed out of the body through **elimination**.
2. Hydrolysis reactions break up macromolecules (protein, fats, nucleic acids, and large polysaccharides) into monomers.
3. The lining of the small intestine is highly folded, increasing its surface area. The higher the surface area, the faster the rate of nutrient absorption through the intestinal wall into the capillaries. The folds are called villi, each covered with cells with hair-like projections called microvilli.

4.

Cell of the gastric gland	Product of the cell	Function
Mucous cell	**Mucous**	**Protects the lining of the stomach**
Chief cell	Pepsinogen	Once pepsinogen is activated by **HCl**, it turns into the protein-digesting enzyme **pepsin**
Parietal cell	**HCl**	Creates **acidic** conditions

5. Mechanical digestion (such as chewing up food and bile's emulsification of fat) is simply breaking up big pieces into little pieces; no covalent bonds are broken and no enzymes are used. Chemical digestion, on the other hand, is driven by enzymatic hydrolysis and covalent bonds are broken.

6.

1. k
2. j
3. e
4. f
5. i
6. g
7. h
8. a
9. c
10. d
11. b

7. The tiger is a carnivore, whose diet consists of mostly meat (and very little cellulose). Since the cecum plays a role in cellulose digestion, the tiger would have a small cecum. The rabbit, on the other hand, is an herbivore with an entirely plant-based diet. In order to help digestion, the rabbit would evolve to have a very large cecum.

8. Interesting poop trivia! About a third of your poop is composed of **bacteria**, and poop is brown because of pigments produced from the breakdown of **red blood cells/hemoglobin/bile/bilirubin (any of these answers work!)**.
9. **c.** Starch begins digestion in the mouth (salivary amylase) and continues as it passes through the small intestine (pancreatic amylase).

Animal Circulation and Respiration

1. The organism's body must have a high surface-area-to-volume ratio.
2. A four-chambered heart has two separate ventricles separated by a septum. This prevents oxygen-poor blood in the right ventricle from mixing with oxygen-rich blood in the left ventricle.
3. The place of gas exchange between the respiratory system and the circulatory system occurs in the **alveoli** of the respiratory system and the surrounding **capillaries** of the circulatory system.
4. The partial pressure of oxygen would be higher than that of carbon dioxide in the bloodstream after it leaves the lungs (having dropped off CO_2 and picked up O_2), before it enters the oxygen-starved tissues.
5. When prokaryotic cells became too large, their SA:VOL ratio decreased. In order to increase their membrane surface area to compensate for their larger volumes, eukaryotic cells evolved to have compartmentalized interiors (organelles).
6. The mammalian **left** ventricle has a thicker wall than the **right** ventricle. The **left** ventricle needs to be stronger because it powers the blood through the entire body (the **systemic** circuit). The **right** ventricle only needs to power the blood out to the **lungs** (the **pulmonary** circuit).

7. The respiratory system brings in the oxygen that is picked up by the circulatory system. After the circulatory system moves the oxygen to the tissues, it picks up the waste carbon dioxide the tissues generated through cellular respiration. The circulatory system moves the carbon dioxide back to the lungs, where the respiratory system expels (exhales) it from the body.
8. Blood coming from the lungs and entering the tissues has a high **oxygen** partial pressure. The tissues have a high **carbon dioxide** partial pressure because the cells have been producing a lot of **carbon dioxide** through the process of cellular respiration. These partial pressures facilitate the movement of oxygen from the **blood** into the **tissues**.
9. **a.** The more surface area exposed to the environment, the more "space" is provided for substances to pass through cell membranes.

Animal Neurons and Signal Transduction

1. A resting neuron has a higher concentration of **K^+** on the inside of the cell and a higher concentration of **Na^+** on the outside of the cell. The net charge on the inside of the cell is **negative**.
2. A neuron transmits a signal by generating an **action potential**, which is a wave of **positive** charge generated when **sodium** ions rush into the cell.
3. **e-b-d-c-a-f**
4. A neuron actively transports 3 Na^+ ions out for every 2 K^+ ions brought in. This creates a net movement of positive charges *out* of the cell. The inside of the neuron also has large proteins that carry an overall negative charge, further contributing to the negative interior.
5. **g-f-b-d-a-e-c**

6. When an action potential reaches the end of the **presynaptic** neuron, it must rely on the chemicals called **neurotransmitters** to cross the space between adjacent neurons (the **synapse**). Once the chemicals bind to the **postsynaptic** neuron, sodium channels will open and the action potential will continue.
7. **d.** The neurotransmitters are contained within vesicles, and for the chemical to be released into the synapse, the vesicles need to migrate toward (and fuse with) the membrane at the synaptic terminal.

The Animal Immune System

1. The first and second lines of defense are nonspecific, meaning the response occurs regardless of the antigen. Conversely, the third line of defense is specific for the particular invading antigen.
2. **1-E, 2-C, 3-H, 4-G, 5-D, 6-A, 7-B, 8-F**
3. The cell-mediated response targets **intracellular** pathogens such as **viruses** and relies on **T cells**.
4. Once the particular lymphocyte (either B or T cell) is selected for by a particular antigen, it will produce two populations of cloned cells: the active cells that take care of the immediate infection, and a second population of memory cells that last for years (even decades) awaiting a second invasion. If the same antigen presents itself again, the memory cells respond immediately to remove the invader before it has a chance to cause problems. Vaccines work in the same way, except the initial "infection" is only a part of the virus/bacterium, just enough to stimulate a lymphocyte response (but not itself able to cause the disease).
5. The other barriers of your body provide the first line of defense. This includes skin, sweat, saliva, and tears.
6. The humoral response targets **extracellular** pathogens and relies on **B cells** to produce **antibodies** to bind to and help remove the invader.

7. Passive immunity is when antibodies are given to the patient (and are not produced by the patient's own B cells, as they are in active immunity). Passive immunity is an important treatment option when the patient does not have the luxury of time to produce their own antibodies (venom from a snake bite, or babies receiving antibodies from their mother).
8. The flu virus mutates every year, and the previous year's vaccine will no longer be close enough of a "match."
9. **a.** When a snake-bite victim receives antivenin, they are given a population of antibodies specific for that particular venom protein. Because the antibodies were generated by a different source (and not the bite-victim themselves), this is an example of passive immunity.

Populations

1. 1-f; 2-d; 3-e; 4-a; 5-c; 6-b
2. A species with a Type III survivorship curve would need to produce a large number of offspring because most will die off very early in life. Plants, insects, and heavily preyed-upon mammals exhibit a Type III survivorship curve.
3.

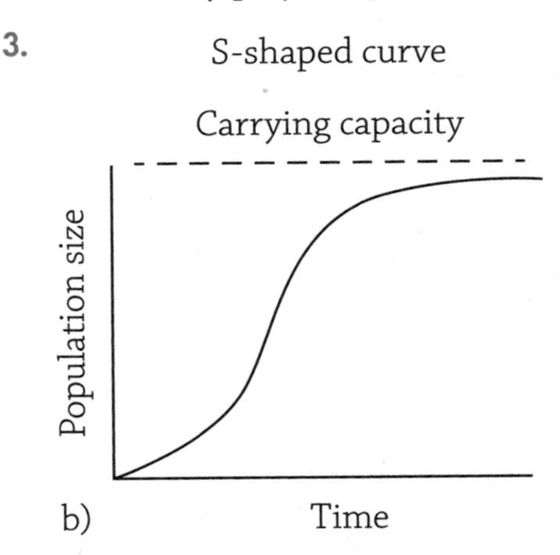

The logistic curve is more realistic because the initial exponential growth is unsustainable. Resources will eventually become limited and population growth will slow and reach the carrying capacity.

4. Mutualism, because both the leafhopper and the bacteria benefit from the arrangement.
5. A population with a **low** death rate and a **high** birthrate would experience an overall increase in the population numbers. This population would most likely have an age structure with mostly **young** individuals.
6. 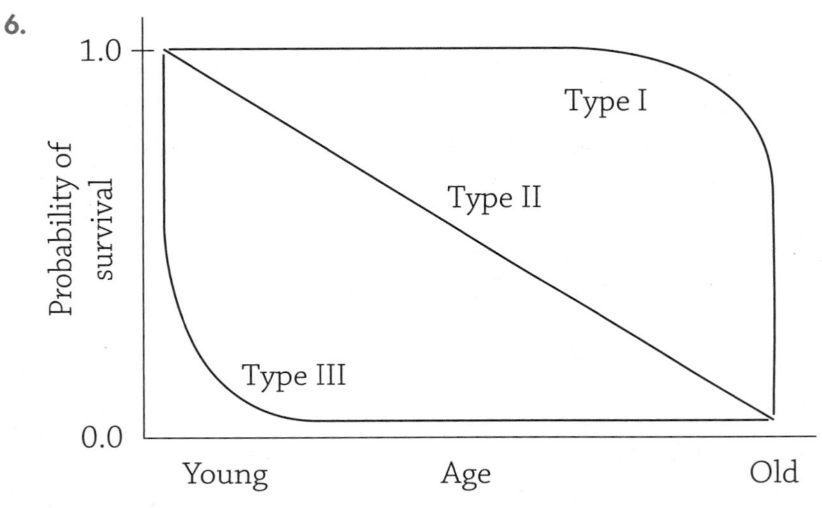

Large mammals exhibit the Type I survivorship curve. The probability of survival remains high throughout their lives (which is good, considering the relatively few offspring produced), and most die after they reach old age.

7.

   a. A viral disease **(DD)**
   b. Competition for mates **(DD)**
   c. Volcanic eruption **(DI)**
   d. Predation **(DD)**
   e. Wildfire **(DI)**
   f. Oil tanker spill in ocean **(DI)**
   g. Clearcutting forest **(DI)**

8. To be commensalism, the louse must benefit, and the fish must be neither harmed nor helped. If the louse perfectly replaces the function of the fish's tongue (and no harm comes to the fish), it would be a commensal relationship.
9. **c.** If there is a decrease in limiting factors, this would release a restriction on growth, and the population would increase.

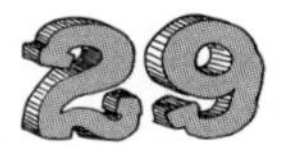

Ecosystems and Their Interconnectivity

1. A habitat only refers to where it lives. A niche, however, also takes into consideration its role in the environment: what it eats, how it reproduces, if it is prey for other organisms, when it sleeps, and so on.
2. The graph depicts the competitive exclusion principle. Both species are initially growing well, but at day 4, resources become limiting and species A outcompetes species B. Species B eventually dies off, leaving only species A to occupy this particular niche.
3. Each step in a food chain depicts the transference of energy as one organism eats another. The second law of thermodynamics states that every time energy is converted from one form to another (such as an herbivore eating plants and then transforming the energy of the plants' tissues into ATP), the process is not 100% efficient. In fact, only a measly 10% of the plants' stored energy is able to be used by the herbivore. That means that, eventually, the available energy will become too small to support another step in the food chain (another trophic level).
4. Photosynthesis is the only process through which carbon dioxide is removed from the atmosphere. Once it is locked into tissues (thanks to food chains), carbon dioxide is released back into the atmosphere through cellular respiration and decomposition. Burning of fossil fuels also releases a massive amount of carbon dioxide.

5. The competitive exclusion principle states that two different **species** cannot occupy the same ecological **niche**. The survivor is the one who was able to **outcompete** the other.
6. The first trophic level must be a producer (autotroph) because every chain is built upon the energy provided by the sun. The only type of organism able to harness photons of sunlight are autotrophs, either photosynthetic plants, protists, or bacteria.
7. Earth is a **closed** system in regards to matter, but an **open** system in regards to energy.
8. The first step of the nitrogen cycle is **nitrogen fixation**, when nitrogen gas is fixed into organic molecules. The two means of nitrogen fixation are **lightning** and **nitrogen-fixing bacteria**.
9. **a.** Denitrifying bacteria remove nitrogenous compounds from the soil. If the numbers of denitrifying bacteria in the soil decrease, the amount of nitrogen that remains in the soil would increase.

Human Impact on Ecosystems

1. Both primary and secondary succession are processes by which ecosystems reestablish their populations after a disturbance. The difference is that primary succession must start from scratch, re-forming soil from the rocks that remain. It takes much longer than secondary succession (which has intact soil).
2. A **renewable** resource can be replaced at the same rate by which it is consumed. Give an example of this type of resource: **solar, wind, hydroelectric, or geothermal energy.**
3. Secondary pollutants are created when primary pollutants undergo chemical reactions in the atmosphere. Primary pollutants are released directly into the atmosphere from burning fossil fuels.

4. When excess **fertilizer** enters a body of water, the **phosphorus** causes an **algal bloom** (phosphorus is normally a **limiting factor** and keeps algal growth in check). The overgrowth of algae soon begins to die, and **decomposers** begin to consume the dead algae. The decomposers are undergoing massive amounts of cellular respiration and using up tons of oxygen in the process, leading to **hypoxic** (low oxygen) conditions in the water. This causes a die-off of aquatic animals and plants.
5. a. Chernobyl nuclear contamination (**SS**), b. California wildfires (**SS**), c. Mount St. Helens volcanic eruption (either **PS or SS**, depending on severity of damage. If soil remains intact, it is SS.), d. Deforestation of Amazon rainforest (**SS**), e. Retreating glaciers in Alaska (**PS**).
6. Lichen is a pioneer species that is able to live on the newly exposed rocks and help break them down into soil (the first step of primary succession).
7. Burning fossil fuels releases carbon dioxide into the atmosphere. The CO_2 in the atmosphere (along with other greenhouse gases) absorbs the infrared radiation released from the heated earth, warming the air.
8. **a.** As natural gas is removed and used as a fuel source, it is not replaced (at least not at a measurable rate). Therefore, it is not a renewable fuel source.

Teacher's Guide

I wrote this book from two different perspectives: as a text to help my students better understand a year's worth of biology topics, and for a teacher who needs a way to distill and simplify the material to better explain it to their class. It was easy to target both audiences because when I teach my kids, I learn the topic through their experience; I want to teach them the same way I would want to learn. There is a large chasm between reading a formal biology textbook and an actual understanding of the material. No matter the class level (Introductory Biology all the way through AP/Advanced Biology), we have a lot of information to cover with a ton of new vocabulary and concepts. This scope can easily become overwhelming to those on both sides of the desk!

This book will help you and help them (your kids). The chapters reflect the most common progression through biology concepts, and the book covers the material needed for almost any level of course. When I wrote each chapter, I focused on the most important topics; these topics would provide an easy scaffold from which you could create a classroom lecture. Furthermore, if your students each had this book alongside their larger biology textbook, it could help them in several ways:

1. **Pre-lecture preview.** Since this book is a more concise and easier read, you could assign a given section/chapter for homework and be confident that your students wouldn't become overwhelmed. It is also much easier for a student to grasp the concepts as laid out here, as opposed to the dense and formal textbook readings. Furthermore, as they read through *Must Know High School Biology*, they can check their understanding with the Bonus Flashcard App. Your students could make note of concepts they found confusing for use as a formative assessment. It would also set them up nicely

with an understanding of the material before you present to them and solidify the information.

2. **Post-lesson review.** After they learn a concept from you, it helps (as you know) to refresh and revisit the material soon thereafter. Their follow-up homework could be to read the appropriate section and complete the chapter review questions (the answers are included for them to check their work). I wrote these questions as I would want my own students to demonstrate their understanding after we finish a unit. The review questions test their basic understanding of the material, but they also delve deeper to assess their ability to see deeper concepts and apply their knowledge (summative assessment).

Regardless of how you use it, I feel strongly that this book will not only help ease the burden of a potentially challenging biology class but also make it very fun and entertaining. I have collected a lot of super-cool examples and facts throughout my 22 years of teaching, and I included them in this text as BTW ("by the way") sidebars and IRL ("in real life") factoids. We are so lucky to biology teachers. It's a lot of work, but it's also a ton of fun. Enjoy the process.

Teacher Resources

If you have some time, please review some of the resources I've supplied here. I've found them both helpful and enlightening in my teaching, and I think you will too.

For starters, if you're looking for learning opportunities mixed in with some entertainment, try these educational YouTube videos and channels:

Stated Clearly
(https://www.youtube.com/c/StatedClearly)

Kurzgesagt – In a Nutshell
(https://www.youtube.com/c/inanutshell)

Amoeba Sisters
(https://www.youtube.com/c/AmoebaSisters)

DNA Learning Center
(https://www.youtube.com/user/DNALearningCenter)

If you're looking for more learning resources to challenge your students:

Khan Academy AP Biology
(https://www.khanacademy.org/science/ap-biology)

Bozeman AP Biology
(http://www.bozemanscience.com/ap-biology)

If you're looking for materials, try these suppliers with a long history of offering a variety of necessary products:

Carolina Biological
(https://www.carolina.com/)

Bio-Rad
(https://www.bio-rad.com/)

Biology Kits and Supplies
(https://www.homesciencetools.com/biology/biology-supplies-kits/)

NOTES

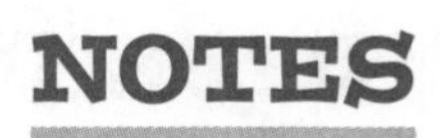

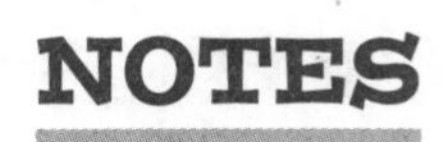

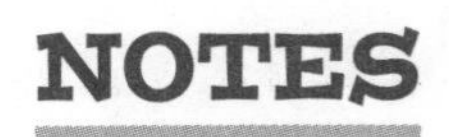